Non-destructive Testing in Civil Engineering

Non-destructive Testing in Civil Engineering

Editors

Jerzy Hoła
Łukasz Sadowski

MDPI • Basel • Beijing • Wuhan • Barcelona • Belgrade • Manchester • Tokyo • Cluj • Tianjin

Editors
Jerzy Hoła
Wrocław University of Science and Technology
Poland

Łukasz Sadowski
Wrocław University of Science and Technology
Poland

Editorial Office
MDPI
St. Alban-Anlage 66
4052 Basel, Switzerland

This is a reprint of articles from the Special Issue published online in the open access journal *Applied Sciences* (ISSN 2076-3417) (available at: https://www.mdpi.com/journal/applsci/special_issues/Non-Destructive_Testing_Civil).

For citation purposes, cite each article independently as indicated on the article page online and as indicated below:

LastName, A.A.; LastName, B.B.; LastName, C.C. Article Title. *Journal Name* **Year**, *Volume Number*, Page Range.

ISBN 978-3-0365-5131-9 (Hbk)
ISBN 978-3-0365-5132-6 (PDF)

© 2022 by the authors. Articles in this book are Open Access and distributed under the Creative Commons Attribution (CC BY) license, which allows users to download, copy and build upon published articles, as long as the author and publisher are properly credited, which ensures maximum dissemination and a wider impact of our publications.

The book as a whole is distributed by MDPI under the terms and conditions of the Creative Commons license CC BY-NC-ND.

Contents

About the Editors vii

Jerzy Hoła and Łukasz Sadowski
Non-Destructive Testing in Civil Engineering
Reprinted from: *Appl. Sci.* **2022**, *12*, 7187, doi:10.3390/app12147187 1

Luis Santana, Diego Rivera and Eric Forcael
Force Measurement with a Strain Gauge Subjected to Pure Bending in the Fluid–Wall Interaction of Open Water Channels
Reprinted from: *Appl. Sci.* **2022**, *12*, 1744, doi:10.3390/app12031744 7

Hoseong Jeong, Baekeun Jeong, Myounghee Han and Dooyong Cho
Analysis of Fine Crack Images Using Image Processing Technique and High-Resolution Camera
Reprinted from: *Appl. Sci.* **2021**, *11*, 9714, doi:10.3390/app11209714 25

Samuel Klein, Henrique Fernandes and Hans-Georg Herrmann
Estimating Thermal Material Properties Using Solar Loading Lock-in Thermography
Reprinted from: *Appl. Sci.* **2021**, *11*, 3097, doi:10.3390/app11073097 49

Jinyoung Hong, Hajin Choi and Tae Keun Oh
Application of Tooth Gear Impact-Echo System for Repeated and Rapid Data Acquisition
Reprinted from: *Appl. Sci.* **2020**, *10*, 4784, doi:10.3390/app10144784 63

Gloria Cosoli, Alessandra Mobili, Francesca Tittarelli, Gian Marco Revel and Paolo Chiariotti
Electrical Resistivity and Electrical Impedance Measurement in Mortar and Concrete Elements: A Systematic Review
Reprinted from: *Appl. Sci.* **2020**, *10*, 9152, doi:10.3390/app10249152 75

Tomáš Húlan, Filip Obert, Ján Ondruška, Igor Štubňa and Anton Trník
The Sonic Resonance Method and the Impulse Excitation Technique: A Comparison Study
Reprinted from: *Appl. Sci.* **2021**, *11*, 10802, doi:10.3390/app112210802 117

Sabine Kruschwitz, Tyler Oesch, Frank Mielentz, Dietmar Meinel and Panagiotis Spyridis
Non-Destructive Multi-Method Assessment of Steel Fiber Orientation in Concrete
Reprinted from: *Appl. Sci.* **2022**, *12*, 697, doi:10.3390/app12020697 127

Blanca Tejedor, Kàtia Gaspar, Miquel Casals and Marta Gangolells
Analysis of the Applicability of Non-Destructive Techniques to Determine In Situ Thermal Transmittance in Passive House Façades
Reprinted from: *Appl. Sci.* **2020**, *10*, 8337, doi:10.3390/app10238337 141

Tim Klewe, Christoph Strangfeld, Tobias Ritzer and Sabine Kruschwitz
Combining Signal Features of Ground-Penetrating Radar to Classify Moisture Damage in Layered Building Floors
Reprinted from: *Appl. Sci.* **2021**, *11*, 8820, doi:10.3390/app11198820 161

Yuqi Jin, Tae-Youl Choi and Arup Neogi
Longitudinal Monostatic Acoustic Effective Bulk Modulus and Effective Density Evaluation of Underground Soil Quality: A Numerical Approach
Reprinted from: *Appl. Sci.* **2021**, *11*, 146, doi:10.3390/app11010146 177

Anna Hoła and Łukasz Sadowski
Verification of a Nondestructive Method for Assessing the Humidity of Saline Brick Walls in Historical Buildings
Reprinted from: *Appl. Sci.* **2020**, *10*, 6926, doi:10.3390/app10196926 . . . . **187**

Clara Pereira, Jorge de Brito, José D. Silvestre and Inês Flores-Colen
Atlas of Defects within a Global Building Inspection System
Reprinted from: *Appl. Sci.* **2020**, *10*, 5879, doi:10.3390/app10175879 . . . . **199**

Anna Rudawska, Radovan Madleňák, Lucia Madleňáková and Paweł Droździel
Investigation of the Effect of Operational Factors on Conveyor Belt Mechanical Properties
Reprinted from: *Appl. Sci.* **2020**, *10*, 4201, doi:10.3390/app10124201 . . . . **221**

Samuel Klein, Tobias Heib and Hans-Georg Herrmann
Estimating Thermal Material Properties Using Step-Heating Thermography Methods in a Solar Loading
Thermography Setup
Reprinted from: *Appl. Sci.* **2021**, *11*, 7456, doi:10.3390/app11167456 . . . . **239**

Mitsuru Uesaka, Jian Yang, Katsuhiro Dobashi, Joichi Kusano, Yuki Mitsuya and Yoshiyuki Iizuka
Quantitative Evaluation of Unfilled Grout in Tendons of Prestressed Concrete Girder Bridges by Portable 950 keV/3.95 MeV X-ray Sources
Reprinted from: *Appl. Sci.* **2021**, *11*, 5525, doi:10.3390/app11125525 . . . . **251**

Katia Gaspar, Miquel Casals and Marta Gangolells
Influence of HFM Thermal Contact on the Accuracy of In Situ Measurements of Façades' U-Value in Operational Stage
Reprinted from: *Appl. Sci.* **2021**, *11*, 979, doi:10.3390/app11030979 . . . . **277**

Marilena Cozzolino, Vincenzo Gentile, Paolo Mauriello and Agni Peditrou
Non-Destructive Techniques for Building Evaluation in Urban Areas: The Case Study of the Redesigning Project of Eleftheria Square (Nicosia, Cyprus)
Reprinted from: *Appl. Sci.* **2020**, *10*, 4296, doi:10.3390/app10124296 . . . . **291**

About the Editors

Jerzy Hoła

Jerzy Hoła is employed at the Faculty of Civil Engineering of the Wrocław University of Science and Technology (WUST). He is a former Dean of the Faculty and a former Senator of the WUST. He is a member of the Civil Engineering Committee and the Chairman of the Construction and Mechanics Committee of the Polish Academy of Sciences. He specializes in general construction and non-destructive testing. He has published over 300 scientific works, including several books and several hundred expertise reports.

Łukasz Sadowski

Łukasz Sadowski is an Associate Professor at the Faculty of Civil Engineering of the Wrocław University of Science and Technology. He is the Head of the Department of Materials Engineering and Construction Processes. He was a fellow of the Foundation for Polish Science and the Ministry of Science and Higher Education. He is a senior member of RILEM and a member of the Polish Academy of Sciences. His achievements include more than 200 publications, 2 books, and tens of expertise reports.

Editorial

Non-Destructive Testing in Civil Engineering

Jerzy Hoła and Łukasz Sadowski *

Faculty of Civil Engineering, Wroclaw University of Science and Technology, 50-370 Wroclaw, Poland; jerzy.hola@pwr.edu.pl
* Correspondence: lukasz.sadowski@pwr.edu.pl

Citation: Hoła, J.; Sadowski, Ł. Non-Destructive Testing in Civil Engineering. *Appl. Sci.* **2022**, *12*, 7187. https://doi.org/10.3390/app12147187

Received: 4 July 2022
Accepted: 15 July 2022
Published: 17 July 2022

Publisher's Note: MDPI stays neutral with regard to jurisdictional claims in published maps and institutional affiliations.

Copyright: © 2022 by the authors. Licensee MDPI, Basel, Switzerland. This article is an open access article distributed under the terms and conditions of the Creative Commons Attribution (CC BY) license (https://creativecommons.org/licenses/by/4.0/).

1. Introduction

The progressive development of civil engineering has forced scientists to improve the known methods and techniques of testing building materials, and also to search for new ones, e.g., non-destructive testing (NDT) methods. These methods usually do not interfere with the tested material and structures during tests. Despite this, NDT methods not only allow for the assessment of many important material properties and parameters, but also for the reliable localization of imperfections, damage, or internal defects in tested elements and structures. This knowledge is needed in many situations in order to e.g., correctly assess operational safety, reliability, durability, and the degree of degradation. It is worth noting that there has recently been significant progress in the development of NDT methods, especially those from the group of acoustic methods. These methods, as is the case in medicine, are very useful for obtaining information about the examined structures and the inside of elements on the basis of recorded acoustic signals. These signals are then processed by appropriate software with the use of complex data analysis, including artificial intelligence.

This Special Issue, entitled "Non-Destructive Testing in Civil Engineering", aims to present to interested researchers and engineers the latest achievements in the field of new research methods, and also the original results of scientific research carried out with their use-not only in laboratory conditions, but also in selected case studies. The articles published in this issue are theoretical-experimental and experimental, and also show the practical nature of the research. They are grouped by topic, and the main content of each article is briefly discussed for your convenience.

These articles extend knowledge in the field of non-destructive testing in civil engineering with regards to new and improved NDT methods, their complementary application, and also the analysis of their results-including the use of sophisticated mathematical algorithms and artificial intelligence, as well as the diagnostics of materials, components, structures, entire buildings, and interesting case studies.

2. Research in the Field of New and Improved Methods and Techniques of Non-Destructive Testing

Ref. [1] presents an experimental method of measuring small forces between the fluid and a wall in open water channels using a strain gauge as a force sensor. For this purpose, six uniaxial strain gauges were used, which were placed throughout the measurement area and subjected to bending tests in order to determine the correlation between the load and the obtained signal. A special data acquisition system was established to record the performance of the strain gauge in relation to the lateral displacement caused by the testing machine. The obtained results indicated a linear relationship between the load and the obtained signal for the situation when the strain gauge was seated in the zone from 30% to 45% from the central axis in the sensor's measuring region. The described sensor can therefore be used to measure small magnitude forces. Additionally, the linear correlation between the load and the obtained signal can be used for calibration, as long as the strain gauge is seated close to the central axis of the detection area.

In turn, article [2] is devoted to the identification of cracks in concrete. The work associated with this kind of research requires effort and equipment, such as articulated ladders. Additionally, there are important health and safety issues, as some structures are not very accessible. To deal with these problems, various studies have used digital imaging to measure cracks in concrete. The purpose of this experimental study is to evaluate the optical limit of digital camera lenses with regard to an increased working distance. Three different lenses and two digital cameras were used to record line images with a thickness of 0.1 to 0.5 mm. Field measurement tests were carried out to verify the measurement parameters identified on the basis of the results of the performed research. The actual crack widths were visually measured, and the obtained values were used for further analysis. Based on the conducted studies, it was confirmed that the number of pixels that corresponds to the working distance had a large impact on the accuracy of the crack width measurement when using image processing. Therefore, the optimal distance and measurement guidelines required to measure the size of some objects were provided for the imaging and optical equipment that was used in this study.

In conventional thermography, a sample is subjected to a periodically changing stream of heat. This heat flux usually enters the sample in one of three ways: by a point source, a line source, or an extended source. Calculations that were conducted on the basis of surface sources are particularly well suited to solar load thermography. This is due to the fact that most natural heat sources and heat sinks can be brought closer in order to be uniformly extended over a certain area of interest. This is especially interesting because the natural thermal phenomena cover large areas, which in turn makes this method suitable for the measurement of large-scale samples. Article [3] describes an investigation of how extended approximation source formulas for determining the properties of thermally thick and thermally thin materials can be used in a naturally excited system. This work also shows the possible sources of errors, and gives quantitative results to estimate the thermal efficiency of a retaining wall structure. It was shown that this method can be used in the case of large-size structures that are subjected to the natural phenomenon of external heating.

Progress in the impact-echo (IE) method is due to the automation of the scanning of concrete bridge decks. The toothed system that was presented in [4] when describing the IE method was developed using gears as impacts and microelectromechanical systems. This system continuously collects a large amount of field test data because a rack generates impacts automatically. The duration of the contact between two gears is assessed, and the contact mechanism is then compared to a steel ball mallet using a high-speed camera. Data were collected based on the measurements of concrete slabs in which artificial voids were embedded at different depths. Based on the experiments, a reduction in pitch, or an increase in the number of teeth, was required to reduce the contact time and to generate the thickness frequency from deep delaminations. Automatically acquired time-domain data was shifted to the frequency-time domain by means of spectrograms in order to identify the dominant frequency of a set of obtained signals. The results display that the developed method enabled to obtain high-quality data during IE tests. In turn, the spectrogram analysis delivered important information about the frequency of the obtained IE signals, and verified the repeatability of the data.

The aim of the article [5] was to analyze the most modern techniques of measuring electrical impedance (and consequently electrical resistance) of mortar/concrete elements. Various measurement methods are described and discussed, with the advantages and disadvantages with regard to their performance, reliability, and degree of maturity being highlighted. The usefulness of electrical resistivity measurements was demonstrated. Due to the fact that electrical resistivity is an important indicator of the health of concrete, and also that it changes whenever there are phenomena modifying the conductivity of the mortar/concrete (e.g., degradation or external influences), the conducted review of measurement techniques was meant to serve as a guide for those interested in these types of measurements.

3. Complementary Applications of the Non-Destructive Testing Methods

Article [6] presents resonance frequency tests of bent concrete samples. These frequencies were determined simultaneously using the sonic resonance method and the impulse excitation technique, with the two methods then being compared. The samples differ in used material and shape. The mean values and corresponding values of standard deviations of the resonance frequencies were compared. The performed tests showed the equivalence of both methods of measuring the resonance frequency. The obtained difference between the values measured using the sonic resonance method and the impulse excitation technique were not significant. The relationship between the resonant frequencies, which is presented graphically, is linear with a slope of 0.9993.

Fibrous reinforcement in high-performance cement materials is widely applied in many areas of the construction industry. One of the most frequently studied features of steel fiber reinforced concrete is the slower development of cracks, which in turn results in the better durability of such concrete. Additional benefits are related to structural properties, as fibers can significantly increase the ductility and tensile strength of concrete. In some applications, it is even possible to completely replace conventional reinforcement with fiber reinforcement, leading to significant logistical and environmental benefits. However, it can have disadvantages, as the fibers can induce anisotropic behavior of the concrete if they are not properly oriented. For the safe use of steel fiber reinforced concrete, non-destructive testing (NDT) methods should be used in order to assess the orientation of the fibers in the hardened concrete. For this purpose, article [7] uses complementary methods of ultrasound, electrical impedance and X-ray computed tomography. The article also demonstrates the capabilities of each of these techniques separately when measuring fiber orientation. Based on these results, conclusions were drawn regarding the most promising areas for future research and development.

Article [8] examines the use of a heat flux meter and quantitative infrared thermography to assess the difference between the predicted and real thermal conductivity of house facades under steady-state conditions in the Mediterranean basin. First, the suitability of NDT techniques was tested experimentally, and then a single-family house was verified in a real environment. The outcomes of this study show that both techniques quantified the difference between the design and real U-value of a house facade. The quantitative infrared thermography method was faster than the heat flux meter method, although the flux meter method has higher accuracy. The presented results will help when choosing the most appropriate test method.

4. Analysis of the Results of Non-Destructive Testing, Including the Use of Sophisticated Mathematical Algorithms and Artificial Intelligence

Until now, the analysis of cores obtained from a structure by drilling is the main option to obtain knowledge about the depth of damage due to water penetrating the floors of buildings. The time-consuming and costly procedure is an additional burden for building insurers, who mention water damage caused by leaking pipelines as the most common claim against insurers. The radar method, due to its high sensitivity to water, can be an important and non-destructive support for this problem. Therefore, the article [9] describes a modular sample, which was developed to obtain the appropriate thickness of screed and insulation material. The resulting dataset was then used to examine the corresponding characteristics of a signal in order to classify three situations: dry insulation, damaged insulation, and damaged screed. It was highlighted that the analysis of the statistical distributions of the scans allows for an accurate identification of damage on floors. Combining the proposed functions with multidimensional data analysis and artificial intelligence was crucial to achieving satisfactory results.

Article [10] shows a numerical analysis of the application of the ultrasonic method for the detection of underground voids in the ground in order to check the properties of the underwater seabed period. The obtained numerical model demonstrates the possibility of detecting (with a spatial resolution of about 0.5 λ) subterranean void airspace. The proposed technique can overcome the limitations of conventional techniques, which use sonar devices that are characterized by a low penetration depth and the leakage of the transverse sound wave propagating in the underground fluid environment.

Article [11] presents the results of the verification of a method of assessing the moisture content of saline brick walls based on non-destructive measurements and artificial intelligence. The method was formerly developed and can be suitable for the non-destructive identification of the moisture content of walls in historic buildings when destructive interference during the research is not possible because of the conservation limitations. However, before its implementation into construction practice, this method requires validation of other historic buildings. The results of testing the dampness of two selected historical objects were used for the experimental verification of the model obtained by the artificial neural network (ANN). These results were different than those used for learning and testing the ANN. The obtained high values of the linear correlation coefficient and low values of mean absolute error confirm that the obtained ANN model is useful in assessing the moisture content of saline brick walls.

5. Diagnostics of Materials, Elements, and Structures, as well as Entire Buildings

The study presented in [12] aims to propose a list of defects that can be found in several types of building elements/materials, which will in turn simplify issues related to the diagnosis of building pathologies. The database was formulated with the use of previously developed elements of the global control system: a fault classification list and the urgency of repairs. This database has been structured using tables ordered according to defect type, building component/material, and urgency to repair (a five-point scale of 0–4). The repair urgency levels are demonstrated with photos and described using brief criteria. Not all the repair urgency levels are applicable to all the combinations of "defect-component/material". Levels 1, 2, and 3 are most frequently taken into account. The proposed list of defects is a novel approach that can be useful to support experts during inspections of buildings, the concept of which can be adapted to further inspection systems.

The aim of the article [13] was to present the influence of temperature and humidity on the mechanical properties of a conveyor belt. The investigations were carried out in both a climatic chamber, which simulates the effects of negative and positive temperatures of −30 °C to 80 °C (243 K to 353 K), and in a thermal shock chamber. The results of the tests in the climatic chamber showed that numerous mechanical parameters have undesirable values at 10 °C (283 K) and 80 °C (353 K) at a relative humidity of 80%. Interestingly, the results showed that the tensile modulus, tensile strength, and yield point are higher at temperatures below 0 °C than at temperatures above 0 °C.

6. Interesting Case Studies in the Field of Non-Destructive Testing in Civil Engineering

In article [14], it was shown that active thermography methods, such as step heating thermography, show a good correlation with the solar load system. Solar load thermography is an approach that has recently gained the attention of scientists. It is beneficial because it is particularly easy to set up and can measure objects on a large scale due to the fact that the sun is the main source of heat. This work also introduces the concept of using a pyranometer as a reference point for evaluation algorithms by providing a direct measurement of the intensity of solar radiation. In addition, a recently introduced method of estimating thermal efficiency is assessed using thermograms of the environment.

Article [15] presents the portable X-ray sources that were developed in the last 10 years and which are based on a linear electron accelerator with a power of 950 keV/3.95 MeV in the X band (9.3 GHz). Moreover, it also describes the inspections of prestressed concrete bridges. A bridge with a T-shaped PC girder, 200–400 mm thick, and a bridge with a box-shaped PC girder, 200–800 mm thick were subjected to tests. X-ray images of the defects of the tendon ducts were observed. An attempt to quantify unfilled mortar was made. This is due to the fact that this is the major defect that causes corrosion. On the taken X-ray images, gray values were obtained, which correspond to the X-ray attenuation coefficients of filled and unfilled mortars in tendon ducts. The gray ratio of the filled/unfilled tendon ducts was then compared in order to determine the degree of filling. For this purpose, data obtained from an actual T-shaped post-tensioned concrete bridge and model samples were used to validate the method.

Precise evidence of the thermal conductivity of walls is essential for selecting the proper energy-saving measures in existing buildings. For reliable testing using the heat meter method (HFM), good thermal contact should be provided between the heat meter plate and the surface of the wall. The aim of the article [16] was to assess the effect of an imperfect thermal contact of the heat meter plates on the accuracy of the in-situ measurements of the U value of the facade after applying foil. The foil was applied in order to avoid damage to the surface of the wall, which is a normal procedure during the operational phase of a building. The results show that the deviations between the measured U-values and the values obtained using the HFM directly on the wall surface and when HFM was installed with a PVC film were significantly different from the theoretical values.

Article [17] concerned the application of geophysical techniques in the urban environment of the city of Nicosia in Cyprus. The main goal of the research, being part of the Eleftheria Square redesign project, was to visualize subsurface properties in order to reduce the impact of threats on old buildings (and thus to preserve the cultural heritage of the site) and new infrastructure under construction. Since 2008, various phases of the project have used electrofusion tomography, radar, and also electromagnetic induction methods to provide an understanding of the geological stratigraphy of buried objects (archaeological and underground structures) and unexpected events (such as water infiltration). The results of geophysical research confirmed the effectiveness of the adopted methods, and added cognitive value with regard to the studied area. The new information gathered helped public administration technicians to plan direct and targeted interventions and to modify the original design in line with the discovery of archaeological finds.

Funding: This research received no external funding.

Conflicts of Interest: The authors declare no conflict of interest.

References

1. Santana, L.; Rivera, D.; Forcael, E. Force Measurement with a Strain Gauge Subjected to Pure Bending in the Fluid–Wall Interaction of Open Water Channels. *Appl. Sci.* **2022**, *12*, 1744. [CrossRef]
2. Jeong, H.; Jeong, B.; Han, M.; Cho, D. Analysis of Fine Crack Images Using Image Processing Technique and High-Resolution Camera. *Appl. Sci.* **2021**, *11*, 9714. [CrossRef]
3. Klein, S.; Fernandes, H.; Herrmann, H.G. Estimating Thermal Material Properties Using Solar Loading Lock-in Thermography. *Appl. Sci.* **2021**, *11*, 3097. [CrossRef]
4. Hong, J.; Choi, H.; Oh, T.K. Application of Tooth Gear Impact-Echo System for Repeated and Rapid Data Acquisition. *Appl. Sci.* **2020**, *10*, 4784. [CrossRef]
5. Cosoli, G.; Mobili, A.; Tittarelli, F.; Revel, G.M.; Chiariotti, P. Electrical Resistivity and Electrical Impedance Measurement in Mortar and Concrete Elements: A Systematic Review. *Appl. Sci.* **2020**, *10*, 9152. [CrossRef]
6. Húlan, T.; Obert, F.; Ondruška, J.; Štubňa, I.; Trník, A. The Sonic Resonance Method and the Impulse Excitation Technique: A Comparison Study. *Appl. Sci.* **2021**, *11*, 10802. [CrossRef]
7. Kruschwitz, S.; Oesch, T.; Mielentz, F.; Meinel, D.; Spyridis, P. Non-Destructive Multi-Method Assessment of Steel Fiber Orientation in Concrete. *Appl. Sci.* **2022**, *12*, 697. [CrossRef]
8. Tejedor, B.; Gaspar, K.; Casals, M.; Gangolells, M. Analysis of the Applicability of Non-Destructive Techniques to Determine In Situ Thermal Transmittance in Passive House Façades. *Appl. Sci.* **2020**, *10*, 8337. [CrossRef]
9. Klewe, T.; Strangfeld, C.; Ritzer, T.; Kruschwitz, S. Combining Signal Features of Ground-Penetrating Radar to Classify Moisture Damage in Layered Building Floors. *Appl. Sci.* **2021**, *11*, 8820. [CrossRef]
10. Jin, Y.; Choi, T.Y.; Neogi, A. Longitudinal Monostatic Acoustic Effective Bulk Modulus and Effective Density Evaluation of Underground Soil Quality: A Numerical Approach. *Appl. Sci.* **2021**, *11*, 146. [CrossRef]
11. Hoła, A.; Sadowski, Ł. Verification of a Nondestructive Method for Assessing the Humidity of Saline Brick Walls in Historical Buildings. *Appl. Sci.* **2020**, *10*, 6926. [CrossRef]
12. Pereira, C.; de Brito, J.; Silvestre, J.D.; Flores-Colen, I. Atlas of Defects within a Global Building Inspection System. *Appl. Sci.* **2020**, *10*, 5879. [CrossRef]

13. Rudawska, A.; Madleňák, R.; Madleňáková, L.; Droździel, P. Investigation of the Effect of Operational Factors on Conveyor Belt Mechanical Properties. *Appl. Sci.* **2020**, *10*, 4201. [CrossRef]
14. Klein, S.; Heib, T.; Herrmann, H.G. Estimating Thermal Material Properties Using Step-Heating Thermography Methods in a Solar Loading Thermography Setup. *Appl. Sci.* **2021**, *11*, 7456. [CrossRef]
15. Uesaka, M.; Yang, J.; Dobashi, K.; Kusano, J.; Mitsuya, Y.; Iizuka, Y. Quantitative Evaluation of Unfilled Grout in Tendons of Prestressed Concrete Girder Bridges by Portable 950 keV/3.95 MeV X-ray Sources. *Appl. Sci.* **2021**, *11*, 5525. [CrossRef]
16. Gaspar, K.; Casals, M.; Gangolells, M. Influence of HFM Thermal Contact on the Accuracy of In Situ Measurements of Façades' U-Value in Operational Stage. *Appl. Sci.* **2021**, *11*, 979. [CrossRef]
17. Cozzolino, M.; Gentile, V.; Mauriello, P.; Peditrou, A. Non-Destructive Techniques for Building Evaluation in Urban Areas: The Case Study of the Redesigning Project of Eleftheria Square (Nicosia, Cyprus). *Appl. Sci.* **2020**, *10*, 4296. [CrossRef]

Article

Force Measurement with a Strain Gauge Subjected to Pure Bending in the Fluid–Wall Interaction of Open Water Channels

Luis Santana [1,2], Diego Rivera [3] and Eric Forcael [1,*]

1. Department of Civil and Environmental Engineering, Universidad del Bío-Bío, Concepción 4081112, Chile; lsantana@ubiobio.cl
2. Department of Water Resources, College of Agricultural Engineering, Universidad de Concepción, Chillán 3812120, Chile
3. Centro de Investigación en Sustentabilidad y Gestión Energética de Recursos (CiSGER), School of Engineering, Universidad del Desarrollo, Las Condes 7610658, Chile; diegorivera@udd.cl
* Correspondence: eforcael@ubiobio.cl; Tel.: +56-41-3111700

Featured Application: This research has specific applications in the measurement of forces of small magnitude, which occur in the fluid–wall interaction, through a non-destructive technique that allows for measuring the shear stress in open water channels.

Abstract: An experimental method to measure forces of small magnitude with a strain gauge as a force sensor in the fluid–wall interaction of open water channels is presented. Six uniaxial strain gauges were employed for this purpose, which were embedded across the entire sensing area and subjected to pure bending, employing two-point bending tests. Sixteen two-point bending tests were performed to determine the existence of a direct relationship between the load and the instrument signal. Furthermore, a regression analysis was used to estimate the parameters of the model. A data acquisition system was developed to register the behavior of the strain gauge relative to the lateral displacement induced by the loading nose of the universal testing machine. The results showed a significant linear relationship between the load and the instrumental signal, provided that the strain gauge was embedded between 30% and 45% of the central axis in the sensing area of the sensor ($R^2 > 0.99$). Thus, the proposed sensor can be employed to measure forces of small magnitude. Additionally, the linear relationship between the load and the instrumental signal can be used as a calibration equation, provided that the strain gauge is embedded close to the central axis of the sensing area.

Keywords: small force; load; instrumental signal; two-point bending test; strain gauge; calibration equation

1. Introduction

The direct measurement of small forces has shown to have great practical significance in many fields such as sciences, engineering, the industrial sector, and medicine—among many others—and its relevance was first recognized in the 1970s [1], continuing its growth to this day.

One of the applications of measurement of small forces can be found when studying the fluid–wall interaction of open water channels. In this sense, shear stress in open water channels—the action exerted by water on the channel bed material—can alter the state of rest or movement of the bed material, either by sedimentation, dragging, scour, or transport of the material. Wall–water stores oppose the movement of the fluid and affect the flow velocity profile. Drag and shear stress due to vegetation reduce flow discharge in channels, but they also allow flood attenuation and sediment deposition [2,3]. Shear stress, dragging, and sediment transport have been widely studied, and there exist limitations on using conventional formulas, such as the Manning equation [2]. Average bed shear stress is difficult to derive from the bulk flow characteristics, such as applying predefined velocity profiles or the shear stress distribution. Currently, there is a wide range of methods for direct and indirect measurement of shear stress in open channels. However, direct measurements of shear stress in the wall–fluid interface are less frequent. Tinoco et al. [4] classify intrusive approaches (acoustic Doppler velocimeters, optical backscatter sensors, ultrasonic velocity profilers, and acoustic Doppler current profilers) and non-intrusive techniques (laser-induced fluorescence, particle image velocimetry, and laser Doppler velocimetry) to detect the magnitude of the flow disturbance by the experimental probe itself. Detailed reviews about methods to measure and estimate shear stress can be found in Huai et al. [5]. Experimental setups range from large-scale measurements (e.g., Errico et al. [6]) to highly controlled laboratory-based experiments (e.g., Duan et al. [7] and Bashirzadeh et al. [8]). The direct measurement of shear stress requires sensitivity to small changes in flow velocity, generates minimal flow disturbance, especially in the vicinity of the bed, and maintains the principle of no sliding of the material in the bed [9–11].

The development of microsystems, nanotechnology, fluid dynamics, aerodynamics, biotechnology, and biomedical technology, among others, involves the application of devices suitable to measure forces from the force scale of millinewtons (mN) to micronewtons (μN). On one side, microelectromechanical sensors (MEMS) such as systems based on interferometry, oil films, and liquid crystal coatings have shown improvements in performance and accuracy relative to conventional techniques (hybrid MEMS) and have been used to measure skin friction [12]. On the other hand, multidimensional F/M MEMS are used to measure force and moment in robotics, electronics, and the development of medical equipment and smart devices [1].

The techniques employed for these measurements can be direct or indirect [13,14]. Despite the usefulness of these sophisticated devices to measure forces induced by external loads, they are limited by their complex calibration and testing procedures, which are not sufficiently structured because of the miniaturization of sensors [12–16].

Electrical resistance strain gauges are one of the simplest and most frequently used passive transducers. They are usually bonded to one of the surfaces of the specimen to be tested to generate only axial tension and compression. They have been used as load or strain transducers of sophisticated devices since their manufacturing technology and encapsulation prevent them from suffering mechanical and environmental damage. Due to their small size, sensitivity, accuracy, simplicity, and low cost compared to sophisticated devices, strain gauges are broadly used in laboratories worldwide as sensing devices [16]. Nonetheless, no research has been found that explores the operating principle of strain gauges subjected to pure bending. This is because the bending effect is considered negligible concerning the tension–compression effect. In the fields of fluid dynamics and aerodynamics, the measurement of small forces involved in the fluid–wall interaction is required to analyze the stability of walls and fluids.

Consequently, there is a lack of simple devices for measuring forces of small magnitude, in the order of mN, that can be employed to measure flows of water in open and closed systems. Due to the foregoing, the strain gauge is proposed to be used as a device for measuring small forces, used in a way different from that for which it was designed; that is to say, the strain gauge itself is subjected to pure bending. Furthermore, lateral displacement employing a two-point bending test (cantilever) induced in a way different from that defined by ASTM D747 [17] is proposed.

The proposed method is justified for the following reasons: Firstly, the size of the proposed device is in the order of millimeters. Therefore, it is smaller than the minimum dimensions required for standardized tests [17]. Secondly, in this type of test, the data analysis is based on the conventional beam theory [18], or the elastic theory [19], provided that there is a direct relationship between the applied load and the resultant lateral displacement within the elastic range of the specimen's material. However, the appropriateness of the conventional beam theory for determining the relationship between the load and lateral displacement of a sample is called into question when the dimensions of the specimen are in the order of millimeters; thus, uncertainties are generated in the bending tests. Thirdly, because the geometric dimensions of the proposed device correspond to the ones of a small plate, it is difficult to obtain an analytical relationship between load and displacement. The numerical solutions (finite elements or differences) and analytical solutions (superposition method or integral transform method) [20] are well known; however, they require complex and time-consuming procedures. Furthermore, thin flats behave differently from beams in terms of strain and internal stress distribution, which is under Love–Kirchhoff's theory.

For these reasons, the objective of this paper is to present the experimental results of two-point bending tests applied directly to strain gauges to measure forces of small magnitude directly. A direct relationship between the load and the instrumental signal is expected to be found because both of them depend on the lateral displacement of the device. Finally, these results are expected to be a practical contribution to the evidence about the mechanical behavior of strain gauges subjected to pure bending.

2. Materials and Methods
2.1. Materials and Experimental Setup

Figure 1 shows the experimental setup for the implementation of the proposed two-point bending test. Consequently, a total of sixteen two-point bending tests were carried out on the six strain gauges used as force sensors. The tests were carried out utilizing an INSTRON 4467 universal testing machine, which has a cell load of 10 N and a precision of 0.0025 N. Through the control panel and the software IX of the INSTRON, the geometric parameters and the characteristics of each test were defined. Furthermore, the reference values for force and displacement were established. All of the tests were performed at a constant speed of 1 mm/min, with a total displacement of the loading nose of 2 mm, which is in agreement with the ASAE Standard method S368.2 [21]. Simultaneously, the instrumental signal (indicated in Section 2.2) was recorded as a result of the data acquisition system, the devices developed, and the proposed strain gauge. In this sense, to subject the strain gauge to pure bending, each sensor was embedded within the sensing area employing two 3 mm thick printed circuit boards (PCB). Thus, these points are mentioned in Table 1, while the PCB plates were fitted to a 10-mm-thick acrylic plate (base of the materials testing machine) through an aluminum device as shown in Figure 1a,c.

Table 1. Dimensions of the strain gauge sensors and characteristics of the bending tests.

Sensor	Sensor Dimensions				Trial Characteristics					
	b (mm)	e (mm)	Lo (mm)	Lo/e	x_0 (mm)	Rep	x_0/e	x_0/Lo	k (mm)	T (°C)
G1	2.54	0.060	2.27	37.8	1.22	5	20.3	0.54	0.30	27.9 ± 0.4
G2	2.58	0.055	3.25	59.1	2.30	2	41.8	0.71	0.60	26.4 ± 1.0
G3	2.54	0.050	2.30	46.0	1.35	2	27.0	0.59	0.35	26.4 ± 0.6
G4	2.58	0.065	2.53	38.9	1.68	2	25.8	0.66	0.69	26.6 ± 0.4
G5	2.54	0.065	2.46	37.8	1.51	2	23.2	0.61	0.44	26.5 ± 0.4
G6	2.52	0.065	1.88	28.9	1.00	3	15.4	0.53	−0.05	19.2 ± 2.6
Average	2.55	0.060	2.45							

In this study, six strain gauges (from G1 to G6 as shown in Figure 1a,c) used as force sensors were assessed. The strain gauges employed were simple, uniaxial, miniature strain gauges fabricated with copper-nickel alloy and completely coated with polyamide [22]. Their average geometric dimensions were 2.56 mm wide, 6.0 mm long and 0.06 mm thick, and their sensing area was 1.6 mm wide and 2.0 mm long, with a nominal resistance of 120 Ω.

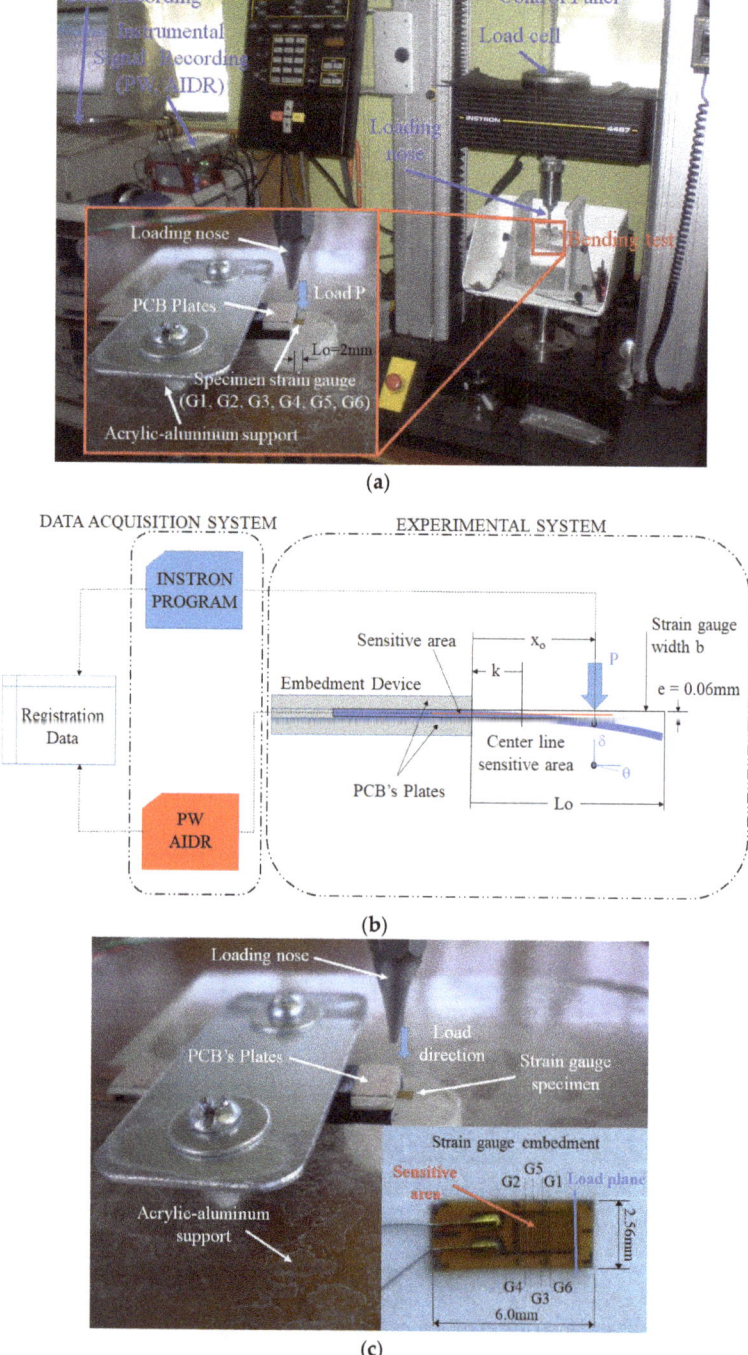

Figure 1. Proposed two-point bending test: (**a**) INSTRON and experimental setup for implementing the bending test (loading nose width 1.0 ± 0.05 mm), (**b**) scheme of the bending test and data recording, (**c**) details of the strain gauge embedment.

Table 1 shows the geometric dimensions and mechanical characteristics of the strain gauges used in the bending tests. Width (b), thickness (e), and the beam span (L_0) were measured using a micrometer with a precision of 0.01 mm. The position of the loading nose concerning the embedment (x_0) and the location of the embedment concerning the central axis of the sensing area (k) were measured using a Vernier caliper with a precision of 0.05 mm and verified using a photographic sample with a precision of 0.01 mm. The number of repetitions (R) was reduced to two due to the low coefficient of variation obtained for the G1 sensor, for which five repetitions were carried out. The tests, except that of the G6 sensor, were performed in a room with controlled temperature, as it is shown in Table 1. It has to be noted that the instrumental signals may be affected by temperature change; however, in this research, the temperature's effect was not considered, since the calibration was carried out at a constant temperature in a controlled environment as previously mentioned.

The repeatability of the responses in the experimental tests was estimated for each sensor to analyze the validity of the data acquisition system and the universal testing machine. For each repetition, the beam span, the position, and the load speed of the loading nose remained constant. After finishing a test, the loading nose was returned to its initial position and the test was repeated.

2.2. Instrumental Signal Measurements in the Two-Point Bending Tests

Figure 2 shows a scheme of the electronic circuits used to measure the instrumental signal in the two-point bending tests. The change in the electrical resistivity (ΔR) of the sensor when the strain gauge was being bent by the loading nose (see Figure 1b) was measured through a Wheatstone bridge (PW) because of its great precision [16,23]. As the signal (or drop in voltage) obtained from the PW was very small, in the range of millivolts, an instrumentation amplifier was required for the signal conditioning. In addition, a fixed-gain differential-input amplifier (AIDR) with high common-mode rejection was developed.

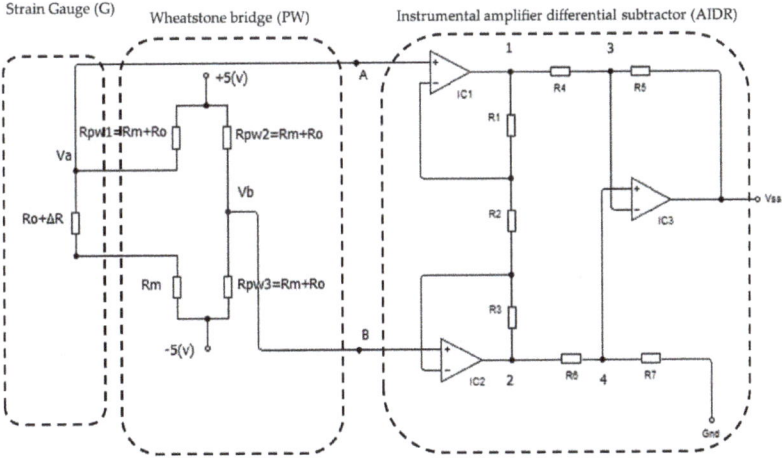

Figure 2. Scheme of the instrumental signal measurement during bending tests.

A Wheatstone bridge circuit (PW) in a quarter-bridge arrangement was used to register the changes in the electrical resistance of the sensor concerning its initial resistance (R_o). The quarter-bridge configuration was balanced ($V_{AB} = 0$) and showed equality of nominal resistance values [8]. Moreover, an inactive gauge was added to prevent changes in resistance for the initial resistance. Resistors in series, Rm, were added to each component of the PW to establish a 10-mA upper limit for the current flow through the sensor. The resistor ratio R_m/R_o = 3.25 was employed for the resistors. In practice, the change in resistivity $\Delta R/R_o$ is in the range of 10^{-2}, which is negligible relative to unity.

The response of the proposed device (sensor) was indicated in the PW through a voltage drop, as Equation (1) illustrates. This voltage drop and the change in resistivity of the sensor are in a proportional relationship because of the lateral displacement of the sensor. The constant of proportionality was 0.588.

$$V_{AB} = \frac{\Delta R/R_o}{[2\Delta R/R_o + 4(1 + R_m/R_o)]} V_r \approx \frac{\Delta R/R_o}{4(1 + R_m/R_o)} V_r = 0.588 \Delta R/R_o \tag{1}$$

The response of the sensor obtained by the PW is amplified by the AIDR circuit, wherein the voltage drops in nodes 3 and 4 must be the same (see Figure 2) to achieve a linear performance of the instrumentation amplifier. Therefore, the instrumental signal was obtained by applying Ohm's law to each loop of the circuit shown in Figure 2, according to the output voltage given by Equation (2).

$$V_s = A_1 V_A - A_2 V_B, \tag{2}$$

where $A_1 = \left[\frac{R_7}{R_4} \frac{(R_4+R_5)}{(R_7+R_6)} \frac{(R_2+R_3)}{R_2} + \frac{R_1 R_5}{R_2 R_4}\right]$ and $A_2 = \left[\frac{R_7}{R_4} \frac{(R_4+R_5)}{(R_7+R_6)} \frac{R_3}{R_2} + \frac{R_5}{R_4} \frac{(R_1+R_2)}{R_2}\right]$ are the equivalent voltage drop percentages in terminals A and B of the Wheatstone bridge, whereas the nominal resistance values of the amplifier are represented by R_i, with i = 1 to 7. To rearrange Equation (2) in terms of the voltage drops in differential mode, $V_{AB} = V_A - V_B$, and in common mode, $V_{mc} = (V_A + V_B)/2$, the voltage drops of the instrumental signal, V_s, can be rewritten in terms of the common-mode gain, G_{mc}, and the gain in differential mode, G_{AB}, as Equation (3) illustrates:

$$V_s = G_{mc} V_{mc} + G_{AB} V_{AB}, \tag{3}$$

where $G_{mc} = \left[\left(1 - \frac{R_1 R_6}{R_7 R_4}\right) / \left(1 + \frac{R_6}{R_7}\right)\right]$ and $G_{AB} = \left[\left(1 + \frac{R_6}{R_7}\right)\left(\frac{R_3}{R_2} + \frac{1}{2}\right) / \left(1 + \frac{R_6}{R_7}\right) + \frac{R_5}{R_4}\left(\frac{R_1}{R_2} + \frac{1}{2}\right)\right]$.

Given that the resistances $R_5 R_6 = R_7 R_4$, the amplifier worked in a high common-mode rejection ratio, wherein G_{mc} is null. Thus, the amplification factor, K, was estimated as K = $R_5/R_4 = R_7/R_6$, and the average gain factor, G, was estimated as $G/2 = R_1/R_2 = R_3/R_2$. Then, the instrumental output signal, Vs, given by Equation (3), can be replaced by Equation (4).

$$V_s = K(1+G)V_{AB} + V_{offsetAI}, \tag{4}$$

where $V_{offsetAI}$ corresponds to the voltage drops inside the AIDR and the parasitic currents, which could be electronically adjusted through a variable resistor. In addition to this, the voltage drops could be avoided by employing the differential output voltage as $\Delta V_s = V_s - V_{so}$ because the initial signal must comply with Equation (4). Hence, the instrumental signal was subjected to the residual voltage of the sensor and became independent from the initial adjustment required by the Wheatstone bridge.

Finally, the signal delivered by the sensor, given by Equation (1), was amplified by the AIDR through Equation (5), where a linear relationship between the instrumental signal and the change in electrical resistance was determined:

$$\Delta V_s = K(1+G)\Delta V_{AB} \approx \frac{K(1+G)V_r}{4(1+R_m/R_o)} (\Delta R - \Delta R_o)/R_o \tag{5}$$

2.3. Load–Instrumental Signal Relationship

When the load system is uniformly distributed across the width of the specimen by the loading nose, Love–Kirchhoff's model of plates is equivalent to the Euler–Bernoulli beam theory, which is developed as a cantilever beam. Moreover, for small displacements or small loads, the classical beam theory and the plate theory can be regarded as valid; hence, when the load is uniformly applied along the width of the sensor, a direct relationship between the load and the lateral displacement of the sensor is expected.

The load and the instrumental signal are related through the lateral strain of the sensor. A linear relationship between load and displacement can be expected for small lateral displacements of the plate, where Hooke's law is valid [20]. On the other hand, if the difference in lateral displacement is proportional to the change in electrical resistance, a linear relationship between the instrumental signal and the force applied on the sensor can be possibly expected.

To achieve that, the data about load and instrumental signal were simultaneously recorded by the universal testing machine (precision 0.0025 N) and the data acquisition system (precision of 0.17 mV). To synchronize both sets of data, the lateral displacement and the measuring time were registered and related to the speed ratio. To minimize synchronization errors between the universal testing machine and the measuring equipment during the data collection and recording, the load values (P) and the instrumental signal (V_s) were recalculated at 1-s time intervals by using their respective time-weighted averages. The data obtained were employed to obtain the relationship between the load and the instrumental signal for each test.

The average performance concerning the load and the instrumental signal of each sensor concerning the displacement of the loading nose was presented in a graphic form, according to the characteristics shown in Table 1 and Figure 2. To find the relationship between the load and the instrumental signal, dimensionless values for load and instrumental signal were employed. The load was divided by the maximum load of the tests and the instrumental signal was divided by the analog-to-digital reference voltage (V_{ref} = 2.56 V). Furthermore, a linear regression analysis was performed to find a linear relationship between the load and the adimensionalized instrumental signal. The parameters of the linear model were calculated using the least-squares approach [24].

The values for the parameters of the model were compared using a 95% confidence level through determining the confidence interval and the probability of the Fisher's value, assuming that the slope of the model did not differ significantly from the zero value.

Finally, the difference between the estimated and the measured values for the adimensionalized load was used as estimation error in the expected linear regression model.

2.4. Sensitivity and Performance of the Device

The device's sensitivity is in the range of 9 to 18 mV/mN, depending on the location of the embedding region (near or within the sensitive area of the strain gauge as shown in Figure 1). Graphically, Figure 3 shows that the change in device output signal explains more than 99% of the load variability for strain gauges G1 to G6. Therefore, there is a high positive correlation between the load and the instrumental signal of the new device, where the variation in sensor sensitivity is attributable to manufacturing differences. There is high linearity within 30% to 80% of the maximum load because the sensors are, by design, within the elastic range.

In general, the sensor was tested for loads between 0 and 100 mN, showing high levels of linearity between load and instrumental signal and of lateral displacement between 0 and 1 mm, without generating gauge delamination or residual deformation effect and with an affordable cost and consumption. However, the experimental setup to measure shear stress in channels' beds is expensive. The order of magnitude of detectable forces is less than 1 N/m². On the other hand, the bed shear stress in open channels is in the order of 1–10 N/m² [25–28]. In this case, the most relevant characteristic is that the dimension of the device is small enough to avoid significant interferences in the flow. It has to be noticed that there is room for improving: (1) the sensor embedding procedure to attain perfect embedding and (2) the capture of the device's output signal.

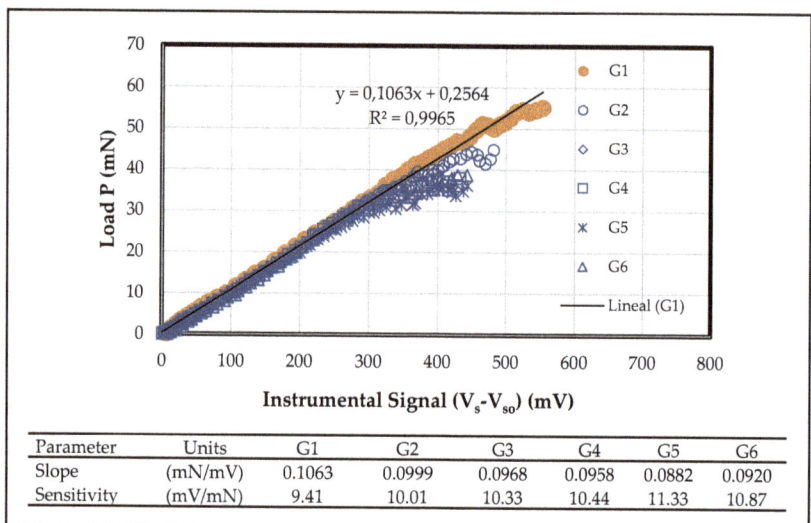

Figure 3. Relationship between instrumental signal and load P.

Currently, it is difficult to provide an accurate estimation of the resolution and the minimum detectable force for the proposed device. However, an appropriate estimate is that the precision for the force is 2.5 mN and for the device's output signal is 0.17 mV; therefore, the shear stress resolution would be 0.1 mN.

3. Results and Discussion

3.1. Load and Instrumental Signal Behavior in the Bending Tests

Figure 4 shows the typical development of the average curves of load and instrumental signal in the two-point bending tests relative to the displacement of the loading nose.

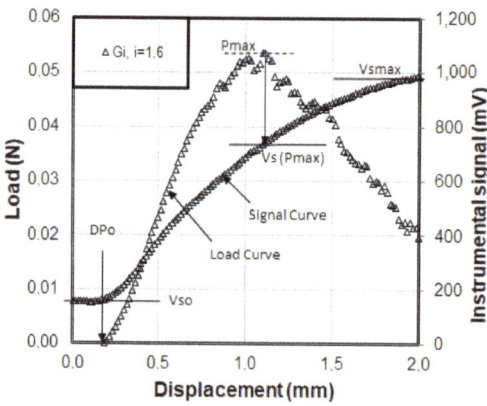

Figure 4. Generic parameters of the average load and instrumental signal curves from the bending tests.

The average and the variation coefficients of the generic parameters for each sensor, such as the maximum instrumental signal (V_{smax}), the instrumental signal at the beginning of the test (V_{so}), the instrumental signal at maximum load (V_s (P_{max})), the maximum load (P_{max}), and the displacement when the loading nose touches the strain gauge (DP_o) are determined and presented in Table 2.

Table 2. Experimental values of generic parameters for load and instrumental signal.

Sensor	V_{smax} (mV)	CV (%)	V_{so} (mV)	CV (%)	$V_s(P_{max})$ (mV)	CV (%)	P_{max} (N)	CV (%)	DP_o (mm)	CV (%)
G1	984.7	0.7	154.4	8.4	707.0	3.3	0.056	4.7	0.203	17.9
G2	703.5	0.3	284.2	0.2	670.4	6.0	0.018	2.0	0.216	32.5
G3	1159.6	0.5	348.0	3.1	872.8	0.5	0.046	5.1	0.092	38.2
G4	906.8	0.2	363.1	1.9	903.8	0.3	0.021	1.7	0.291	12.1
G5	1317.9	0.3	239.1	22.9	1052.3	3.7	0.037	7.0	0.233	30.4
G6	1746.4	0.0	876.5	7.5	1530.8	3.4	0.070	4.5	0.095	94.6
Average	1014.5	0.4	277.7	7.3	841.3	2.8	0.035	4.1	0.207	26.2

Different mean DP_o and V_{so} values were observed for each sensor and throughout the test repetitions, which can be attributed to experimental errors. These errors occurred due to difficulties in determining the moment when the surface of the strain gauge was touched by the loading nose and due to the initial settling of the measuring system of the strain gauge. Consequently, new uncertainties that have a significant impact on the behavior of the instrumental signal are posed in the estimation of V_{so} (see Table 3). Furthermore, the value for the maximum load (P_{max}) was consistent for each strain gauge in all of the repetitions (see Table 2). However, the magnitude of P_{max} differed significantly for each strain gauge, which is attributed to the position of the embedment and an inverse relationship between load and position of the loading nose [18]. Then, P_{max} diminished as the embedment moved away from the central axis of the sensing area of the strain gauge. Hence, when the distance of x_o increases, a lower total applied load is required to maintain the maximum bending moment constant in the area of embedment. Therefore, due to the dimensions of the tested material, new uncertainties concerning the measuring of x_o are introduced.

Table 3. Statistical parameters of the linear model between load and instrumental signal ($p < 0.8P_{max}$).

Sensor	Intercept (mN)				Slope (mN/mV)				R^2	Vsoe (mV)	Diff (%)
	Lower Limit	Value	Upper Limit	Prob-F	Lower Limit	Value	Upper Limit	Prob-F			
G1	−19.57	−18.98	−18.40	0.0000	0.1110	0.1124	0.1138	0.0000	0.999	168.96	9.4
G2	−15.40	−13.45	−11.50	0.0000	0.0529	0.0578	0.0627	0.0000	0.935	232.78	−18.1
G3	−42.91	−41.19	−39.48	0.0000	0.1083	0.1113	0.1143	0.0000	0.997	370.15	6.4
G4	−11.68	−10.78	−9.87	0.0000	0.0380	0.0397	0.0414	0.0000	0.977	271.46	−25.2
G5	−13.46	−12.65	−11.84	0.0000	0.0533	0.0546	0.0559	0.0000	0.997	231.62	−3.1
G6	−188.05	−170.55	−153.06	0.0000	0.1754	0.1917	0.2079	0.0000	0.975	889.82	1.5

In Table 2, the average values of the generic parameters are different for each sensor due to the characteristics of the tests presented in Table 1. This is attributed to the position of the embedment in the sensing area of the strain gauge and to the absence of a calibration process to adjust the display of the data acquisition system to zero.

The repeatability of the tests was determined due to the low value of the coefficients of variation (CV) observed for the instrumental signal (less than 1%) and the load (less than 7%), which validated the measuring systems. Thus, there is no influence of the measuring systems on the trends shown in the tests for each sensor, which implies that the variations or differences between them were only due to experimental errors.

The average performance of the load and the instrumental signal for each sensor in the two-point bending tests is shown in Figure 5a,b, respectively. In Figure 5, the displacement value considered as small-displacement limit (SDL) is shown by a vertical dashed line [29].

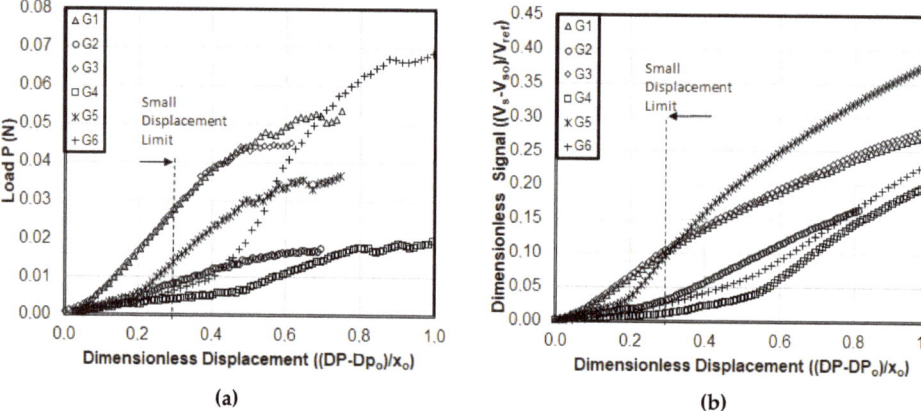

Figure 5. Average load and instrumental signal curves obtained from the two-point bending tests: (**a**) dimensionless load–displacement curve, (**b**) dimensionless instrumental signal–displacement curve.

In Figures 4 and 5, it is observed that the sensors were subjected to large lateral displacements through the two-point bending tests. Similarities and differences between the average load–displacement and instrumental signal–displacement curves were found. A non-linear relationship between both the load and the instrumental signal and the lateral displacement was observed in all tests performed. The test repetitions showed a consistent trend for the load and the instrumental signal, namely, a third-order polynomial trend line for the whole range of the lateral displacement of the sensor. Furthermore, it is observed that the slope of the load–displacement curve changes sign when the lateral displacement of the sensor exceeds the maximum load value (P_{max}). In contrast, the slope of the instrumental signal–displacement curve diminishes when the lateral displacement reaches its maximum value (V_{smax}) (see Figure 5b and Table 2).

Within the SDL, a linear relationship can be observed in almost all of the responses of the average load–displacement and instrumental signal–displacement curves. This linear relationship can be extended beyond the SDL value up to lateral displacements induced by loads close to 80% of the maximum load, owing to the flexibility of the material used (strain gauges). This is consistent with the classical beam theory or the plate theory, as well as with the hypothesis proposed in this study. In addition, minimum slopes for the load–displacement and instrumental signal–displacement curves were observed for loads lower than 20% of the maximum load. This is attributed to the initial settling of the material when the surface of the sensing area was touched by the loading nose (G1, G3, and G5) or when the sensor was turned from its initial position owing to the absence of a perfect embedment (G4 and G6). Moreover, for lateral displacements larger than 0.5 times the x_o value and that exceeded 80% of the maximum load, a loss of linearity was observed. This can be attributed to the loss of friction between the loading nose and the surface of the sensor. Additionally, the variation on the slopes of the curves can be attributed to the fact that the greater the lateral displacements to which the strain gauges were subjected, the lower the load required to maintain the maximum bending moment.

Finally, in Figure 5a,b, two types of behavior can be observed from the relationship between load–displacement and instrumental signal–displacement. This can be explained according to the position of the loading nose concerning the embedment of the strain gauge. The first type of behavior corresponded to an immediate response of the instrumental signal and the load on the sensor caused by the lateral displacement induced by the loading nose on those strain gauges embedded between 30% and 45% of the central axis of their sensing areas (sensors G1 and G3). The second type of behavior corresponded to a slow response of the instrumental signal and the load on the sensor when the embedment was located in the center or on the edge of the sensing area of the sensor (sensors G2, G4, and G6). Moreover, large lateral displacements were required to induce changes in the electrical resistance of the sensing area due to the absence of a perfect embedding of the strain gauge.

3.2. Relationship between Load and Instrumental Signal

Figure 6 shows the dimensionless average load–instrumental signal curve obtained in the two-point bending tests for the dataset about load and instrumental signal lower than the maximum load (P_{max}) at 1-s time intervals. Additionally, the figure shows that the SDL as a vertical dashed line, which corresponds to the ratio between the lateral displacement of the strain gauge and the position of application of the loading nose concerning the embedment, δ_B/x_o, equals 0.3 [29].

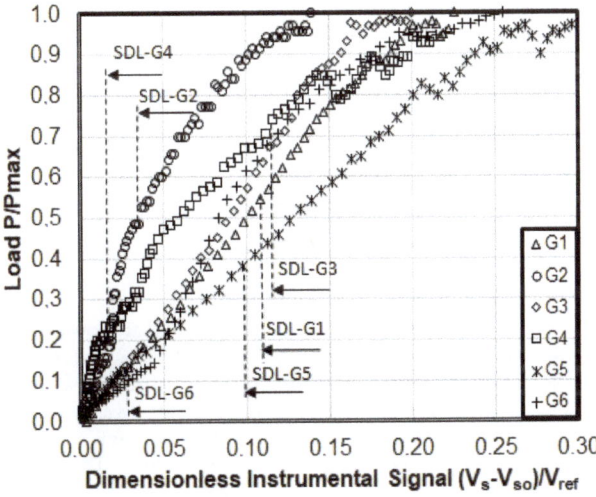

Figure 6. Dimensionless average load–instrumental signal curve from the bending tests using 1-s time-weighted average values.

For the loading range between 20 and 80 percent of maximum load, a linear model between the load (mN) and the instrumental signal (mV) was established. The parameters of the model, intercept (mN) and slope (mN/mV), are determined and presented in Table 3. The loading range was proposed to avoid the influence of the initial settling of the sensor, the effect of the loss of friction of the loading nose for large displacements, and the effect of the instrumental signal at the beginning of the test (V_{so}). In addition to this, in Table 3, the confidence interval and the probability of the F value of Fisher, obtained through the regression analysis, are presented for each parameter.

In Figure 6, a positive, dimensionless relationship between the load and the instrumental signal on each strain gauge is observed for the whole range of load values between zero and the maximum load [0, P_{max}]. In the bending tests, two trends in the load–instrumental signal relationship were identified, which are explained by the position of the embedment relative to the strain gauge. These trends coincide with the results obtained in Figure 4.

The first trend was a significant linear relationship that showed a coefficient of determination greater than 0.99 for all loading ranges, including the lateral displacements greater than the SDL. It corresponded to the embedment ratios between 0.3 and 0.45 (G1, G3, and G5). For these trends, the estimation error did not exceed 5% of the measured value (see Figure 7).

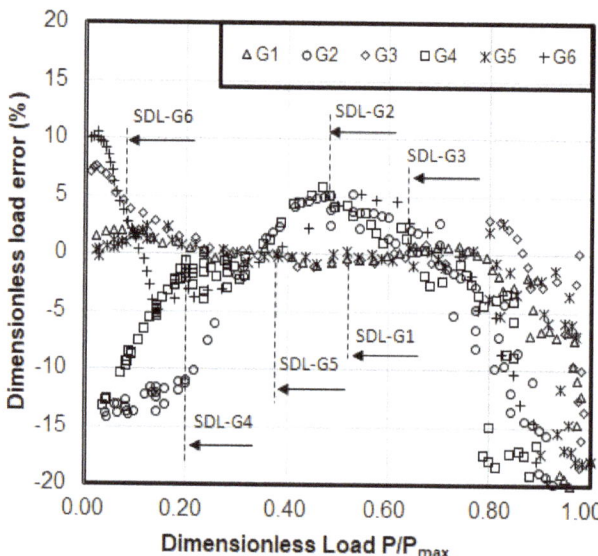

Figure 7. Estimation error of the dimensionless load, P/P_{max}.

The second trend corresponded to the strain gauges embedded on the edge or in the center of the sensing area (G2, G4, and G6) that showed a positive, non-linear relationship between load and instrumental signal. For small lateral displacements of the sensor, the linear trend between the load and the instrumental signal was significant. Nevertheless, the slope of the load–instrumental signal curve was two to four times greater than the slope of the first trend. This is due to the narrow range of variation of the instrumental signal compared to the range of variation of the load. This linear trend was significant for the whole loading range, but the coefficient of determination decreased to 0.93 (see Table 3) and the estimation errors exceeded 20% of the measured value (see Figure 7).

It is worthwhile to highlight that, in Figure 6, none of the sensors showed the initial settling of the measuring system in the bending tests illustrated in Figure 5. This can be attributed to the fact that the performance of the load sensor of the INSTRON is compared with the performance of the strain gauge itself, thereby annihilating the effects of the initial settling of the sensor and obtaining a linear relationship between the load and the instrumental signal from the beginning of the test. Thus, the loss of linearity of the load–instrumental signal curve for loads lower than 20% of the maximum load can be attributed to the absence of a perfect embedment of the sensor.

Additionally, for displacements greater than the SDL and close to 80% of the maximum load, an abrupt variation in the linear trend line, which has an error of overestimation greater than 15% of the measured value of the load, was observed (see Figure 7). This can be associated with the loss of friction between the loading nose and the surface of the strain gauge.

In Table 3, it is observed that all the regression coefficients (slopes) of the linear model rejected the null hypothesis. Furthermore, significant differences in the magnitude of the slope of the linear model were identified between sensors, which implies that the regression coefficient was individual for each one. The mean values for the slope of the linear model were lower for the sensors G1, G3, and G5, which had an embedment ratio of 0.30–0.45 from the axis of the sensing area of the strain gauge. The confidence intervals for these sensors were narrower than those of G2, G4, and G6. On the other hand, the intercept values for G2, G4, G5, and G6 were significantly different from zero. This is attributed to a shift in the initial voltage of the instrumental signal, which was different from the initial value of the instrumental signal at the beginning of the test only for sensor G1 (see

Tables 2 and 3). Although the difference between the values of the instrumental signal, V_{so}, was greater than 18% for sensors G2 and G4, it was not considered significant because of the broad confidence interval of the intercept due to the absence of a perfect embedment. Thus, the V_{so} value was an important source of uncertainty for the linear model that only influenced the intercept of the model. This is due to the settling of the sensor before taking the load and determining the moment when the loading nose touched the surface of the sensor. Nevertheless, the uncertainties regarding the coupling time between the load and the instrumental signal were reduced by using 1-s time-weighted average intervals.

On the other hand, Figure 7 shows the approximation error concerning the maximum load of the linear model between load and instrumental signal, according to the dimensionless load (P/P_{max}). In Figure 7, it can be noticed that the greatest relative errors of approximation of the linear model were presented for the strain gauges embedded on the edge or in the center of the sensing area (G2, G4, and G6), which implied an error in the specification of the regression model used. On the other hand, for the embedment ratios close to 0.30–0.45 from the central axis of the sensing area (G1, G2, and G4), the linear model adequately approached the data obtained between the values 0.2 and 0.8 of the dimensionless load and showed a relative error inferior to 5%. This can lead to the conclusion that the location of the embedment relative to the strain gauge was the main factor affecting the regression model. Provided that this embedment was located within 30% to 45% of the sensing area, a linear response between the load and the instrumental signal was obtained.

Alternatively, Figure 8 shows an inverse relationship between the slope of the load–instrumental signal curve and the location of the embedment concerning the central axis of the sensing area (embedment ratio). Figure 8 also shows an inverse relationship between the maximum load on each sensor and the embedment ratio. The foregoing is because both the maximum load and the slope of the load–instrumental signal curve were inversely related to the distance of application of the loading nose. This is because the location of the embedment (k) on the sensor was a fraction of the distance of application of the loading nose (x_o). Thus, both P_{max} and x_o influenced the values of the parameters of the linear regression model, intercept, and slope; however, they did not affect the specification of the model, which depended solely on the location of the embedment within the sensing area.

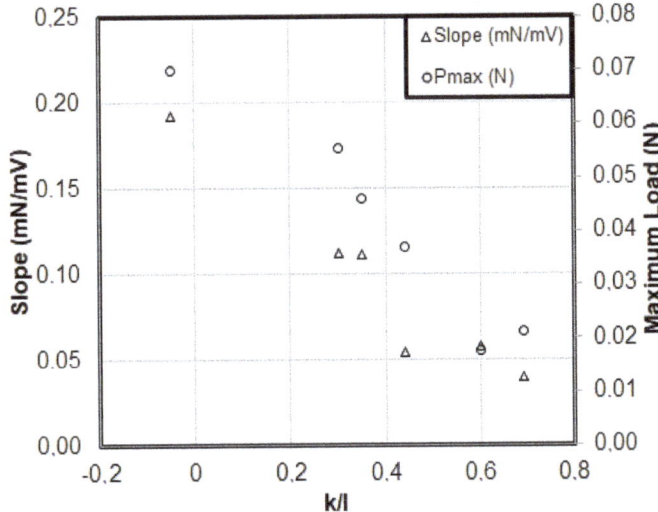

Figure 8. Slope and maximum load versus embedment ratio.

In summary, the sensors G1, G3, and G5 showed better performances concerning the linear model between the load and the instrumental signal, for both great and small displacements of the strain gauge when the sensor was embedded between 30% and 45% of the axis of the sensing area, and approximation errors were inferior to 5%. The experimental load–displacement and instrumental signal–displacement curves (Figure 5a,b, respectively) showed a third-order polynomial trend for loads between zero and the maximum load. To obtain a linear relationship between the load and the instrumental signal, it is required that the load–displacement and instrumental signal–displacement curves be collinear; that is to say, the coefficients of the third-order polynomial model need to be perfectly scaled; and thus, both the load–displacement curve and the instrumental signal–displacement curve can measure the same phenomenon proportionally.

3.3. Measurements in an Experimental Open Channel

The proposed method shows a new setup to use a widely studied piezoresistive force micro-sensor based on a strain gauge as it derives the shear stress from bending (e.g., Pommois et al. [30], Allen et al. [31], Wang et al. [11], and Hua et al. [32]) instead of pure axial deformation. Preliminary results in controlled laboratory experiments were able to measure small forces due to local flow changes, as we observed a significant linear relationship between the change in strain gauge signal and the relative fluid-sensor velocity for flow depths greater than 5 cm, while flow depth less than 2 cm did not show changes in the strain gauges [33]. The experimental setup considered a granular bed and sensors located in the vicinity of the bed to relate flow depth to dynamic pressure inside the boundary layer and the pressure variation of the surface flow. The channel used was a rectangular prismatic section 0.48 m wide and 0.45 m high with an acrylic bed and walls. The laboratory flume is of variable slope, 12 m long, and was left with zero slope for experimental activities. The experimental test protocol proposed by Sepúlveda [33] was used, which adapted the ISO 3455:2007 standard for windlass calibration to the characteristics of operation and functioning of the sensor. Sepúlveda [33] assumed valid the principle of action and reaction of the force sensor on liquid water at rest, where it is expected that the constant movement of the sensor along the channel is equivalent to the movement of the fluid with the sensor at rest.

Changes in the coefficient of determination between flow velocity and the signal from the strain gauge were independent of temperature changes, but a high scattering of the data between tests is attributable to experimental errors such as orthogonality and/or twisting of the sensor in the direction of motion. In other experimental runs considering a moving fluid, the proposed sensor detected changes in flow depth, flow magnitude, and flow surface oscillations. Sensors located at the bed and near the bed (5.0 mm) showed a sudden decrease in the signal and then increased exponentially until the signal became constant. In contrast, for the sensor located 35 mm from the bed, the signal did not show a sudden decrease but an exponential and oscillatory increase consistent with the flow pressure gradient, until it reached a plateau. The magnitude of the instrumental signal change correlated positively with flow magnitude only for the sensor located 35 mm from the bed in all tests. Differences in response related to the location are due to the presence and influence of the boundary layer at the bed that affects the response of the signal. The sensor located outside the boundary layer (free zone) showed a coherent development with the changes of the flow variables. For sensors located inside the boundary layer, the hydraulic load was the most relevant variable measured by the instrumental signal. Figure 9 shows some pictures of the experimental test in an open channel with water in motion.

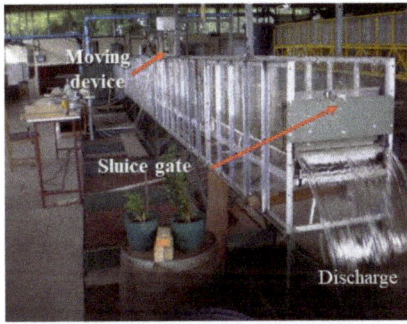  

Figure 9. Experimental test in an open channel.

4. Conclusions

The linear model between the load and the instrumental signal was significant for all the sensors in the two-point bending tests, even for displacements above the SDL value. Moreover, the results were under the classical theory only for small displacements.

The sensors embedded within 30 to 45 percent of the axis of the sensing area showed a linear trend between the load and the instrumental signal, an approximation error of the load inferior to 5 percent, and a coefficient of determination greater than 0.99. The sensors embedded on the edge and in the center of the sensing area of the strain gauge presented a significant linear trend, an approximation error of the load greater than 15 percent, and a coefficient of determination between 0.93 and 0.98. This was due to the error of specification of the model for these cases.

For loads inferior to 20 percent of the maximum load, a loss of linearity was observed in the bending tests, which is attributed to an initial shift because of a lack of perfect embedment; whereas for loads superior to 20 percent of the maximum load, a loss of linearity was observed that can be attributed to the loss of friction between the loading nose and the surface of the sensor.

The parameters of the linear model, intercept, and slope were observed to be unique for each sensor. The value of the slope was inversely proportional to the location of the embedment on the sensor. In contrast, the intercept was directly affected by the value of the initial instrumental signal, V_{so}, and constituted a source of error. Therefore, it is advisable to consider the V_{so} value as a parameter of adjustment of the regression equation to reduce the uncertainties in the linear model.

Finally, the strain gauge itself, subjected to pure bending, can be employed to measure forces of small magnitude, in the order of mN, through a linear relationship between the load and the instrumental signal, provided that it is embedded within 30 to 45 percent of the sensing area. However, to the best of the authors' knowledge, the present study is an isolated attempt to use strain gauges for the direct measurement of shear stress in open channels. Thus, it was not possible to compare the sensitivity results shown in the present research against other studies.

Future studies may consider a FEM study that may determine pressure sensors' sensitivity in terms of device dimensions and linearity or improve the understanding of the dependence of various geometrical parameters on the overall sensor performance. In addition, for future research, a set of simulations to prove the capability of the proposed sensors may be conducted.

Author Contributions: Conceptualization, L.S. and D.R.; methodology, L.S. and D.R.; validation, L.S., D.R., and E.F.; formal analysis, L.S. and D.R.; investigation, L.S., D.R., and E.F.; resources, L.S.; data curation, L.S. and D.R.; writing—original draft preparation, L.S. and D.R.; writing—review and editing, E.F. and L.S.; visualization, L.S.; supervision, L.S. and D.R.; project administration, L.S.; funding acquisition, L.S. and E.F. All authors have read and agreed to the published version of the manuscript.

Funding: The APC was funded by the Universidad del Bío-Bío, Chile (Funding Number VRAPROYE-UBB2055).

Institutional Review Board Statement: Not applicable.

Informed Consent Statement: Not applicable.

Data Availability Statement: Not applicable.

Acknowledgments: The authors want to acknowledge the support provided by the Department of Agroindustry of the College of Agricultural Engineering at the Universidad de Concepción, Chile and by the Department of Civil and Environmental Engineering at the Universidad del Bío-Bío, Chile. In addition, the first author would like to acknowledge the scholarship provided by the National Commission for Scientific and Technological Research (CONICYT) to carry out his doctoral studies. Finally, Diego Rivera thanks the support from the Water Research Center for Agriculture and Mining, CRHIAM (ANID/FONDAP/15130015).

Conflicts of Interest: The authors declare no conflict of interest.

References

1. Liang, Q.; Zhang, D.; Coppola, G.; Wang, Y.; Wei, S.; Ge, Y. Multi-Dimensional MEMS/Micro Sensor for Force and Moment Sensing: A Review. *IEEE Sens. J.* **2014**, *14*, 2643–2657. [CrossRef]
2. Cheng, N.-S.; Nguyen, H.T. Hydraulic Radius for Evaluating Resistance Induced by Simulated Emergent Vegetation in Open-Channel Flows. *J. Hydraul. Eng.* **2011**, *137*, 995–1004. [CrossRef]
3. D'Ippolito, A.; Calomino, F.; Alfonsi, G.; Lauria, A. Flow Resistance in Open Channel Due to Vegetation at Reach Scale: A Review. *Water* **2021**, *13*, 116. [CrossRef]
4. Tinoco, R.O.; San Juan, J.E.; Mullarney, J.C. Simplification bias: Lessons from laboratory and field experiments on flow through aquatic vegetation. *Earth Surf. Process. Landf.* **2020**, *45*, 121–143. [CrossRef]
5. Huai, W.; Li, S.; Katul, G.G.; Liu, M.; Yang, Z. Flow dynamics and sediment transport in vegetated rivers: A review. *J. Hydrodyn.* **2021**, *33*, 400–420. [CrossRef]
6. Errico, A.; Lama, G.F.C.; Francalanci, S.; Chirico, G.B.; Solari, L.; Preti, F. Flow dynamics and turbulence patterns in a drainage channel colonized by common reed (*Phragmites australis*) under different scenarios of vegetation management. *Ecol. Eng.* **2019**, *133*, 39–52. [CrossRef]
7. Duan, Y.; Zhong, Q.; Wang, G.; Zhang, P.; Li, D. Contributions of different scales of turbulent motions to the mean wall-shear stress in open channel flows at low-to-moderate Reynolds numbers. *J. Fluid Mech.* **2021**, *918*, A40. [CrossRef]
8. Bashirzadeh, Y.; Qian, S.; Maruthamuthu, V. Non-intrusive measurement of wall shear stress in flow channels. *Sens. Actuators A Phys.* **2018**, *271*, 118–123. [CrossRef]
9. Naughton, J.; Sheplak, M. Modern skin friction measurement techniques—Description, use, and what to do with the data. In Proceedings of the 21st Aerodynamic Measurement Technology and Ground Testing Conference, Denver, CO, USA, 19–22 June 2000; American Institute of Aeronautics and Astronautics: Reston, VI, USA, 2000; pp. 1–27.
10. Örlü, R.; Vinuesa, R. Instantaneous wall-shear-stress measurements: Advances and application to near-wall extreme events. *Meas. Sci. Technol.* **2020**, *31*, 1–19. [CrossRef]
11. Wang, J.; Pan, C.; Wang, J. Characteristics of fluctuating wall-shear stress in a turbulent boundary layer at low-to-moderate Reynolds number. *Phys. Rev. Fluids* **2020**, *5*, 074605. [CrossRef]
12. Sheplak, M.; Cattafesta, L.; Nishida, T.; McGinley, C. MEMS Shear Stress Sensors: Promise and Progress. In Proceedings of the 24th AIAA Aerodynamic Measurement Technology and Ground Testing Conference, Portland, OR, USA, 28 June–1 July 2004; American Institute of Aeronautics and Astronautics: Reston, VI, USA, 2004; p. 13.
13. Wei, J. Silicon MEMS for Detection of Liquid and Solid Fronts. Ph.D. Thesis, Delft University of Technology, Delft, The Netherlands, 2010.
14. Haritonidis, J.H. The Measurement of Wall Shear Stress. In *Advances in Fluid Mechanics Measurement*; Springer: Berlin/Heidelberg, German, 1989; pp. 229–261.
15. Kolitawong, C.; Giacomin, A.J.; Johnson, L.M. Invited Article: Local shear stress transduction. *Rev. Sci. Instrum.* **2010**, *81*, 1–20. [CrossRef] [PubMed]
16. Patel, B.; Srinivas, A.R. Validation of experimental strain measurement technique and development of force transducer. *Int. J. Sci. Eng. Res.* **2012**, *3*, 1–4.
17. ASTM D747-10; Standard Test Method for Apparent Bending Modulus of Plastics by Means of a Cantilever Beam. ASTM International: West Conshohocken, PA, USA, 2019.
18. Timoshenko, S.; Gere, J.M. *Mechanics of Materials*; Van Nostrand Reinhold, Co.: New York, NY, USA, 1972.
19. Ortiz Berrocal, L. *Resistencia de Materiales*, 3rd ed.; McGraw-Hill Interamericana de España S.L.: Madrid, Spain, 2007; ISBN 844-815-6351.

20. Tian, B.; Zhong, Y.; Li, R. Analytic bending solutions of rectangular cantilever thin plates. *Arch. Civ. Mech. Eng.* **2011**, *11*, 1043–1052. [CrossRef]
21. Hahn, R.; Rosentreter, E.E. *ASAE Standards 1991: Standards, Engineering Practices and Data*; American Society of Agricultural Engineers: St. Joseph, MI, USA, 1991; ISBN 092-9355-13X.
22. Müller, I.; Machado, R.; Pereira, C.; Brusamarello, V. Load cells in force sensing analysis—Theory and a novel application. *IEEE Instrum. Meas. Mag.* **2010**, *13*, 15–19. [CrossRef]
23. Palomo, F.; Vega-Leal, A.; Galván, E. *Problemas Resueltos de Instrumentación Electrónica*, 3rd ed.; Ingeniería–Universidad de Sevilla: Sevilla, Spain, 2013; ISBN 978-84-472-1061-9.
24. Gujarati, D.; Porter, D.C. *Basic Econometrics*, 5th ed.; McGraw-Hill Education: New York, NY, USA, 2008; ISBN 007-3375-772.
25. Knight, D.W.; Sterling, M. Boundary Shear in Circular Pipes Running Partially Full. *J. Hydraul. Eng.* **2000**, *126*, 263–275. [CrossRef]
26. Lashkar-Ar, B.; Fathi-Mogh, M. Wall and Bed Shear Forces in Open Channels. *Res. J. Phys.* **2010**, *4*, 1–10. [CrossRef]
27. Pan, J.; Shen, H.T.; Cheng, N.-S. Bed and wall shear stresses in smooth rectangular channels. *J. Hydraul. Res.* **2021**, *59*, 847–857. [CrossRef]
28. Singh, P.K.; Banerjee, S.; Naik, B.; Kumar, A.; Khatua, K.K. Lateral distribution of depth average velocity & boundary shear stress in a gravel bed open channel flow. *ISH J. Hydraul. Eng.* **2021**, *27*, 23–37. [CrossRef]
29. ISO 14125:1998. *Fibre-Reinforced Plastic Composites: Determination of Flexural Properties*; ISO: Geneva, Switzerland, 2013; p. 18.
30. Pommois, R.; Furusawa, G.; Kosuge, T.; Yasunaga, S.; Hanawa, H.; Takahashi, H.; Kan, T.; Aoyama, H. Micro Water Flow Measurement Using a Temperature-Compensated MEMS Piezoresistive Cantilever. *Micromachines* **2020**, *11*, 647. [CrossRef]
31. Allen, N.; Wood, D.; Rosamond, M.; Sims-Williams, D. Fabrication of an in-plane SU-8 cantilever with integrated strain gauge for wall shear stress measurements in fluid flows. *Procedia Chem.* **2009**, *1*, 923–926. [CrossRef]
32. Hua, D.; Suzuki, H.; Mochizuki, S. Local wall shear stress measurements with a thin plate submerged in the sublayer in wall turbulent flows. *Exp. Fluids* **2017**, *58*, 124. [CrossRef]
33. Sepúlveda, R. *Calibración de Sensores de Fuerza Inmersos en Agua en Un Canal de Laboratorio*; University of Bío-Bío: Concepción, Chile, 2013.

Article

Analysis of Fine Crack Images Using Image Processing Technique and High-Resolution Camera

Hoseong Jeong [1], Baekeun Jeong [2], Myounghee Han [2] and Dooyong Cho [2,*]

[1] Institution of Agricultural Science, Chungnam National University, 99 Daehak-ro, Daejeon 34134, Korea; hsjeong@cnu.ac.kr

[2] Department of Convergence System Engineering, Chungnam National University, 99 Daehak-ro, Daejeon 34134, Korea; bridgeworld@naver.com (B.J.); mhhan73@gmail.com (M.H.)

* Correspondence: dooyongcho@cnu.ac.kr; Tel.: +82-42-821-5693

Featured Application: Analysis of Fine Cracks.

Abstract: Visual inspections are performed to investigate cracks in concrete infrastructure. These activities require manpower or equipment such as articulated ladders. Additionally, there are health and safety issues because some structures have low accessibility. To deal with these problems, crack measurement with digital images and digital image processing (DIP) techniques have been adopted in various studies. The objective of this experimental study is to evaluate the optical limit of digital camera lenses as working distance increases. Three different lenses and two digital cameras were used to capture images of lines ranging from 0.1 to 0.5 mm in thickness. As a result of the experiments, it was found that many elements affect width measurement. However, crack width measurement is dependent on the measured pixel values. To accurately measure width, the measured pixel values must be in decimal units, but that is theoretically impossible. According to the results, in the case of 0.3 mm wide or wider cracks, a working distance of 1 m was secured when the focal length was 50 mm, and working distances of 3 m and 4 m were secured when the focal length was 100 mm and 135 mm, respectively. However, for cracks not wider than 0.1 mm, focal lengths of 100 mm and 135 mm showed measurability within 1 m, but a focal length of 50 mm was judged to hardly enable measurement except for certain working positions. Field measurement tests were conducted to verify measurement parameters identified by the results of the indoor experiment. The widths of actual cracks were measured through visual inspection and used for the analysis. From the evaluation, it was confirmed that the number of pixels corresponding to the working distance had a great influence on crack width measurement accuracy when using image processing. Therefore, the optimal distance and measurement guidelines required for the measurement of the size of certain objects was presented for the imaging equipment and optical equipment applied in this study.

Keywords: crack measurement; image processing; high resolution; working distance; crack width; visual inspection; structure state assessment; ground sample distance (GSD)

1. Introduction

Structures undergo deterioration and damage over time due to various external forces and environmental factors. Among this damage, cracks are indicators that inform us about the current state of structures well. Cracks are usually measured by visual inspection using manpower.

Structures that cannot be easily accessed, such as civil engineering structures, are inspected by manpower using equipment (such as articulated ladders). The difficulty in finding cracks during inspections and the input of professional manpower increases costs and requires large amounts of time. Furthermore, in the case of the lower part of a bridge, nuclear power plant facilities, and dams, which are poorly accessible, it is hard to secure stability because workers have to deal with poor work environments. Moreover, there are problems such as poor objectivity and errors occurring in the detection of defects due to the worker's subjective evaluation depending on their working hours, experience, personal ability, and environmental factors. In addition, for the efficient maintenance of structures, preventive maintenance is required with timely repair and reinforcement rather than one-time, large-scale reinforcement. However, frequent inspections are difficult due to limited maintenance budgets, and this makes preventive maintenance difficult.

To remedy the above-mentioned problems, digital image processing (DIP) technologies using imaging equipment have been introduced, and various crack detection algorithms have been developed over the past several decades [1–3]. These technologies are expected to minimize the subjective evaluation of inspectors, thereby securing the objectivity of inspection results, and to relieve constraints on inspection timing and location, as well as solve problems such as the large manpower and cost required for existing inspections and investigations [4]. However, early studies using imaging equipment and image processing techniques used simple image processing technologies or were limited to crack information extraction techniques and had limitations in the measurement of the characteristic information (width, length, shape, etc.) of cracks due to low resolution. Early studies on image processing were mainly those that installed charge-coupled device (CCD) cameras [5], ground penetration radar (GPR), laser systems (LS), or hybrid systems (HS) [6,7], which are classified as non-contact evaluation equipment, on a vehicle such as a van to develop image recognition equipment and equipment that inspects and measures cracks on road surfaces [8]. In related studies on image processing techniques such as ones that applied the Otsu method [9], which is based on shooting variables and threshold value settings, to concrete structures such as roads, tunnels, and bridges to detect and measure cracks [10–13], image preprocessing studies have been conducted using histograms, mean values, median values, and Gaussian filters. Studies on feature extraction using morphology techniques [14–16] and studies on contour detection through primary and secondary differentiation [17,18] have also been conducted and repeatedly developed. Thereafter, the field of applicable target structures in studies conducted on crack detection was expanded to include bridges, tunnels, and dams that cannot be easily accessed.

The enhancement of the resolution of imaging equipment and the development of optical equipment has enabled the detection of cracks from long distances, which has led to studies on crack measurement. In 2012, Lee detected cracks in structures and measured the widths, lengths, and directions of the cracks using an 18-megapixel camera and a 600 mm lens, and concluded that 0.2 mm cracks can be detected at a working distance of 40 m [19]. In 2014, Li secured a working distance of 60 m in a study that applied a focal length of 1000 mm using a 20-megapixel video camera, a 500 mm lens, and a 2X converter, and showed an error rate of 92.6% [20]. Jahanshahi measured virtual images of 0.4 to 2.0 mm wide cracks using a 17-megapixel imaging device while changing the working distance from 725 to 1760 mm. He analyzed the filmed images and reported that increases in the working distance led to increases in the ground sample distance (GSD) of the images, that the accuracy of crack width measurement decreased as the number of pixels representing cracks decreased, and that to enhance the accuracy of crack measurement, the working distance should be reduced, the focal length should be increased, and the resolution should be enhanced [21,22]. Lins proposed a crack detection and measurement algorithm to automatically measure cracks and measured the crack widths of experimental subjects and actual structures. The measured crack sizes ranged from 1.22 mm to 30 mm and error rates of 7–8% were shown [23]. Khalili used 12-megapixel imaging equipment to measure 15.1 mm long cracks. He proposed a morphological operator-based algorithm and made efforts to reduce measurement errors [24]. Most crack measurement studies in the past dealt with cracks greater than 0.5 mm, and studies on microcracks smaller than 0.3 mm have been limited.

Recently, studies that linked unmanned aerial vehicles (UAV) [25,26] with machine-learning-based image processing methods such as neural networks [27,28] to detect or measure cracks in concrete structures have been steadily increasing. Various kinds of imaging equipment are applied according to the operational limits of unmanned aerial vehicles. To acquire crack information, the pixel size of the video image according to the working distance must be known. However, the resolution of imaging equipment and optical equipment mounted on general unmanned aerial vehicles and the precision of GPS cause limitations in acquiring crack information, and related studies are insufficient.

In this paper, an indoor measurement experiment was conducted to measure cracks in concrete structures and present guidelines for the image acquisition and processing of imaging equipment. Two units of DSLR video equipment, which is applied in various fields, were selected, and both were equipped with full-frame image sensors. In addition, the focal length was changed using 50 mm, 100 mm, and 135 mm lenses. In order to derive the crack measurement limits and optimal working distances of the selected imaging equipment and optical equipment to analyze precision according to differences in resolution, shooting was conducted while changing the working distance at intervals of 10 mm. In the experiment, virtual images equivalent to microcracks not larger than 0.3 mm were fabricated, and the images were acquired. The pixel values of these images were measured with Otsu's image processing technique through a preprocessing process, and the areas of the virtual images were obtained by applying the *GSD* values according to the working distance. The effects of the pixel values measured through the respective imaging equipment and optical equipment on the measurement of crack width were analyzed with theoretical and experimental approaches.

2. Geometric Relationships of Image Acquisition

Measurement methods using imaging equipment measure the sizes of objects in an image coordinate system by analyzing the correlation between the size of the real object and the image. Therefore, every measurement unit has the characteristic of having pixel units constituting an image. Measurement methods using acquired images require a process of converting the measured pixel value into an engineering unit.

As can be seen in Figure 1, image acquisition applies the general lens principle. A digital camera uses an image sensor such as a charge-coupled device (CCD) or a complementary metal-oxide-semiconductor (CMOS). It shows the relationship between the object plane that shows the object and the image plane that shows the image based on the image sensor in the camera. Here, the relationship between the captured image and the real object is determined using the ratio of the distance (*O*) from the measured object to the center point of the lens to the distance (*i*) from the camera lens to the surface of the image sensor, or the measured value in pixel units can be converted into actual length units by using the number of pixels (*l*) of the image that corresponds to the actual size (*L*) of the object.

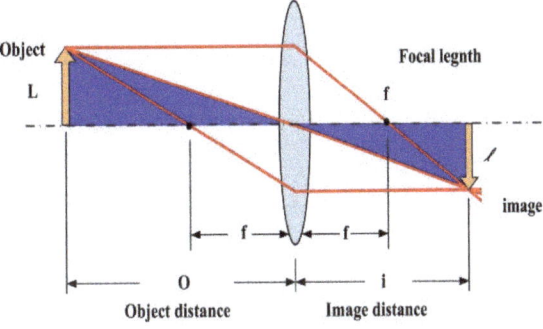

Figure 1. The principle of a thin lens.

Ground sample distance (*GSD*) or ground resolution refers to the size of one pixel corresponding to the actual size of the object. The unit used for the foregoing is mm/pixel, and it can be obtained using the size and resolution of the camera sensor, the focal length of the lens, and the working distance. The distance conversion coefficient *GSD* can be generalized through the lens principle from previous researchers and the Gaussian imaging equation.

The Gaussian imaging equation is as shown in Equation (1), and Equation (2) below can be obtained through the geometric relationship between the size of the object, the number of pixels in the image, the distance from the lens to the object, and the distance from the lens to the image.

$$\frac{1}{O} + \frac{1}{i} = \frac{1}{f} \quad (1)$$

$$L = \frac{O-f}{f} \times l \quad (2)$$

where O is the distance (mm) from the lens to the object, which is the working distance; i is the distance (mm) from the lens to the image sensor; f is the length (mm); L is the size (mm) of the object; and l is the number of pixels in the image corresponding to the object.

When an object is photographed using a camera, an image of the object is formed on the image sensor. In this case, the size of the object varies according to the size and resolution of the image sensor. This can be defined as the image-forming principle and is illustrated in Figure 2. The size of the image sensor is shown as $w \times h$ (mm), and the resolution is expressed as $W_{re} \times H_{re}$ (pixels).

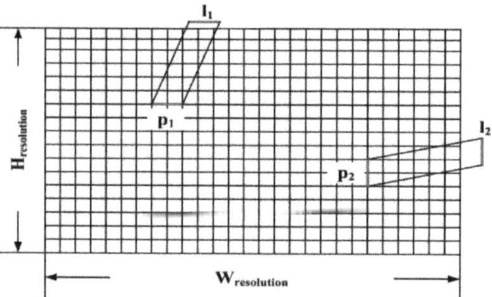

Figure 2. Image-forming principle of the sensor.

In this case, when the number of measured pixels is expressed as p_1, and the corresponding image length is denoted as l_1, the following Equation (3) can be obtained from the correlation between the horizontal and vertical sizes of the image sensor and the resolution.

$$\frac{W}{w} = \frac{p_1}{l_1} \text{ or } \frac{H}{h} = \frac{p_2}{l_2} \quad (3)$$

where W and H are the horizontal and vertical resolution values (pixels) of the sensor and w and h are the horizontal and vertical dimensions (mm) of the sensor. If Equation (3) is substituted into Equation (2), vertical or horizontal length is defined as Equation (4) below.

$$L = \frac{O-f}{f} \times \frac{w \times p_1}{W} \quad (4)$$

Basically, since one pixel is a square, the value of the size does not change according to the direction. Therefore, there is no need to use the values for different directions separately. Equation (4) is an equation for crack measurement and can be changed and expressed as shown by the following Equation (5).

$$GSD = \frac{L-f}{f} \times \frac{S_s}{S_R}, \ CW(crack\ width) = GSD \times N_m \quad (5)$$

where L is working distance, f is the focal length, S_s is the horizontal or vertical dimensions (mm) of the sensor, and S_R is the horizontal or vertical resolution values (pixels) of the sensor. *CW* is the crack width, *GSD* is the ground sample distance, and N_m is the number of measured pixels.

3. Width Measurement Using Image Processing

3.1. Overview of Indoor Measurement Experiments

3.1.1. Virtual Image Production and Measurement Experimental Variables

Cracks occur in structures over time due to various external forces and structural and environmental factors, and vary widely in width, length, and shape. In order to properly measure a crack, the pixel value of the video image must correspond to the value of the actual crack width. Therefore, in this experiment, linear virtual images were produced to substitute for the sizes of microcracks. As can be seen in Figure 3, the images were produced to improve the contrast between the lines and the background, and the lines were produced in 0.1, 0.3, and 0.5 mm widths. As experimental measurement variables, the pixel values and widths of each virtual image were measured according to the differences in the resolution of the imaging equipment at the same focal length and changes in the focal length at the same resolution of the imaging equipment, as shown in Table 1.

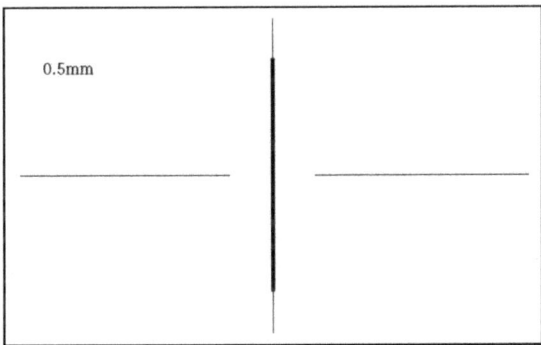

Figure 3. Test image.

Table 1. Measurement experimental variables.

Trial	Manufacturer	Focal Length (mm)	Width (mm)	Resolution (million)
CC50	Canon	50		22.1
SC50		50	0.1, 0.3, and 0.5	
SC100	Sony	100		61.0
SC135		135		

OC OO - OO
① ② ③

① CC: Canon Camera; SC: Sony camera
② Focal length
③ Value of width

3.1.2. Crack Measurement Limits

In many countries, the standard for the assessment of states regarding crack widths is stipulated as 0.5 mm or smaller for the prevention of corrosion, the improvement of durability, and aesthetic reasons. ACI-318 regulations stipulate the standard as 0.41 mm for indoor exposure, 0.3 mm for outdoor exposure, and 0.2 mm for extreme environments. British standards (BS) 8110 stipulate that cracks should not exceed 0.3 mm in a general environment, and South Korea stipulates five levels of standards for the assessment of states regarding the maximum crack widths in visual inspections of various structures. Among them, in the case of grade B, the standards are presented as 0.1 to 0.3 mm for reinforced concrete and as smaller than 0.2 mm for prestressed concrete. In this study, the allowable limit for crack width detection and measurement was set to 0.1 mm to 0.3 mm, and an experimental plan was established for the selection of optimum image measurement equipment and the determination of shooting conditions and image processing techniques, such as the degrees of recognition of cracks according to working distances and the extraction of cracks with limits.

3.1.3. Image Equipment and Optical Equipment (Lens)

The most important thing in the use of imaging equipment and the crack measurement system using image processing techniques is selecting a high-definition digital camera. Once the crack measurement limit range has been set, the imaging equipment must be selected in consideration of the camera resolution, sensor size, and shutter speed. The imaging equipment used in this study was Sony's mirrorless A7RIV and Canon's EOS 5D Mark III, which were installed with image sensors in full-frame 35.7×23.8 mm^2 and 36×24 mm^2 sizes with resolutions of 61 million (9504×6336) and 22.1 million (5760×3840) pixels, respectively.

The lens collects light and sends it to the image sensor, and the working distance is secured according to the specifications of the lens. The types of lenses can be divided into wide-angle, telephoto, standard, and prime. Wide-angle lenses enable shooting wider background ranges compared to standard lenses and shooting large pictures of subjects that cannot be approached from distant locations. Prime lenses are only suitable for either long distances or short distances, thereby minimizing the problems above. The focal length of the lens necessary to satisfy accuracy in the determined working distance range can be obtained by using Equation (5). Figure 4 below shows the focal lengths according to the working distances necessary to acquire the size of the subject clearly and accurately. In this study, lenses that can optimize the performance of digital high-definition cameras were selected. The applied lenses consisted of 50, 100, and 135 mm single vision lenses selected according to manufacturer, and the specifications of the lenses are shown in Table 2. As for the range of the final working distances, the distances to the points where the values of the size of one pixel corresponding to individual line widths become the same were set as the measurement limit distances.

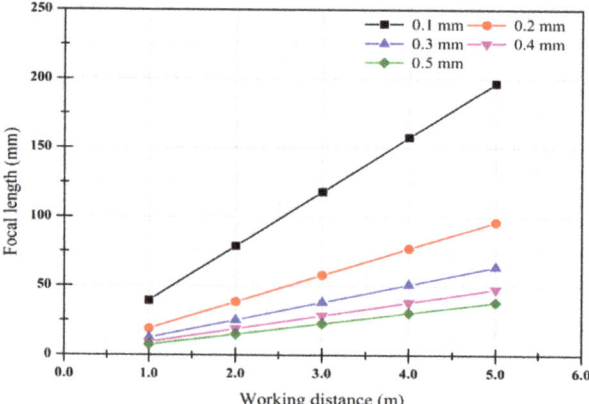

Figure 4. Focal length satisfying width measurement.

Table 2. Lens specification.

Classification	Sony			Canon
Focal length	50 mm	100 mm	135 mm	50 mm
Min. focus distance	0.45 m	0.57 m (0.57–1 m) 0.85 m (0.85–∞ m)	0.7 m	0.35 m
Weight	778 g	700 g	950 g	160 g
Angle of view	47°	24°	18°	46°

3.2. Image Acquisition

Imaging equipment, such as a camera, reacts sensitively to the amount of light. Good images can be obtained by adjusting the amount of light to the optimum by changing the aperture value or shutter speed value. However, with regard to filming outdoors, it is very difficult to secure images of consistent quality because the amount of light changes from moment to moment due to the weather, environmental conditions, and sunlight. In addition, not only does the brightness of a video image vary depending on the direction of light, but shadows may also be misprocessed as cracks during image processing because of geographic conditions present in the image. To remove the factors that cause errors in advance and keep the image quality consistent, an indoor experiment was conducted under LED lighting. The average illuminance of the LED lights in the laboratory was measured using a portable illuminance meter and was determined to be 1455 lux.

There were five image shooting methods: fully automatic shooting, shutter priority, aperture priority, program auto, and manual exposure modes. In the case of the fully automatic shooting mode, everything in the camera is set to automatic, and as for shutter and aperture priority-based modes, the shutter and aperture are adjusted to obtain an exposure value that corresponds to the brightness of the subject. In general, whereas the shutter speed is adjusted when shooting dynamic images, an aperture priority-based shooting method is used when shooting static images. In this experiment, an aperture priority-based shooting method was used to obtain clear video images. As can be seen in Figure 5, when the aperture value is low, the foreground part in focus is clear, but the background is blurred, and when the aperture value is high, the overall video image is clear.

(a) Value of aperture: 2.8   (b) Value of aperture: 22

Figure 5. Experiment condition.

The working distance was measured by measuring the distance from the virtual image plane to the position of the camera sensor using an aluminum staff used for level measuring. As can be seen in Figure 6a, the position of the camera sensor is indicated by the focal plane mark, a Saturn-shaped symbol on the top of the camera. Since it was difficult to measure the sensor position, a three-axis self-leveling laser was used to match the position of the sensor with the position of the staff scale, and the distance between the virtual image plane and the sensor plane was measured as shown in Figure 6b. The equipment and working distance measuring method used in the experiment are shown in Figure 6c,d. Feng et al. (2015) conducted numerical studies to quantify the error resulting from camera non-perpendicularity. The range of the optical axis tilt angle was from 0° to 30°, and they determined that the error increases as the tilt angle increases. However, they discovered that a measurement error of less than 1% was determined at a 30 degree optical axis tilt angle, and this measurement error is acceptable [29]. Therefore, the orthogonality between the image plane and object surface was not considered in this study. Figure 7 shows an overall view of the indoor measurement experiment, and image shooting was carried out until the final working distance of each linear virtual image was reached, while changing the working distance in units of 10 mm to acquire the images.

(**a**) Position of sensor

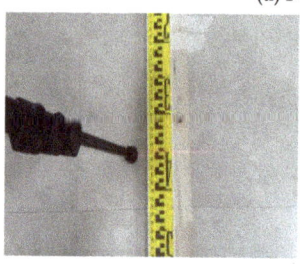 

(**b**) Method of working distance measurement

 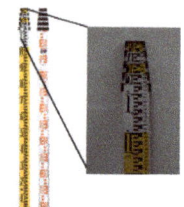

(**c**) Self-leveling 3-beam laser (**d**) Level staff

Figure 6. Experimental methods and apparatus.

3.3. Image Processing

First, for local rather than global image processing of acquired images, regions of interest (ROI) were set, and to make the images in the ROIs clearer, preprocessing processes such as image enhancement and sharpening were carried out. For image processing, the image processing toolbox (IPT) of the Matlab program of Mathworks Co. was used.

The IPT provides a comprehensive set of standard algorithms, functions, applications for image processing, and image processing techniques. Since the linear virtual images used in the indoor experiment have clear contrasts between the linear images and the background, a simple image processing technique was applied to measure the pixel value of the line width. As for the applied image processing techniques, color images were first converted into grayscale images, and then the grayscale images were converted into binary images using Otsu's method. Then, pixel values corresponding to the width of the linear virtual images were obtained.

Figure 7. Indoor test setting.

4. Measurement Results and Analysis

4.1. Pixel Values According to Working Distance

4.1.1. Measured Pixel Values According to Changes in Resolution

To verify the measurement limits of imaging equipment according to differences in resolution, lenses with the same focal length were applied, and lenses suitable for individual manufacturers were applied. As presented above, the names of imaging equipment units were written as SC (Sony camera) and CC (Canon camera) to improve understanding of the study findings and analysis values. The pixel values from the video images acquired using each imaging equipment unit, measured using the Matlab program, are shown in Figures 8 and 9. According to the physical specifications of the applied single-vision lens with a focal length of 50 mm, working distances ranging from the minimum distances of 300 mm (CC) and 400 mm (SC) to the working distances at which the pixel sizes become the same as the sizes of the virtual images are shown. A relative comparison was carried out at a working distance of 400 mm, which is the minimum distance of the SC equipment.

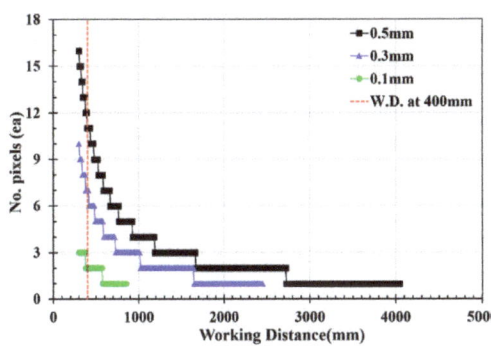
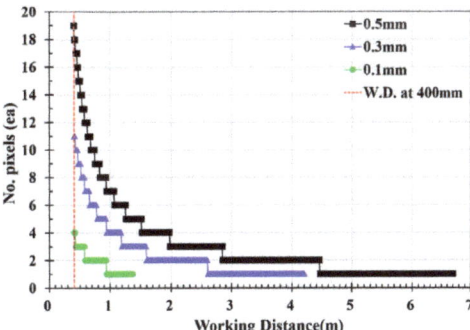

(**a**) CC using focal length of 50 mm   (**b**) SC using focal length of 50 mm

Figure 8. Results of pixel measurement by each manufacturer.

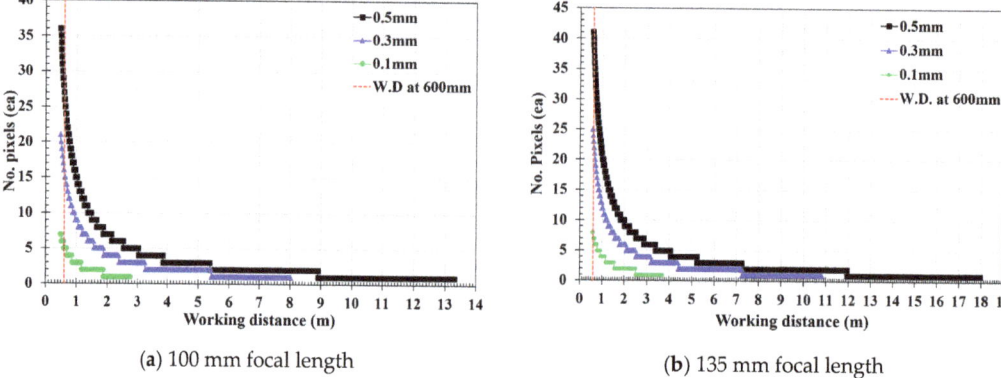

(a) 100 mm focal length　　　　　(b) 135 mm focal length

Figure 9. Results of pixel measurement applying different focal lengths using SC.

The pixel values measured using CC50 and SC50 at a working distance of 400 mm are shown in Table 3 and Figure 8. The pixel values measured while the width value increased by five times from 0.1 mm to 0.5 mm show a resulting increase by 5.5 times from 2 to 11 in the case of CC, and by 4.75 times from 4 to 19 in the case of SC, at a working distance of 400 mm. When the width value was 0.1 mm, the relevant working distance was 850 mm in the case of CC and 1370 mm in the case of SC, with an increase of 1.61 times, and when the width value was 0.3 mm, the relevant working distance increased by 1.63 times from 2450 to 4010 mm, and when the width value was 0.5 mm, the relevant working distance increased by 1.64 times from 4050 to 6650 mm. Therefore, it was confirmed that the number of pixels measured and the limitation of working distance increased dependence on the resolution of the image sensor and the size of the object. Regardless of the proportional relationship for all the variables, it was found that the number of pixels acquired exponentially decreases as the working distance increases. Additionally, the number of pixels at the minimum working distance does not proportionally increase as the physical dimensions of an object increase.

Table 3. The number of pixels measured by resolution differences at 400 mm working distance.

Classification	Equivalent Focal Length	
	Number of Pixels (Each)	
	CC50	SC50
0.1 mm	2	4
0.3 mm	8	11
0.5 mm	11	19

4.1.2. Measured Pixel Values According to Changes in Focal Length

In order to identify the measured pixel values according to changes in the focal length, 100 and 135 mm were applied to the SC equipment as focal lengths, and the results are shown in Figure 9 below. With regard to SC100-0.1, a pixel value of 7 was obtained through measurement at the minimum working distance of 470 mm, and pixel values of 21 and 36 were identified for SC100-0.3 and SC100-0.5, respectively. In the case of SC135-0.1, 0.3, and 0.5, pixel values of 8, 25, and 41, respectively, were obtained when measured at the minimum working distance of 565 mm. Relative comparisons with the results for SC50 were carried out. Since the minimum working distance varied depending on the physical characteristics of the lens, the comparisons were carried out at the same working distance of 600 mm. Compared to SC50-0.1, a 2.5 times higher value was obtained for SC100-0.1, and a four times higher value was obtained for SC135-0.1. In the case of a 0.3 mm virtual image, approximately 2.29 times and 3.57 times higher values were obtained, respectively, and with regard to a 0.5 mm virtual image, approximately 2.16 times and 3.16 times higher values were obtained, as can be seen in Table 4.

Table 4. The number of pixels measured by focal length changes at a working distance of 600 mm.

Classification	Equivalent Resolution		
	Number of Pixels (Each)		
	SC50	SC100	SC135
0.1 mm	2(4)	5(7)	8(8)
0.3 mm	7(11)	16(21)	23(25)
0.5 mm	12(19)	26(36)	38(41)

( ) The number of measurement pixels according to each focal length at min. working distance.

4.2. Measured Values of Line Widths According to Working Distance

4.2.1. Measured Values of Line Widths according to Differences in the Resolution of Imaging Equipment

As written above, the *GSD* of the pixel value measured by each variable through image processing was calculated according to the working distance to measure the width value of the virtual image. Figure 10 shows the width values of individual virtual images with the same focal length of 50 mm applied, measured by working distance using individual units of imaging equipment. Here, the ground truth represents the width value of the actual virtual image, and the shaded part of the red box is the percentage within the limit of error, which is less than 10% (PWL-10). In addition, the working position where the size of one measured pixel has the same value as the width of the virtual image was expressed as the measurement limit distance of the imaging equipment.

As a result of measurement of the width of CC50-0.1 in Figure 10a, there were 56 working positions within the measurement limit distance of 850 mm, and the number of working positions that had values within PWL-10 was 22, accounting for 39.29%. However, in the case of SC50-0.1, the number of working positions that had values within PWL-10 was 38, accounting for 38.77% of the 98 working positions within the measurement limit distance.

With regard to CC50-0.3, the number of working positions within the measurement limit distance of 2450 mm was 216, and the number of working positions that had values within PWL-10 was 105, which is 48.6% in ratio. However, for SC50-0.3, the number of working positions that had values within PWL-10 within the measurement limit distance of 4010 mm was 178, accounting for 49.2% of the entire 362 working positions within the measurement limit distance. Here, all working positions were within the working distance of 1 m except for those at 930 and 940 mm, which had values within PWL-10, and the ratio was 96.7%. Within the working distance of 1.5 m, there were 99 working positions, and the ratio of those working positions that had values within PWL-10 was 89.1%, as can be seen in Figure 10b. The results of the measurement of the width of a 0.5 mm virtual image are shown in Figure 10c. In the case of CC50-0.5, the ratio of working positions within 1 m that had values within PWL-10 was 95%, and it was seen that errors increased as the working distance increased.

For the accumulated shooting section, 376 working positions were identified within the measurement limit distance of 4050 mm, and the number of working positions that had values within PWL-10 was 188, with a ratio of 50.26%. In the case of SC50-0.5, it was found that the values measured at all working positions in the entire section within a working distance of 1.5 m corresponded to PWL-10. The number of working positions that had values within PWL-10 within a working distance of 2 m was 144 out of 161 locations, with a ratio of 89.4%. It was confirmed that the precision showed a tendency to decrease after that, since the shooting section grew as the working distance increased. Within the measurement limit distance of 6650 mm, 321 out of 624 working positions showed values within PWL-10, with an accuracy of 51.3%. Table 5 shows the resulting measurement values of each image with different resolutions for the same focal length. The measurement accuracy increased slightly, but this seems to be a result of the increase in the measurement limit distance. For instance, as can be seen in Table 5, in the case of CC50-0.1, the ratio of working positions that had values within PWL-10 was 39.29% within the limit distance of 850 mm, but in the case of SC50-0.1, the ratio was the same when the same limit distance was applied. In conclusion, it was demonstrated that the SC equipment had 2.76 times higher resolution corresponding to the sensor area and 1.65 times larger horizontal and vertical dimensions than the CC equipment, but no significant difference in measurement accuracy was shown between the two equipment units within their respective limit distances.

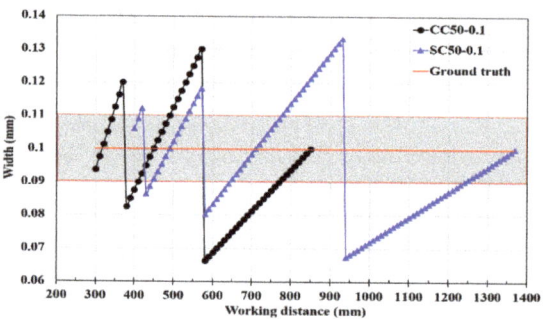

(a) 0.1 mm line width

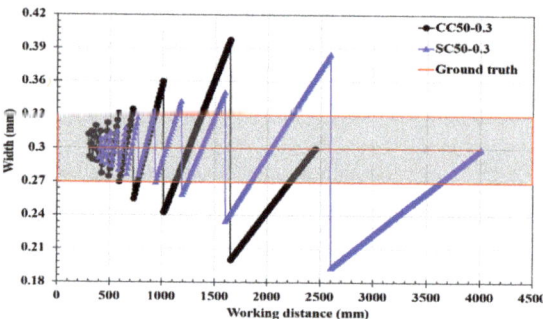

(b) 0.3 mm line width

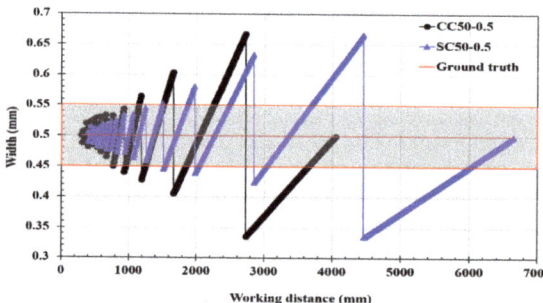

(c) 0.5 mm line width

Figure 10. Width measurement results of Canon and Sony cameras.

Table 5. Value of PWL-10 within working distance by resolution difference.

Classification		PWL-10					Unit: %
		SC50			CC50		
		0.1 mm	0.3 mm	0.5 mm	0.1 mm	0.3 mm	0.5 mm
Working distance (m)	0.85	52.17	100	100	39.29	87.71	100
	1.00	39.34 (38.77) *	96.72	100		78.87	95.77
	1.50		89.18	100		72.07	88.28
	2.00		72.04	89.44		46.11 (48.61) *	72.51
	2.50		65.40	89.57			65.40
	3.00		52.87	72.41			52.87
	4.00		49.03 (49.17) *	70.63 (51.44) *			47.92 (50.26) *

( ) * Value of PWL-10 at the measurement limit distance.

However, relative comparisons by limit distance and shooting section of the CC equipment showed that the SC equipment had somewhat higher precision, and it was confirmed that the difference in resolution increased working distance as well as accuracy. In addition, it was judged that the size measurement of a specific object is based on the numbers of horizontal and vertical pixels, not the pixel value in the space corresponding to the sensor. At the current level of imaging equipment, when measuring a 0.1 mm crack by applying a focal length of 50 mm, a minimum working distance from the crack surface is required, and it is judged that crack measurement is somewhat difficult without increasing the resolution and focal length.

4.2.2. Measured Values of Line Widths According to Increases in Focal Length Using the Same Imaging Equipment

Earlier, the resulting measurement values of the widths of virtual images according to differences in resolution at the same focal length of 50 mm were checked. In this chapter, the widths of virtual images were measured while increasing the focal length at the same resolution, and the resulting values were analyzed.

Figure 11 shows the resulting measurement values of a 0.1 mm wide virtual image. With regard to SC100-0.1, the number of working positions that had values within PWL-10 was 96 out of 218 total working positions within the measurement limit distance of 2740 mm, representing a ratio of 46.49%. As for SC135-0.1, the ratio of working positions that had values within PWL-10 was 95.55% within a working distance of 1 m, and the ratio decreased as the working distance increased to show values lower than 82.10% at a working distance of 1.5 m. The ratio showed a tendency to decrease rapidly after that. Within the measurement limit distance of 3700 mm, 136 out of 301 working positions had values within PWL-10, representing a ratio of 47.94%.

In Figure 12, with regard to SC100-0.3, 379 out of 746 working positions within the measurement limit distance of 8020 mm had values within PWL-10, representing a ratio of 51.45%. In this case, when checked by the shooting section, the measurement accuracy at which all working positions in the entire section, ranging from the minimum working distance to 1.5 m, had values within PWL-10, as shown. However, in the case of SC135-0.3, 100% measurement accuracy was shown over the entire shooting section within 2 m, and the measurement accuracy gradually decreased afterward as the distance increased so that 97.95% measurement accuracy was secured within 3 m. The measurement accuracy was 91.01% at the 4 m working position and then showed a tendency to decrease from the working position, so the number of working positions that had values within PWL-10 became 517 out of 1014 working positions within the measurement limit distance of 10,830 mm, representing a ratio of 51.94%.

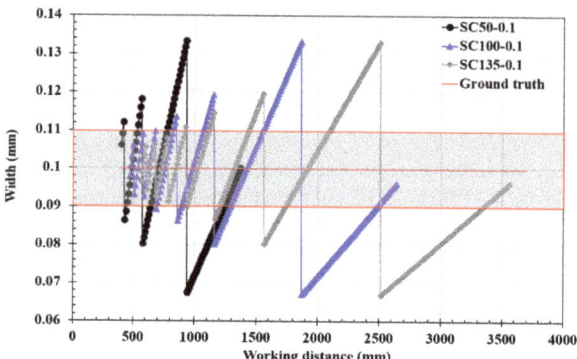

Figure 11. Results of 0.1 mm width measurement.

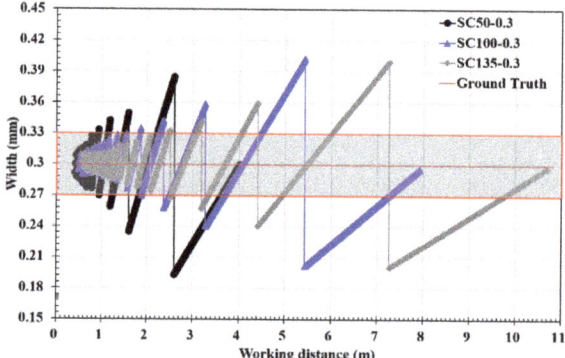

Figure 12. Results of 0.3 mm width measurement.

In the results of measurement of a 0.5 mm wide virtual image, in the case of SC100-0.5, 663 out of 1274 working positions within the measurement limit distance of 13,300 mm had values within PWL-10, representing a ratio of 52.41%, as seen in Figure 13. When checked by the shooting section, it was found that measurement precision was secured over the entire section, from the minimum working distance to 3 m. As for SC135-0.5, all working positions within 4 m had values within PWL-10, and the precision decreased afterward so that 98.20% of all working positions within 5 m had values within PWL-10. The resulting value was 89.49% in the shooting section within 5.5 m, but the precision increased slightly after that so that 91.26% of all working positions in the shooting section within 6.5 m had values within PWL-10. The reason for the slight increase after decreasing is that the number of working positions that had values within PWL-10 remained the same, while the working distance continued to increase. In the shooting section within 7 m, the value was 86.66%, and as the working distance increased, 900 out of 1727 sections within the measurement limit distance of 17,960 mm had values within PWL-10, representing a ratio of 52.70%. The overall precision seems to have increased slightly compared to SC100-0.5, and the reason is judged to be the fact that not only the maximum shooting section but also the number of working positions that had values within PWL-10 increased.

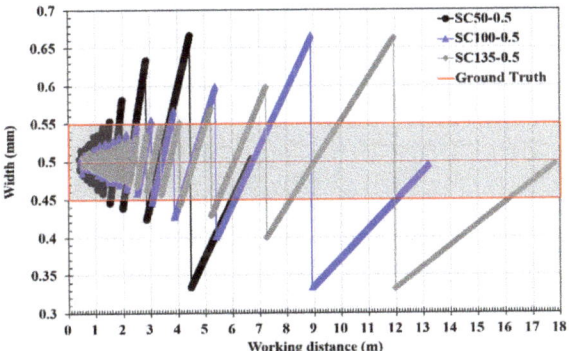

Figure 13. Results of 0.5 mm width measurement.

The levels of measurement accuracy by shooting section according to changes in focal length are summarized in Figure 14 and Table 6. The resulting values of SC135-0.1 and SC50-0.5, SC100-0.3 and SC50-0.5, and SC135-0.3 and SC100-0.5 show similar trends. It was found that in the case of SC135-0.5, at least 95% measurement accuracy was secured up to 5 m. For widths smaller than 0.3 mm, which is the crack measurement limit, but not smaller than 0.1 mm, which is the purpose of this study, SC50-0.3, SC100-0.3, and SC135-0.3 secured at least 90% measurement accuracy within working distances of 1 m, 3 m, and 4 m. However, in the case of 0.1 mm width, at least 85% and 95% measurement accuracy levels were secured within a 1 m working distance at focal lengths of 100 mm and 135 mm, respectively. However, as mentioned above, it is judged that when measuring 0.1 mm cracks, if a focal length of 50 mm is applied, the cracks can hardly be measured except for certain working positions. If microcracks not wider than 0.1 mm are measured by applying imaging equipment with the same resolution as the one applied in this experiment and a focal length of 50 mm, it is judged that the measurement should be carried out at distances shorter than the minimum working distance from the crack surface. In addition, to secure at least 90% measurement accuracy within the working distance of 1 m, 135 mm should be applied as the minimum focal length. Furthermore, on estimating the resolution value required when the focal length is 50 mm using the *GSD* value with the 135 mm focal length applied, it is thought that the resolution range of the imaging equipment should have a performance of 400 million pixels.

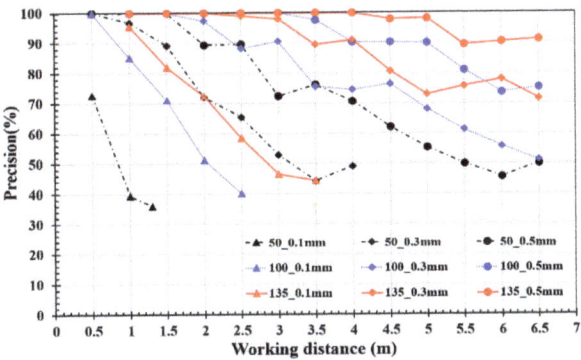

Figure 14. Value of PWL-10 according to focal length difference.

Table 6. Value of PWL-10 within working distance by focal length difference.

Classification		PWL-10					Unit: %
		SC100			SC135		
		0.1 mm	0.3 mm	0.5 mm	0.1 mm	0.3 mm	0.5 mm
Working distance (m)	1.00	85.18	100	100	95.56	100	100
	1.50	71.15	100	100	82.10	100	100
	2.00	51.29 (46.49) *	97.40	89.44	72.41	100	100
	3.00		90.55	100	46.53 (47.94) *	97.95	100
	4.00		74.57	90.39		91.01	100
	4.50		76.48	90.34		80.75	97.97
	5.00		68.06	90.08		73.03	98.20
	5.50		61.30	81.15		75.75	89.49
	6.00		55.77	73.82		77.98	90.45
	6.50		51.15 (51.45) *	75.33 (52.41) *		71.59 (51.94) *	91.26 (52.70) *

( ) * Value of PWL-10 at the measurement limit distance.

4.3. Analysis of the Causes of Crack Measurement Errors

Previous researchers pointed out that the causes of errors in crack measurement are in the resolution, image processing method, thresholding process during image processing, and mixed-pixel phenomenon, in which the background is recognized as a crack or a crack is recognized as a background [30]. Nevertheless, based on the results obtained in this experiment, it is judged that the errors occur due to the resolution of the imaging equipment and the range of the applied focal length.

Given the basic image processing theory, the smallest unit of digital images is expressed as integers in pixel units [31]. However, in order to accurately measure crack widths, the measured pixel values should be in decimal units, but this is not the case in practice. To help understanding, the measured pixel counts and width values of SC100-0.5 and SC50-0.5 within the same shooting section range among the resulting values derived using SC equipment are shown in Figure 15, where Pm is a pixel value measured through image processing and Pr is a value obtained by dividing the ground truth value by the GSD value according to the working distance applied to each focal length, which can be calculated using Equation (6).

$$Number\ of\ real\ pixels = \frac{Ground\ truth}{GSD} \quad (6)$$

$$GSD = \frac{L-f}{f} \times \frac{s_S}{s_R} \quad (7)$$

Wm is an area value in consideration of the *GSD* value for the number of pixels acquired through image processing. The ground truth is the width value of the virtual image, and it is the area value considering the *GSD* value for the actual pixel value. In addition, for ease of reading the figure, only the inflection point parts were expressed using markers.

In Figure 15a, it can be seen that PWL-10 values are more precisely measured in the case of SC100-0.5 than SC50-0.5 when the width is measured. Moreover, the position where the measured pixel value and the theoretical actual pixel value intersect was found to be a point with 0% error, as shown in Table 7.

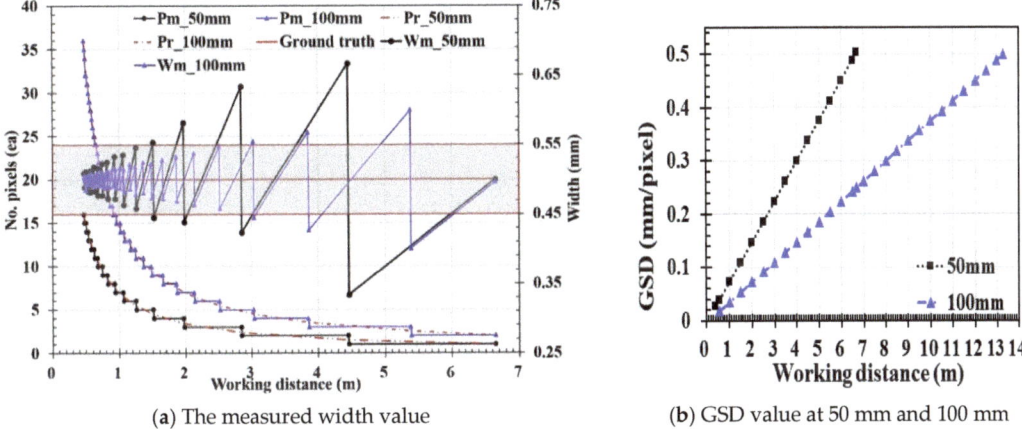

(a) The measured width value (b) GSD value at 50 mm and 100 mm

Figure 15. (a) The actual number of pixels of a 0.5 mm wide virtual image and the measured width value based on the measured pixel value; (b) *GSD* value at 50 mm and 100 mm focal lengths according to working distance.

Table 7. Position of shooting with zero error of measurement width value.

Classification		Working Position (Error = 0)			Unit: mm
		Width (mm)			
		0.1 mm	0.3 mm	0.5 mm	
Focal length	50 mm	490, 710, 1370	410, 490, 710, 1370, 2030, 4010	490, 600, 650, 710, 1150, 1700, 2250, 3350, 6650	
	100 mm	540, 760, 980, 1420, 2740	540, 760, 820, 980, 1090, 1420, 2080, 2740, 4060, 8020	540, 650, 700, 760, 980, 1200, 1300, 1420, 1750, 2300, 2740, 3400, 4500, 6700, 13,300	

Given the measurement results shown in Table 8, when a focal length of 50 mm is applied at a 400 mm point, the pixel value should be 18.85 to become a value of 0.5 mm width, but in reality, the pixel value measured through image processing was 19. At a working distance of 470 mm, the actual pixel values at focal lengths of 50 mm and 100 mm should be 12.69 and 28.08, respectively, but the values obtained through measurement were 16 and 36, respectively. For example, in the case of a certain section that is from 3000 mm to 4000 mm, two-pixel values were measured at 50 mm focal length in the entire section. As can be seen in Figure 15b, the number of pixel values was maintained at two even though the *GSD* continued. This is judged to be the cause of the changes in the crack value measured in the section. Given these results, it is considered that the errors occur because of the crack value measured in the form of a positive integer rounded to the nearest integer of the actual crack value, the *GSD* that increases proportionally as the working distance increases, and the repeating number of pixels. Although the resolution, thresholding process during image processing, and mixed pixels are important elements in crack measurement, measured pixel values are considered to have the greatest effects as they affect the entire section, and they are judged to be a limitation in measurement using imaging equipment.

Table 8. Measured line width (0.5 mm).

WD (mm)	GSD (mm/Pixel)		Real Pixels (ea)		Measured Pixel (ea)		Measured Width (mm)		Error (%)		Ground Truth (mm)
	||||||||| Focal Length (mm) |||||||||										
	50	100	50	100	50	100	50	100	50	100	
400	0.0265	-	18.857	-	19	-	0.504	-	0.75	-	
410	0.0273	-	18.333	-	18	-	0.491	-	−1.81	-	
470	0.0318	0.0145	15.714	35.675	16	36	0.509	0.504	1.18	0.90	
480	0.0325	0.0143	15.349	34.736	15	35	0.488	0.503	−2.27	0.75	
1000	0.0719	0.0341	6.947	14.667	7	15	0.504	0.511	0.75	2.27	
2000	0.1477	0.0719	3.385	6.947	3	7	0.443	0.503	−11.36	0.75	0.5
3000	0.2234	0.1098	2.237	4.552	2	5	0.447	0.549	−10.61	9.85	
4000	0.2992	0.1477	1.671	3.385	2	3	0.598	0.443	19.70	−11.36	
5000	0.375	0.1856	1.333	2.694	1	3	0.375	0.557	−25.00	11.36	
6650	0.5000	0.2481	1.000	2.015	1	2	0.5	0.496	0.00	−0.75	

5. Width Measurement: Real Crack (Concrete Bridge)

In this chapter, we investigate the effect of the measured pixels when measuring an actual crack. The cracks at the abutment of a pedestrian bridge were selected as the ground truth due to safety issues during visual inspection. Crack measurement was carried out through visual inspection, and a scale loupe with a magnifying glass of 10X magnification was used to check the width of fine concrete cracks, as shown in Figure 16. Concrete crack images were obtained from a minimum working distance to 2000 mm, at 200 mm increments. For the analysis of the acquired crack images, the same image processing method as the method used in the indoor experiment was implemented. However, unlike the linear virtual images, the crack images had bad conditions and a texture on the target surface, so preprocessing was performed to remove noise. The preprocessing procedure was to separate the foreground (cracks and noise) from the background by applying adjacency and connectivity conditions on neighboring pixels. The area of the crack and noise was computed and indexed in descending order. As shown in Figure 17, the noise regions were removed, excluding those with the largest area.

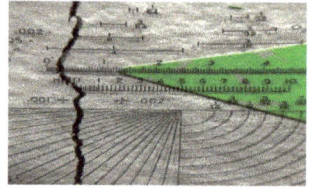

(a) Scale loupe   (b) Method of crack measurement

Figure 16. Concrete crack measurement apparatus and method.

 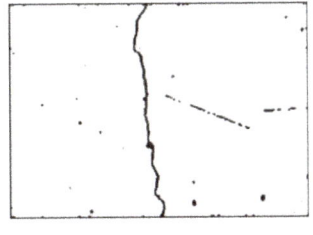 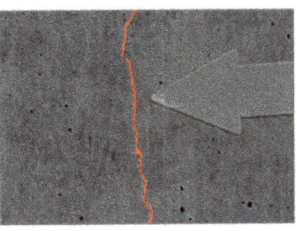

(**a**) Original image (**b**) Binary image (**c**) Processed image

Figure 17. Image processing results of concrete crack.

Table 9, and Figures 18 and 19 indicate the number of pixels (Pm), measured crack value (Wm), and errors (Em) according to the width value of the target crack through image processing. In addition, the estimated pixels in Table 9 were derived using the approximation curve in Figure 18 and crack measurement using the estimated pixels illustrated in Figure 19. Regression analysis using the power law was carried out, and the coefficient of determination was 0.9829. The approximated pixels (Pa), crack width (Wa), and error (Ea) were the calculated crack width and errors according to the working distance, respectively. Here, Pa showed a similar tendency to the pixel Pr corresponding to the actual crack. Moreover, the Wa calculated by the estimated pixel indicated an error of up to 5% and a great agreement with the ground truth. However, the biggest Em was 19.32% at the working distance of 1800 mm, while the rest of them were less than 10%. As the working distance increased, the Wm did not show a clear trend in error change and repeated the absolute increase or decrease, because the Pm was measured somewhat higher or lower than the value of Pa or Pr. As mentioned in Section 4.3, in order to accurately measure crack width, the Pm must be in decimal units. However, the Pm was computed as an integer, and it was found that the difference between Pm and Pr led to an error in the crack measurement. In addition, it can be seen that Pr becomes 3 at the position of 1450 mm where the Pm, Pr, and Pa intersect in Figure 18. Moreover, it can be seen that the measurement error at the same working distance represents positive or negative values, as shown in Table 9. As shown in Table 9 and Figure 18, it was indicated that the error increased as the distance increased, moving away from the reference position. In summary, it was judged that the pixel measured as an integer and the same pixel within a specific working distance range was the cause of the error in the crack measurement. From this evaluation, it was confirmed that the number of pixels corresponding to the working distance had a great influence on the measurement accuracy of crack width when using image processing. Once the working distance is accurately defined, it is judged that the number of pixels measured at each shooting location can be identified through the approximation method.

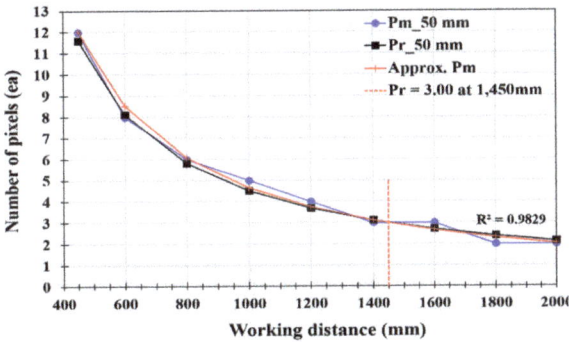

Figure 18. Measured pixels and approximation curves of number of pixels.

Table 9. *GSD* and crack width (0.308 mm).

WD (mm)	GSD (mm/pixel)	Measured			Approx.			Real	
		Pm (ea)	Wm (mm)	Em (%)	Pa (ea)	Wa (mm)	Ea (%)	Pr (ea)	Ground Truth (mm)
450	0.0265	12	0.318	3.40	12.03	0.319	3.65	11.59	
600	0.0379	8	0.303	−1.42	8.54	0.324	5.00	8.11	
800	0.0531	6	0.318	3.40	6.06	0.322	4.41	5.80	
1000	0.0683	5	0.341	9.84	4.65	0.317	3.02	4.51	
1200	0.0835	4	0.334	7.79	3.74	0.312	1.41	3.69	0.308
1400	0.0987	3	0.296	−4.02	3.11	0.307	0.24	3.12	
1600	0.1139	3	0.341	9.84	2.66	0.302	1.85	2.70	
1800	0.1291	2	0.258	−19.32	2.31	0.297	3.40	2.39	
2000	0.1442	2	0.288	−6.76	2.04	0.293	4.88	2.14	

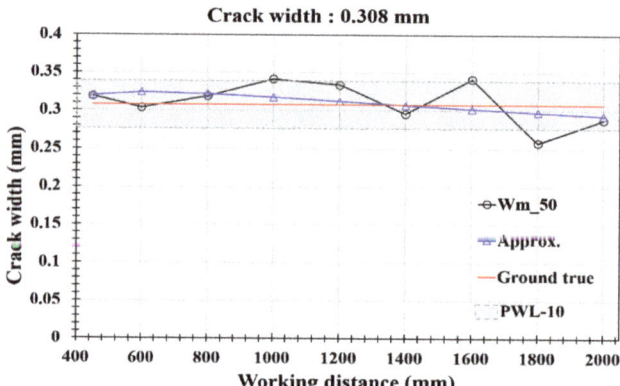

Figure 19. Measurement results of concrete crack.

6. Conclusions

In this paper, indoor measurement experiments were conducted to measure cracks in concrete structures and to present guidelines for image acquisition and processing using imaging equipment. Two units of DSLR imaging equipment, which is applied in various fields, were selected, and lenses with focal lengths of 50, 100, and 135 mm were applied. Linear virtual images with widths of 0.1, 0.3, and 0.5 mm were applied to the indoor measurement experiments, and video images were obtained at 10 mm intervals from the minimum working distance to the measurement limit distance to measure the line width values of the virtual images through image processing. The crack measurement limits, as well as optimal working distances of the selected imaging equipment and optical equipment, were derived to analyze measurement accuracy according to the difference in resolution and focal lengths. Field measurement tests were conducted to verify measurement parameters identified by the results of the indoor experiment. The conclusions drawn from this study are as follows.

- The pixel values were measured using the CC (Canon camera) and SC (Sony camera) equipment units with performances of 22.1 and 61 million pixels, respectively, and applying the same focal lengths. The SC equipment had 2.76 times higher resolution corresponding to the sensor area and 1.65 times larger horizontal and vertical dimensions than the CC equipment, respectively. As a result of applying the same

- focal length, it was found that the measured number of pixels and the measurement limit distance increased because the resolution and size of the virtual image increased. However, despite an increase in resolution by 1.65 times, the measured pixel values of the SC50 were estimated to be 1.33, 1.10, and 1.19 times higher, respectively, than the CC50 at the minimum working distance. In addition, when the virtual image size increased by three times from 0.1 to 0.3 mm, the pixel value of CC50 at the minimum working distance increased by 3.3 times, and when it increased by five times from 0.1 mm to 0.5 mm, the pixel value increased by 5.5 times. As for SC50, the pixel value increased by 2.75 times and 4.75 times, respectively.
- As for increasing the focal length at the same resolution, the value showed a similar tendency with the results of increasing the resolution. However, the pixel value of SC100 was 1.85 times larger at average than that of SC50 at the minimum working distance. When the size of the virtual image increased by three times and by five times, the pixel value of SC100 was estimated at the same rate and by 5.13 times. As a result of experiments, the pixel value was more affected by the increase in focal length than the increase in resolution. However, even if the resolution, focal length, and size of the object increased, the pixel values did not increase at the same rate. It is considered that the difference between the pixel value and the rate could occur because of the minimum working distance, in accordance with the physical features of an optical lens.
- According to the results, in the case of CC50-0.1 and SC50-0.1 mm, 39.29% and 38.77% of the working positions, respectively, had values within PWL-10 within the respective measurement limit distances. With regard to the 0.3 mm wide virtual image, 48.61 and 49.17%, respectively, of the working positions had values within PWL-10. For the 0.5 mm virtual image, 50.26 and 51.44%, respectively, of the working positions had values within PWL-10 within the limit distance. There was no significant difference in measurement accuracy within the limit distance of the imaging equipment. However, relative comparisons by shooting section showed that the SC equipment had somewhat higher precision than the CC equipment.
- Increasing the focal length enables more precise measurement by increasing the measurement limit distance and pixel measurement values, as well as reducing the *GSD* value of imaging equipment. When measurement accuracy was checked by section while increasing the working distance from the minimum working distance, it was found that within a 1 m working distance, with regard to SC50-0.1 and SC100-0.1, 39.34 and 85.18% of working positions, respectively, had values within PWL-10, and as for SC135-0.1, 95.55% of working positions had values within PWL-10. In the case of SC50-0.3, the ratio was 96.72% within 1 m of the working distance, and for SC100-0.3 and SC135-0.3, all working positions within working distances of 1.5 m and 2 m, respectively, had values within PWL-10. In the case of SC50-0.5, all working positions within a working distance of 1.5 m had values within PWL-10, and as for SC100-0.5 and SC135-0.5, all working positions within working distances of 3 m and 4 m, respectively, had values within PWL-10. Based on the results of analysis of PWL-10 in this study, cracks were measured using a unit of 61-megapixel imaging equipment, and according to the results, in the case of 0.3 mm wide or wider cracks, a working distance of 1 m was secured when the focal length was 50 mm, and working distances of 3 and 4 m were secured when the focal lengths were 100 and 135 mm, respectively. However, with regard to cracks not wider than 0.1 mm, focal lengths of 100 and 135 mm showed measurability within 1 m, but a focal length of 50 mm was judged to hardly enable measurement, except for certain working positions.
- Many elements affect width measurement. However, based on the results of this study, width measurement is highly dependent on the measured pixel values. In order to accurately measure width, the measured pixel values must be in decimal units, but that is theoretically impossible. Field measurement tests were conducted to verify measurement parameters identified by the results of the indoor experiment. An

approximate curve was predicted based on the pixel value according to the working distance, and an estimated pixel value was derived using power law distribution. As the working distance increased, the crack measurement value (Wm) did not show a clear trend in error change and fluctuated with ground truth. However, the crack width (Wa) calculated by the estimated pixel indicated an error of up to 5% and showed great agreement with the ground truth. From the evaluation, it was confirmed that the number of pixels corresponding to the working distance had a great influence on the measurement accuracy of crack width when using image processing. Once the working distance is accurately defined, it is judged that the number of pixels measured at each shooting location can be identified through the approximation method.

- If microcracks not wider than 0.1 mm are measured by applying imaging equipment with the same resolution as applied in this experiment and a focal length of 50 mm, the measurement should be carried out at distances shorter than the minimum working distance from the crack surface. To secure at least 90% measurement accuracy for widths not larger than 0.1 mm at working distances not longer than 1 m, a focal length of 135 mm should be applied. In addition, the resolution value required when a focal length of 50 mm is applied was estimated using the *GSD* value calculated by applying a focal length of 135 mm, and according to the results, it is thought that the resolution range of the imaging equipment should be at least about 400 million pixels.

Author Contributions: Conceptualization, H.J. and D.C.; methodology, H.J., B.J., M.H. and D.C.; software, H.J. and M.H.; validation, H.J. and D.C.; formal analysis, H.J. and D.C.; investigation, H.J. and M.H.; resources, D.C.; data curation, B.J. and M.H.; writing—original draft preparation, H.J.; writing—review and editing, H.J. and D.C.; supervision, D.C.; project administration, D.C. All authors have read and agreed to the published version of the manuscript.

Funding: This research received no external funding.

Institutional Review Board Statement: Not applicable.

Informed Consent Statement: Not applicable.

Data Availability Statement: The data presented in this study are included within the article.

Conflicts of Interest: The authors declare no conflict of interest.

References

1. Mohan, A.; Poobal, S. Crack detection using image processing: A critical review and analysis. *Alex. Eng. J.* **2018**, *57*, 787–798. [CrossRef]
2. Zakeri, H.; Nejad, F.M.; Fahimifar, A. Image Based Techniques for Crack Detection, Classification and Quantification in Asphalt Pavement: A Review. *Arch. Comput. Methods Eng.* **2016**, *24*, 935–977. [CrossRef]
3. Koch, C.; Georgieva, K.; Kasireddy, V.; Akinci, B.; Fieguth, P. A review on computer vision based defect detection and condition assessment of concrete and asphalt civil infrastructure. *Adv. Eng. Inform.* **2015**, *29*, 196–210. [CrossRef]
4. Lee, B.Y.; Kim, Y.Y.; Yi, S.T.; Kim, J.K. Automated image processing technique for detecting and analyzing concrete surface cracks. *Struct. Infrast. Eng.* **2013**, *9*, 567–577. [CrossRef]
5. Naidoo, T.; Joubert, D.; Chiweme, T.; Tyatyantsi, A.; Rancati, B.; Mbizeni, A. Visual surveying platform for the automated detection of road surface distresses. In Proceedings of the SPIE The International Society for Optical Engineering, Kruger National Park, Skukuza, South Africa, 23 June 2014.
6. Huang, J.; Liu, W.; Sun, X. A Pavement Crack Detection Method Combining 2D with 3D Information Based on Dempster-Shafer Theory. *Comp. Aided Civil Infrast. Eng.* **2014**, *29*, 299–313. [CrossRef]
7. Mathavan, S.; Kamal, K.; Rajman, M. A Review of Three-Dimensional Imaging Technologies for Pavement Distress Detection and Measurements. *IEEE Trans. Int. Trans. Syst.* **2015**, *16*, 2353–2362. [CrossRef]
8. Guan, H.; Li, J.; Yu, Y.; Chapman, M.; Wang, C. Automated Road Information Extraction from Mobile Laser Scanning Data. *IEEE Trans. Int. Trans. Syst.* **2015**, *16*, 194–205. [CrossRef]
9. Otsu, N. A Threshold Selection Method from Gray-Level Histograms. *IEEE Trans. Syst. Man Cybern.* **1979**, *9*, 62–66. [CrossRef]
10. Ito, A.; Aoki, Y.; Hashimoto, S. Accurate extraction and measurement of fine cracks from concrete block surface image. In Proceedings of the IEEE 2002 28th Annual Conference of the Industrial Electronics Society, Seville, Spain, 5–8 November 2002.
11. Fujita, Y.; Mitani, Y.; Hamanoto, Y. A Method for Crack Detection on a Concrete Structure. In Proceedings of the 18th International Conference on Pattern Recognition, Hong Kong, China, 20–24 August 2006.

12. Oliveira, H.; Correia, P.L. Automatic crack detection on road imagery using anisotropic diffusion and region linkage. In Proceedings of the 2010 18th European Signal Processing Conference, Aalborg, Denmark, 23–27 August 2010.
13. Zhang, W.; Zhang, Z.; Qi, D.; Liu, Y. Automatic Crack Detection and Classification Method for subway Tunnel Safety Monitoring. *Sensors* **2014**, *14*, 19307–19328. [CrossRef]
14. Sinha, S.K.; Fieguth, P.W. Automated detection of cracks in buried concrete pipe images. *Autom. Constr.* **2006**, *15*, 58–72. [CrossRef]
15. Merazi-Meksen, T.; Boudraa, M.; Boudraa, B. Mathematical morphology for TOFD image analysis and automatic crack detection. *Ultrasonics* **2014**, *54*, 1642–1648. [CrossRef]
16. Su, T.C. Application of Computer Vision to Crack Detection of Concrete Structure. *IACSIT Int. J. Eng. Technol.* **2013**, *5*, 457–461. [CrossRef]
17. Zhao, H.; Qin, G.; Wang, X. Improvement of canny algorithm based on pavement edge detection. In Proceedings of the 2010 3rd International Congress on Image and Signal Processing, Yantai, China, 16–18 October 2010.
18. Wang, G.; Tse, P.W.; Yuan, M. Automatic internal crack detection from a sequence of infrared images with a triple-threshold Canny edge detector. *Meas. Sci. Technol.* **2017**, *29*, 025403. [CrossRef]
19. Lee, H.B.; Kim, J.W.; Jang, I.Y. Development of Automatic Crack Detection System for Concrete Structure Using Image Processing Method. *J. Kor. Inst. Struct. Maint. Insp.* **2012**, *16*, 64–77.
20. Li, G.; He, S.; Ju, Y.; Du, K. Long-distance precision inspection method for bridge cracks with image processing. *Autom. Constr.* **2014**, *41*, 83–95. [CrossRef]
21. Jahanshahi, M.R.; Masri, S.F. Adaptive vision-based crack detection using 3D scene reconstruction for condition assessment of structures. *Autom. Constr.* **2012**, *22*, 567–576. [CrossRef]
22. Jahanshahi, M.R.; Masri, S.F.; Padgett, C.W.; Sukhantme, G.S. An innovative methodology for detection and quantification of cracks through incorporation of depth perception. *Mach. Vis. Appl.* **2013**, *24*, 227–241. [CrossRef]
23. Lins, R.G.; Givigi, S.N. Automatic crack detection and measurement based on image analysis. *IEEE Trans. Instrum. Meas.* **2016**, *65*, 583–590. [CrossRef]
24. Khalili, K.; Vahidnia, M. Improving the accuracy of crack length measurement using machine vision. In Proceedings of the 8th International Conference Interdisciplinarity in Engineering, INTER-ENG 2014, Tigru Mures, Romania, 9–10 October 2014.
25. Bhowmick, S.; Nagarajaisa, S.; Veeraraghavan, A. Vision and deep learning-based algorithms to detect and quantify cracks on concrete surfaces from UAV videos. *Sensors* **2020**, *20*, 6299. [CrossRef] [PubMed]
26. Kim, J.W.; Kim, S.B.; Park, J.C.; Nam, J.W. Development of crack detection system with unmanned aerial vehicles and digital image processing. In Proceedings of the 2015 World Congress on Advances in Structural Engineering and Mechanics (ASEM15), Incheon, Korea, 25–29 August 2015.
27. Li, S.; Zhao, X. Automatic crack detection and measurement of concrete structure using convolutional encoder-decoder network. *IEEE Access* **2020**, *8*, 134602–134618. [CrossRef]
28. Feng, C.; Zhang, U.; Wang, H.; Wang, S.; Li, Y. Automatic pixel-level crack detection on dam surface using deep convolutional network. *Sensors* **2020**, *20*, 2069. [CrossRef] [PubMed]
29. Feng, D.; Feng, M.Q.; Ozer, E.; Fukuda, Y. A Vision-Based Sensor for Noncontact Structural Displacement Measurement. *Sensors* **2015**, *15*, 16557–16575. [CrossRef] [PubMed]
30. Cho, H.W. A Study on Image Based Crack Measurement Using Crack Width Transform Method. Ph.D. Thesis, University of Science and Technology, Deaejon, Korea, 2018.
31. Gonzalez, R.C.; Woods, R.E. *Digital Image Processing*, 3rd ed.; Pearson Education: Upper Saddle River, NJ, USA, 2008.

Article
Estimating Thermal Material Properties Using Solar Loading Lock-in Thermography

Samuel Klein [1,*], Henrique Fernandes [2] and Hans-Georg Herrmann [1,3]

1. Chair for Lightweight Systems, Saarland University, Campus E3 1, 66123 Saarbrücken, Germany; hans-georg.herrmann@izfp.fraunhofer.de
2. Faculty of Computing, Federal University of Uberlandia, Uberlandia 38408-100, Brazil; henrique.fernandes@ufu.br
3. Fraunhofer Institute for Nondestructive Testing IZFP, Campus E3 1, 66123 Saarbrücken, Germany
* Correspondence: samuel.klein@uni-saarland.de

Featured Application: Thermographic monitoring of a retaining wall structure.

Abstract: This work investigates the application of lock-in thermography approach for solar loading thermography applications. In conventional lock-in thermography, a specimen is subjected to a periodically changing heat flux. This heat flux usually enters the specimen in one of three ways: by a point source, a line source or an extended source (area source). Calculations based on area sources are particularly well suited to adapt to solar loading thermography, because most natural heat sources and heat sinks can be approximated to be homogenously extended over a certain region of interest. This is of particular interest because natural heat phenomena cover a large area, which makes this method suitable for measuring large-scale samples. This work investigates how the extended source approximation formulas for determining thermally thick and thermally thin material properties can be used in a naturally excited setup, shows possible error sources, and gives quantitative results for estimating thermal effusivity of a retaining wall structure. It shows that this method can be used on large-scale structures that are subject to natural outside heating phenomena.

Keywords: infrared thermography; solar loading thermography; lock-in thermography; passive thermography; thermal thickness; thermal effusivity; infrastructure; NDT

1. Introduction

The estimation of thermal material properties like effusivity of large structures poses a challenge to conventional non-destructive testing (NDT) methods. Especially active thermography reaches a technical limit when faced with measuring large-scale structures because exciting such a structure thermally is logistically hard. This article shows a method of how a quantitative estimation algorithm can be constructed and shows first results using only naturally occurring heat sources.

Currently thermal material properties like diffusivity are measured using an active setup, where the experiment conductor actively controls the excitation source. Small-scale samples can be measured via a simple heating plate setup [1]. For bigger specimen a laser-excited setup is described in [2]. Both these approaches are infeasible for a large-scale application like the retaining wall investigated in this work.

Solar loading lock-in thermography is a recently introduced measuring method and is preferable because it is particularly easy to set up. As the primary excitation period is the day-night-cycle, the measurement depth is large compared to conventional, lab-scale lock-in thermography. The thermal wavelength for a periodically excited specimen Λ is proportional to the square root of the excitation period. This thermal wavelength is the depth in which a full temperature cycle is completed. It is generally considered as the measurement depth. In addition, the measurement region can be larger than ordinary active thermography setups would be able to excite properly. Measuring very large structures at once is possible. This is particularly interesting for monitoring buildings and large infrastructure alike. This work focuses on large vertical, thermally thick (d > Λ) structures. Multiple assumptions and approximations were made to use the theoretical lock-in thermography formulas on solar loading thermography data. These assumptions result in systematic error. Later a calibration method will be shown, that eliminates any linear errors that were made, by means of a calibration against known material property values.

Related work shows that this method can be successfully used for detecting defects and characterize their depths [3]. Other related work focusses on the time when the sun is obstructed, and a shadow is cast, triggering the thermal camera to begin a pulse-phase-thermography (PPT) measurement [4]. For this work these exact requirements of shadow cast are not needed. Instead, this work specifically choses the 24 h period to suppress transient effects (like cloud shadows or short-term weather effects like gusts of wind) and generate results dependent only of the much slower 24 h excitation intensity. Furthermore, only qualitative results have been shown (i.e., cracks or delaminations) and not quantitative measurements of material properties. This has the benefit of not only being able to detect defects, but also characterize the material under test.

2. Materials and Methods

This work uses two reference plates made out of EN AW 5083 aluminum sized $300 \times 300 \times 10$ mm and $300 \times 300 \times 20$ mm respectively. Using literature values for the thermal material properties described by [5] $\varrho = 2660$ kg/m^3, $\lambda = 125$ W/(m·K), $c_p = 900$ J/(kg·K) results in thermal masses of 2106 J/K and 4212 J/K respectively.

A concrete block was fabricated out of premixed dry mortar with a size of $300 \times 300 \times 300$ mm to be used as a thermally thick reference. Material properties were assumed to be [6] $\varrho = 2240$ kg/m^3, $\lambda = 2$ W/(m·K), $c_p = 900$ J/(kg·K), resulting in a thermal effusivity of $b_{conc} = 2008$ J/(K·m^2·$\sqrt{s}$).

The brickwork measured in the field experiment is assumed to have the following thermal properties [6]: $\varrho = 1920$ kg/m^3, $\lambda = 0.9$ W/(m·K), $c_p = 800$ J/(kg·K), resulting in a thermal effusivity of $b_{brick} = 1176$ J/(K·m^2·$\sqrt{s}$). The retaining wall itself is approx. 8 m tall and extends laterally over 100 m, of which a single measurement site of approx. 4 m by 4 m was chosen.

A custom measurement platform was developed, powered by a rechargeable battery and consisting of an embedded Linux single board computer (RaspberryPi 3B) and the FLIR Boson 640 in USB Video (UVC) mode. The FLIR Boson LWIR camera can directly send raw image data over USB, which was used in all further processing. Thermal simulations were carried out using GNU Octave and MATLAB.

3. Theoretical Framework

Solar loading lock-in thermography uses no artificial heat source, like heat lamps, but only naturally occurring heat sources. These consist of the following: Heat transfer within the structure (conduction). Heat transfer with the surrounding air, which is free to move (convection) and heat transfer via radiation (solar irradiation, radiative heat loss). The fourth heat source, evaporative cooling, is neglected in this work.

Conduction within a static material is a linear process $\sim \Delta T$, governed by the heat equation. Natural convection is a non-linear process that is commonly linearized by defining an "overall heat transfer coefficient" [6], convection in a natural setting, where ambient temperatures, humidity and air speeds are always changing is very hard to accurately represent in a single coefficient. Thermal radiation is a major non-linear process $\sim T^4$ that may be linearized for a single temperature region. The accuracy of thermal radiation linearization is discussed later in more detail.

To estimate thermal material properties the value of the heat input density amplitude (irradiance) of a periodic excitation is needed. The heat input is also assumed to be periodic and sinusoidal.

Estimating the heat input that occurs in a specific frequency content of the total excitation is inaccurate. Especially for overcast days, and frequencies higher than 1/24 h direct measurements are necessary. One way of measuring the heat input is a measuring solar irradiance (e.g., via a pyranometer). Another way of determining heat input, presented in this work, is done by placing thermally known reference samples into the measurement setup and using the thermal response of the reference samples to infer the preceding irradiation amplitude in frequency domain. Subsequently using formulas from lock-in thermography to determine the heat flow amplitude, which in turn arise from the heat equation using various assumptions and simplifications.

The thermal imaging long wave infrared (LWIR) camera is set up in a fixed manner to capture a set of thermal images that are regularly spaced in time. These images represent a 2D temperature field that is assumed to accurately depict the surface temperature of the captured structure. Thermal imaging cameras can be calibrated and adjusted to radiation to measure more accurate results [7]. A two-point calibration with a calibration radiator is performed within the approximate temperature range to get temperature estimates as close to the real temperature as possible. Other effects, like emissivity/reflectivity can be corrected as well to further improve absolute measuring accuracy for a given surface. These kinds of corrections were not performed in the scope of this work but are hypothesized to further increase measurement accuracy.

A time series matrix is arranged from the temperature field T_{xy}, which is called T_{xyt} with the third dimension being time. The time series for each pixel is then transformed into frequency domain by multiplying with the complex frequency ω in time. This is similar to using a lock-in amplifier. The main difference is that the reference oscillator is generated inside the Fast Fourier Transform (FFT) and never subject to any feedback from the signal itself. Therefore, two main conditions have to be met to ensure the error being as low as possible.

(1) For the fundamental frequency and its respective harmonics amplitudes to result in a single bin, the measurement time shall be an exact multiple of the fundamental interval (in this work 24 h).
(2) The local start time of the measurement is captured because it defines the reference Phase value, each calculated phase is related to.

The resulting matrix $S_{xy}(\omega)$ contains the complex coefficients describing either amplitude/phase or real/imaginary (often called 0°/90°) components for each pixel of the image [8].

The reference signal of a conventional lock-in amplification method is generated inherently and is available for the lock-in transform algorithm by design [8]. However, in ambient thermography, the reference signal is generated artificially. A pure sinusoidal signal $ref_t(\omega) = e^{(-i\omega t)}$ is used as the reference. Note that this is a set of vectors in time for each frequency considered:

$$S_{xy}(\omega) = \sum_t T_{xyt} \cdot ref_t(\omega)$$

The amplitude of S_{xy} is of particular interest because it depicts the amount of temperature change the surface has undergone at a particular frequency. Under simplified conditions, these amplitudes are directly related to material properties. Table 1 summarizes these relations in a table.

Table 1. Real and imaginary part of the temperature signal from [8] (p.120).

Thermal Thickness	$\Re\{S_{xy}\}$	$\Im\{S_{xy}\}$
Thermally thin	0	$\frac{p}{\rho d c_p \omega}$
Thermally thick	$\frac{p}{\sqrt{2\omega \lambda \rho c_p}}$	$\frac{p}{\sqrt{2\omega \lambda \rho c_p}}$

Excerpt of the table: p Thermal flux density; λ Thermal conductivity; ω Angular excitation frequency; c_p Specific thermal capacity; ρ Mass density.

Thermal effusivity is defined as $b = \sqrt{\lambda \rho c_p}$. Substitution leads to:

$$b = \frac{p}{|S_{xy}| \cdot \sqrt{\omega}} \quad (1)$$

A three-step method is devised, that estimates the thermal effusivity of every pixel in the scene:

(1) Place a reference sample inside the measured frame with known material properties to get an approximation of p. From Table 1 we get:

$$p = |S_{xy}| \cdot \rho d c_p \omega \quad (2)$$

Note that this is only valid for thermally thin samples.

(2) Empirically correct p to better approximate the input power density amplitude that the actual body is subjected to:

$$p_{eq} = p \cdot c(\omega) \quad (3)$$

where $c(\omega)$ is an empirical function that is determined by linear regression.

(3) Approximate thermal effusivity b by inserting p_{eq} into Equation (1). This results in the final formula:

$$b_{corr} = \frac{p \cdot c(\omega)}{|S_{xy}| \cdot \sqrt{\omega}} \quad (4)$$

The approximation of p via the reference body can only be applied to estimate b if it is used on a part of the surface that is indeed thermally thick. Additionally, approximating heat flow to be linear in relation to input power is detrimental to the accuracy if thermal radiation to the surroundings contributes to a significant part of heat transferred. Thermal radiation heat transfer is in general of the form $p_{rad} \propto T_{obj}^4 - T_{amb}^4$ and can be estimated linearly only for small amplitudes in object and ambient temperature. Simple thermal mass point simulations show that the radiative heat loss accounts for an approximatively constant part of the overall heat loss in respect with excitation strength. Linearizing radiative heat loss within naturally occurring temperatures (0 °C to 40 °C) results in a maximum error of 2.4% percent, which is less than many other approximating factors that were assumed earlier.

On empirically correcting the results: One of the biggest error terms is that the real heat input density in the field experiments is unknown. Especially heat loss via convection and radiation is hard to determine directly as it is dependent on multiple significant factors that were listed above. In a first approach to measuring thermal material properties using solar loading thermography, a regression method is used to match the measured results to the literature value of a specific material (in this case brick and concrete).

Other error sources include but are not limited to:

(1) Temperature distribution in thick materials is dispersive, because thermal material properties like thermal conductivity often depend on the temperature itself.
(2) Heat radiation estimations should consider the temperature distribution of the half space and not estimate the ambient radiation temperature to be constant, especially at nighttime with clear skies, the sky heat sink is significant. The black body temperature T_{amb} of the night sky ranges between -40 °C and -10 °C [9].

To validate the proposed algorithm the following steps were carried out:

(1) Thermal simulations with an idealized setup to validate the theoretical formulas and assess the impact of convective and radiative heat loss.

(2) Laboratory experiments with an artificial infrared heating source and thermally thin (two aluminum plates) and thick (one concrete block) samples being heated from one side in a periodic manner.
(3) A field experiment on a vertical retaining wall. It consists of a combination of brick, mortar, plastering and concrete with varying depths of each of these materials.

4. Results and Discussion

4.1. Thermal Simulations

Multiple transient thermal simulations were carried out to validate the theoretical formulas. The aluminum reference plates were simulated by a mass point with a thermal mass of $m_{th} = \rho V c_p$ each, subjected to a periodic heat input, as well as to convective and radiative cooling.

Convective cooling is assumed to be linear in respect to temperature difference and uses a free convection rate of 5 W/(m^{2}K) for all specimen. Radiative cooling was simulated using a constant ambient temperature of 21 °C and emissivity of $\varepsilon = 1.00$ for all specimen, to simplify radiation calculations.

The simulation results, shown in Figure 1, correlate well with the analytical results according to Equations (1) and (2). For greater heat input amplitudes, the simulation temperature amplitude is lower than expected, as the theoretical formulas do not take the cooling effects into account. Overall, a good correlation is shown for common solar irradiance values (<1200 W/m^2). Moreover, non-linear effects are negligible.

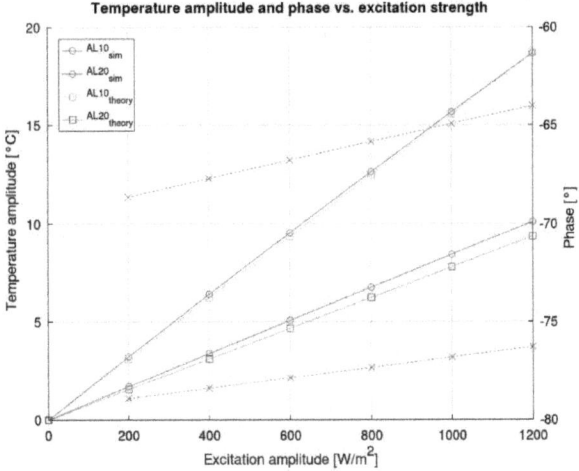

Figure 1. Amplitude and phase angle of the temperature signal from the thermal simulation in comparison to the analytical formula (cf. Equations (1) and (2)). The dotted line shows the phase angle and the solid line the temperature amplitude. Depicted here are temperature amplitudes and phase angles calculated by the FFT in relation to heat input density.

4.2. Laboratory Experiments

In the indoor laboratory experiments, the two reference plates, described in Section 1, were set up vertically on a desk, together with the concrete cube and subjected to an infrared heating lamp, controlled by a computer. The laboratory setup and a sample thermogram is shown in Figure 2.

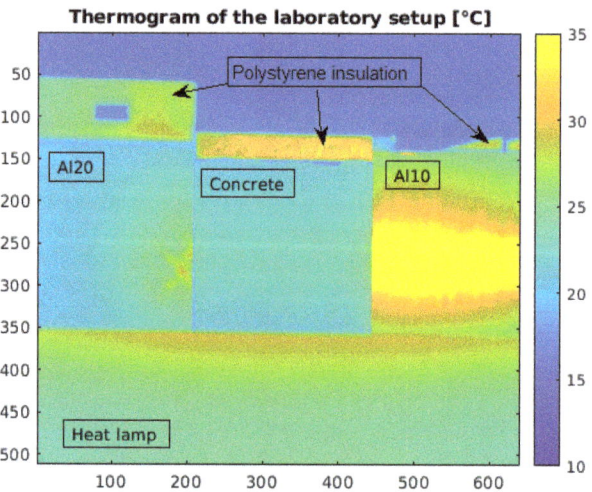

Figure 2. Overview of the laboratory setup. Note the polystyrene insulation in the back of each sample to ensure the backside only experiences minor heat loss.

The periodic heating signal was varied so that a periodic square wave excitation was generated, which in turn could be varied in period. The heat lamp is a 3000 W infrared lamp for sauna application laid horizontal on a desk with a distance of approx. 60 cm to the subjects under test. This results in a maximum irradiation of 800 W/m^2.

This setup was specifically chosen to test the proposed algorithm by having both thermally thick and thermally thin test samples in one setup, to infer information from the reference plates to the concrete block. The excitation periods were: 90, 45, 22.5, 11.25, 5.625 and 2.8125 min.

The upper limit of the excitation period is determined by the thickness of the thermally thick reference body (concrete cube block) that has a side length of 30 cm. The resulting maximum excitation period is ~90 min, as the corresponding thermal wavelength Λ is almost 30 cm. Reflections from the backside of the specimen will become non-negligible if the excitation period were to be extended (the thermally thick reference block would no longer be thermally thick).

Further work could either test outside on a thermally thick wall, construct a larger reference body, or choose a material with a lower effusivity. The latter effectively shortens the thermal wave length. However, due to the increasing effect of environmental effects for materials with lower effusivity, this has the drawback of higher error.

Additionally, in the laboratory setup concrete was chosen as a reference material, because it is much easier to fabricate homogenous block of any size than with brickwork. Using fabricated brickwork here however, would allow a direct comparison of the laboratory experiments and field experiment effusivity value.

The laboratory experiments, depicted in Figure 3, show good correlation in the lower excitation period range with less correlation on longer periods. These result in longer timespans, in which cooling effects are active. In addition, non-linear cooling effects (like radiative cooling) are more pronounced because the specimen heats up more than in experiments with shorter periods. Furthermore, the square root dependency of the thermally thick concrete block temperature amplitude is depicted well (green line with approx. half the slope of the other curves).

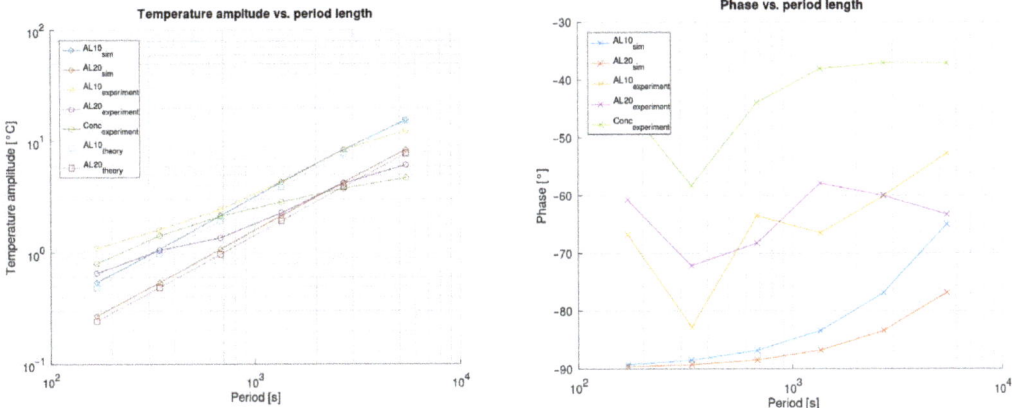

Figure 3. Amplitude and phase angle of the temperature signal from the thermal simulation and laboratory experiments and analytical solution. Depicted here are temperature amplitudes and phase angles calculated by the FFT in relation to excitation period. Note the theoretical phase angle for thermally thin and thick materials is −90° and −45° respectively.

The experimental laboratory results point to greater inaccuracies with longer periods, indicating a poor correlation between theoretical formula and real value in 24 h periods (ultimately used in this work). Therefore, an empirically determined correction factor was introduced to correct these aforementioned errors.

This approximation factor was calculated using a regression function on the various data points collected by means of laboratory experiments and field experiments. In the experiments, the correction factor $c(\omega)$ was calculated using the following equation:

$$c(\omega) = \frac{b_{literature}}{b(\omega)} \qquad (5)$$

where b is calculated from Equation (3). The literature value for the concrete sample block is 2000 J/$\left(K \cdot m^2 \cdot s^{1/2}\right)$ [6] (p. 717). The effusivity of the wall structure was assumed to be 1000 J/$\left(K \cdot m^2 \cdot s^{1/2}\right)$ (Effusivity of brick [6] (p. 716)). Determining the correction factor for each excitation period results in a point plot, which was fitted by a linear function shown in Figure 4.

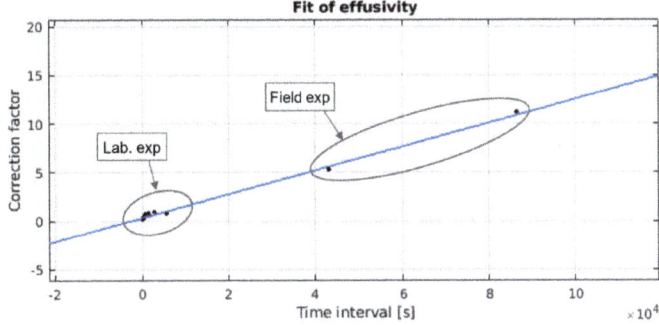

Figure 4. Plot of the correction factors determined by the method depicted above. Note that the shorter time periods were determined in the laboratory experiments and the 24 h and 12 h correction factor was determined in the field experiment.

This correction factor compensates for all systematic errors that rose during simplification but has the limitation that it cannot correct for non-linear errors.

4.3. Field Experiments

The LWIR camera was set up in a fixed manner with approximately 9.5 m distance to the wall. Thermal images were captured for 96 h. The wall under investigation, positioning of the ROIs and a typical thermogram are depicted in Figures 5–7.

Figure 5. Overview of the wall structure under test. The rectangle shows the approximate view of the LWIR camera.

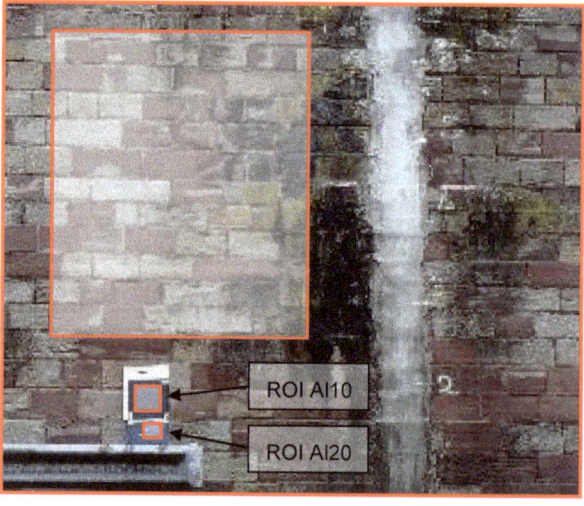

Figure 6. ROI location and size. In the lower left are the two reference plates (aluminum plates, each 300 × 300 mm size, one 10 mm, the other 20 mm thick). On the right side, a plaster strip is visible from earlier experiments on the wall.

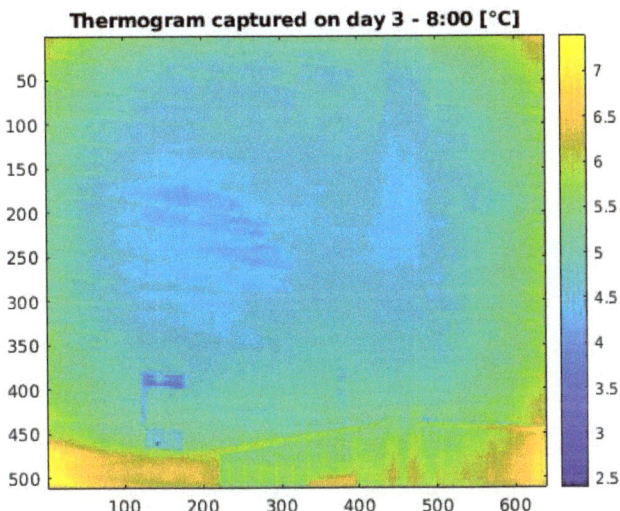

Figure 7. Typical thermogram showing the retaining wall. In the lower left are the two reference plates stacked on top of each other. Note the vignetting effect on the cameras lens has not been fully corrected. However, for the proposed algorithm, this correction is not necessary.

Figure 8 shows the time series values for the three evaluated ROIs over the measurement period. This data shows very poor 24 h period amplitude. This is due to the weather being overcast for almost the whole time (see Table 2). The algorithm (Equation (4)) was evaluated on this data set. Even better results are expected on days with strong sun, and no overcasts to disturb the periodic heat input, also advised by [10]. The amplitude image, shown in Figure 9, is calculated and from these amplitudes the effusivity is estimated and depicted in Figure 10.

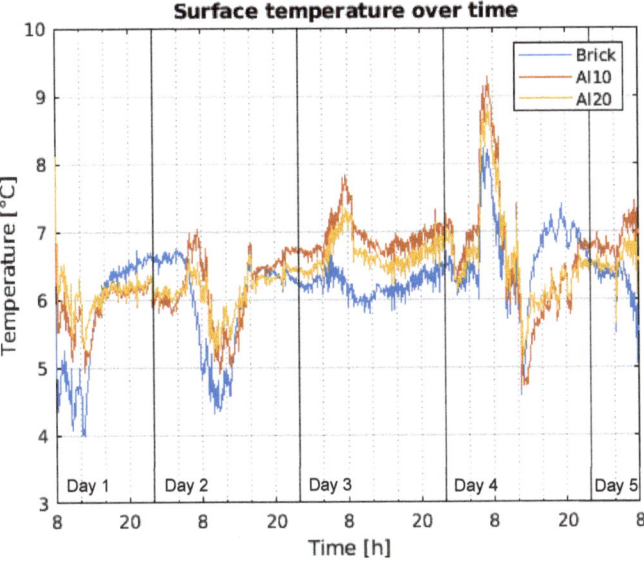

Figure 8. Average surface temperature for the ROIs depicted above. Noticeable is the greater temperature difference of the Al10 plate vs the Al20 plate, caused by the lower thermal mass.

Table 2. Weather table for the 96 h experiment. Weather data is collected by [11] for the weather station "Saarbrücken-Ensheim" which is located ~11 km from the experiment site. Note the high average cloud coverage and low total sunshine per day.

Day	Date	Min. Temp. [°C]	Avg. Temp. [°C]	Max. Temp. [°C]	Total Sunshine [h]	Avg. Humidity [%RH]	Total Rain [mm]	Cloud Coverage [%]
1	9 November 2020	−0.3	3.2	7.0	0.5	93.2	1.1	70
2	10 November 2020	−1.2	0.7	3.0	0.6	96.3	0.0	70
3	11 November 2020	0.6	2.8	4.5	0.0	94.3	2.9	98
4	12 November 2020	0.9	4.2	7.8	1.5	85.1	0.3	78
5	13 November 2020	0.8	2.7	4.2	0.0	94.6	0.5	98

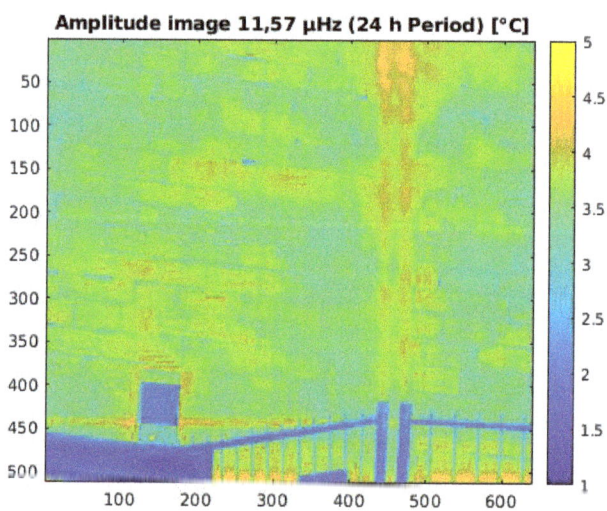

Figure 9. Amplitude image of the fundamental 24 h period. This image depicts the 24 h temperature amplitude average in every pixel over the 96 h measurement period $|S_{xy}|$.

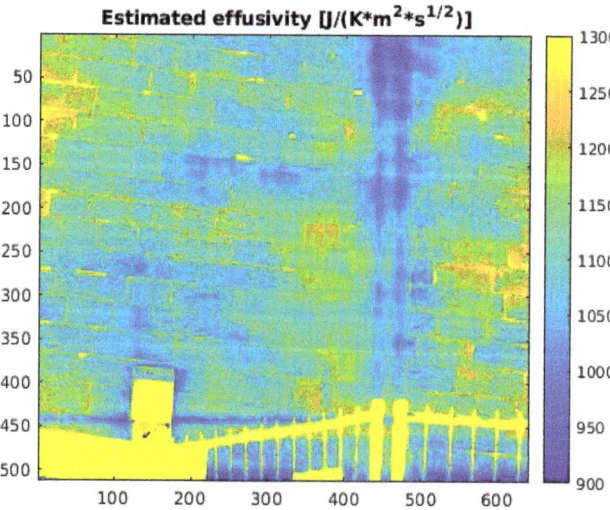

Figure 10. Estimated thermal effusivity for each pixel. Note that these results are only valid for regions that are thermally thick. Regions like the reference plates or the steel fence result in meaningless effusivity values.

There are two regions with higher and lower apparent effusivity of particular interest:
- The gaps between bricks (made out of an unknown type of mortar) correctly measure higher values than the bricks themselves
- The plaster strip on the right correctly measures lower effusivity than brick (plaster ~900 vs. brick ~1100 $J/\left(K \cdot m^2 \cdot s^{1/2}\right)$). Note that the plaster strip is strictly speaking not thermally thick. However, as the sensitivity of the measured effusivity is greatest in the front and less sensitive to the back layers the result correlate well to literature values for these materials. This sensitivity drop along the z-axis stems from the fact that the temperature amplitude of the structure drops exponentially over the depth [8]. Experiencing less amplitude, deep layers cannot influence the surface temperature as much as the front layers. Exact estimations on which layer has how much impact is not fully examined yet.

Phase Evaluation and Thermal Thickness

In theory, thermally thin materials show a negative 90-degree angle relative to the excitation. This makes intuitive sense, because the temperature of the material will rise until the excitation reaches zero again. After which the temperature of the object begins to fall. Thermally thick materials should show negative 45-degrees. This is because internal heat transfer is not negligible and changes how the surface reacts to heat input. These phase angle differences between thermally thick and thin materials are present in laboratory as well as field experiments. The experimental phase angle difference in the laboratory setup is ~30 degrees and in the field ~60 degrees. In both cases using the fundamental excitation frequency, based on phase evaluation the correct objects in frame are identifiable as thermally thick or thin. It is hypothesized that the phase angle information can be used to determine "thermal thickness" or at the very least to distinguish thick from thin in thermally adequate circumstances. Former research shows that phase evaluation is a valid method to identify subsurface defects or material changes [11,12]. This confirms the hypothesis and may contribute a factor to explaining their findings.

In the field experiments, thermal thickness can be inferred correctly from phase angle images, depicted in Figure 11. Figure 5 shows how the reference plates, the guardrail in the lower left corner and the construction fence in the lower part of the image significantly differ in their phase angle compared to the thermally thick retaining wall structure.

To what extent this method generally can be used to determine apparent thermal thickness is subject to ongoing research. What certainly is known that determining other thermal material properties using phase information, like thermal effusivity, thermal conductivity and alike, is much harder, because the phase angle depends on other variables more strongly than solely thermal material properties [11].

Especially excitation variation (over multiple days) will result in phase angle shifts.

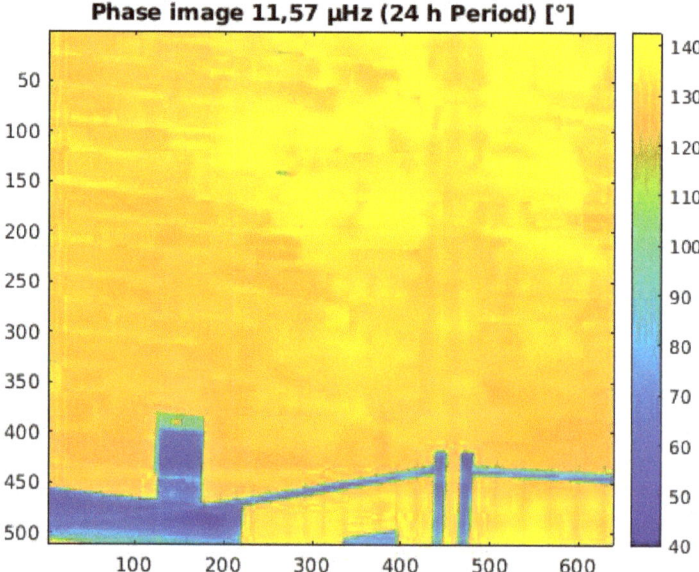

Figure 11. Phase image of the fundamental 24 h period. This image depicts the 24 h phase information every pixel over the 96 h measurement period $arctan(\Im m\{S_{xy}\}/\Re e\{S_{xy}\})$. Note that 90° phase would correlate to an angle of a sine wave, which has its origin at the time of the experiment start (8:00) and maximum 6 h later at 14:00. Note that higher angle values mean 'earlier' thermal responses to the input. If the phase angle value is lower, that corresponds to a 'slower' response to the periodic heat input.

5. Conclusions and Further Work

The presented algorithm has shown good measurement results, albeit non-optimal measurement conditions in the field experiment. It can not only obtain qualitative measurement results on the retaining wall structure but also identify thermal thickness differences using the phase angle information. The application of the presented algorithm can be extended to other kinds of infrastructure for inspection and monitoring purposes.

Further work has to be done to bring the measured results closer to the actual thermal material properties of the structure; this involves separating thermal effects to get a better understanding on the exact heat flow acting on the test subject. How phase angle information can be used to estimate thermal thickness and do quantitative analysis in specific setups is subject to ongoing research as well.

The calibration approximation has to verified under other conditions like other materials or other excitation conditions (i.e., more Sun, another season). If the approximation and regression fit holds for the other conditions, it can be shown to be sufficiently robust to investigate further on how this exact finding can be explained considering the physical effects acting on the test subjects.

Furthermore, increasing accuracy by isolating singular systematic error sources and developing a more sophisticated correction term based on physical properties can be done. For example, measuring apparent ambient temperature with a pyrgeometer and estimating radiative heat flux directly. Using a pyranometer to measure radiative heat input the remaining heat in-/output can be inferred to be convective. Finally, all heat in- and outputs are known and a more accurate model can be set up.

Lastly, a novel method could be devised to include amplitude information from various other frequencies. This work focusses solely on the fundamental 1/24 h frequency, where as other frequency bins are sure to contain additional information.

Author Contributions: Conceptualization, S.K. and H.-G.H.; methodology, S.K.; software, S.K.; validation, S.K., H.F., H.-G.H.; writing—original draft preparation, S.K.; writing—review and editing, S.K., H.F., H.-G.H.; visualization, S.K.; supervision, H.-G.H.; project administration, H.-G.H.; funding acquisition, H.-G.H. All authors have read and agreed to the published version of the manuscript.

Funding: We acknowledge support by the Deutsche Forschungsgemeinschaft (DFG, German Research Foundation) and Saarland University within the funding programme Open Access Publishing; This research was funded by the European Regional Development Fund (ERDF), especially grant number 14.2.1.4-2016-1.

Informed Consent Statement: Informed consent was obtained from all subjects involved in the study.

Data Availability Statement: Most data presented in this study are available on request from the corresponding author, including scripts and embedded software. Some data are not publicly available due to privacy reasons of the property owners.

Acknowledgments: Gratitude goes to the RAG Montan Immobilien for access to the experiment site. Furthermore, the authors thank Rouven Schweizer, Tobias Heib, Stephan Bechtel and Jessica Jacob for their help with experimental setup and conduction.

Conflicts of Interest: The authors declare no conflict of interest. The funders had no role in the design of the study; in the collection, analyses, or interpretation of data; in the writing of the manuscript, or in the decision to publish the results.

References

1. Boué, C.; Holé, S. Infrared thermography protocol for simple measurements of thermal diffusivity and conductivity. *Infrared Phys. Technol.* **2012**, *55*, 376–379. [CrossRef]
2. Ishizaki, T.; Nagano, H. Measurement of 3D thermal diffusivity distribution with lock-in thermography and application for high thermal conductivity CFRPs. *Infrared Phys. Technol.* **2019**, *99*, 248–256. [CrossRef]
3. Sfarra, S.; Marcucci, E.; Ambrosini, D. Infrared exploration of the architectural heritage: From passive infrared thermography to hybrid infrared thermography (HIRT) approach. *Mater. Constr.* **2016**, *66*, 94. [CrossRef]
4. Krankenhagen, R.; Maierhofer, C. Pulse phase thermography for characterising large historical building façades after solar heating and shadow cast—A case study. *Q. InfraRed Thermogr. J.* **2013**, *11*, 10–28. [CrossRef]
5. EN AW 5083 Data Sheet. Available online: https://gleich.de/en/wp-content/uploads/sites/4/2016/10/en_en_aw_5083.pdf (accessed on 5 February 2021).
6. Lienhard, J.H., IV; Lienhard, J.H., V. *A Heat Transfer Textbook*; Phlogiston Press: Cambridge, MA, USA, 2020.
7. Quirin, S.; Herrmann, H.-G. Combining the spectral information of Dual-Band images to enhance contrast and reveal details. QIRT **2018**. [CrossRef]
8. Breitenstein, O.; Warta, W.; Langenkamp, M. *Lock-in Thermography*; Springer: Berlin/Heidelberg, Germany, 2010. [CrossRef]
9. Goforth, M.; Gilchrist, G.; Sirianni, J. Cloud effects on thermal downwelling sky radiance. *AeroSense Int. Soc. Opt. Photonics* **2002**, *4710*, 203–213. [CrossRef]
10. Bortolin, A.; Cadelano, G.; Ferrarini, G.; Bison, P.; Peron, F.; Maldague, X. High-resolution survey of buildings by lock-in IR thermography. In Proceedings of the SPIE 8705, Thermosense: Thermal Infrared Applications XXXV, 870503, Baltimore, MD, USA, 22 May 2013. [CrossRef]
11. Wetter und Klima—Deutscher Wetterdienst. Available online: https://www.dwd.de/DE/leistungen/klimadatendeutschland/klimadatendeutschland.html (accessed on 5 February 2021).
12. Ibarra-Castanedo, C.; Sfarra, S.; Klein, M.; Maldague, X. Solar loading thermography: Timelapsed thermographic survey and advanced thermographic signal processing for the inspection of civil engineering and cultural heritage structures. *Infrared Phys. Technol.* **2017**, *82*, 56–74. [CrossRef]

Article

Application of Tooth Gear Impact-Echo System for Repeated and Rapid Data Acquisition

Jinyoung Hong [1], Hajin Choi [1,*] and Tae Keun Oh [2,3,*]

1 School of Architecture, Soongsil University, Seoul 06978, Korea; jinyoung23@soongsil.ac.kr
2 Department of Safety Engineering, Incheon National University, Incheon 22012, Korea
3 Research Institute for Engineering and Technology, Incheon National University, Incheon 22012, Korea
* Correspondence: hjchoi@ssu.ac.kr (H.C.); tkoh@inu.ac.kr (T.K.O.); Tel.: +82-032-835-8294 (T.K.O.)

Received: 23 June 2020; Accepted: 10 July 2020; Published: 12 July 2020

Abstract: Developments in air-coupled testing hardware in impact-echo (IE) tests have enabled new levels of scanning tests for concrete bridge decks. A tooth gear IE system has been developed using tooth gears as impactors and microelectromechanical systems (MEMS). Since the tooth gear moves and generates impacts itself, this system collects a large amount of test data across the field continuously. The contact duration of two different tooth gears is evaluated and the contact mechanism is compared to a conventional steel ball impactor by a high-speed camera. The data measurements were carried out on concrete slabs, where artificial delaminations were embedded at different depths. Based on our IE experiments, reducing the pitch or increasing the number of teeth was required to decrease the contact duration and generate the thickness mode frequency from deep delaminations. Rapidly obtained time domain data were transferred to the frequency-time domain using spectrograms to identify the dominant frequency band of the signal set. The results show that the developed system enabled us to acquire high-quality data during air-coupled IE tests and spectrogram analysis provided meaningful frequency information and verified its repeatability.

Keywords: impact-echo; tooth gear impactor; contact duration; delamination; rapid scanning; non-destructive testing; concrete slab

1. Introduction

Delamination-like defects are a major concrete bridge-deck deterioration concern. The internal delamination causes degradation of the structural elements in bridges as well as problems of serviceability by exposing the surface. Therefore, early detection of the internal defects is required to repair bridge decks. Impact-echo (IE) tests are a well-used non-destructive testing (NDT) method to evaluate delamination-like defects in concrete. Compared to other NDT methods, the IE test is a relatively simple testing configuration, which includes a hammer-type source and a vibration sensor. The hammer-type source literally makes an impact onto the surface of the concrete and the corresponding echo is measured. The IE test identifies the presence of delamination with depth information, and so it has been applied to bridge-deck inspection [1].

Since the IE test has proven its potential for internal damage detection [2,3], various studies have been conducted to understand the damage mechanism as well as to improve measuring systems. Gibson and Popovics discovered the physical relationship between the measured frequency and the depth of a plate-like structure, suggesting the effect of Poisson's ratio [4]. Zhu and Popovics proposed adding air-coupled sensing to the IE test, demonstrating the leakage of the vibration mode into the air [5]. Oh et al. investigated the application of the air-coupled IE test in the case of shallow delamination and the visualization of its vibration modes [6]. Kee et al. also explained the boundary condition effect on shallow delamination [7]. Further efforts have been made on the application of air-coupled sensing through the combination of different automated systems [8,9], their research reaching into efficient impact sources with air-coupled sensing. Mazzeo et al. showed chain dragging as an impact source while Sun et al. used a ball-chain system to filter out dragging noise [10,11].

Studies on impact sources are crucial for rapid examination of the field. Mazzeo et al. presented an interesting ice ball as an impactor [10]. The disposal of ice into concrete generated mechanical vibrations indicating that shallow delamination had been detected. Evani and Popovics also demonstrated the potential of a rolling impactor for the air-coupled IE test [12]. As established from previous studies, impact sources for the IE test could be found from surrounding products. In this study, a commercially available tooth gear was used as an impactor, generating continuous vibrational forces. A hand-controlled IE system was developed, including microelectromechanical systems (MEMS) as receivers. The aim of this paper is to demonstrate that tooth gears can be applied to air-coupled IE tests and to propose a rapid scanning test configuration including acoustic noise reduction by IE system development, measurement of thickness mode frequency and data visualization. To generate IE vibration modes, the contact duration onto the surface of concrete was experimentally evaluated and excited modes were measured from intact and delamination cases in concrete slabs. Consequently, system designs will improve and extend the understanding of field applications of the IE method.

2. Theoretical Background

2.1. Impact-Echo Vibration

The IE method provides the thickness information of a concrete slab or its delamination by measuring the dominant frequency of vibration. After applying an impact source onto the surface of the concrete, the measured vibration is analyzed in the frequency domain where the frequency at the highest amplitude corresponds to the thickness mode. The relationship between the measured frequency ($f_{thickness}$) and the thickness (h) is

$$f_{thickness} = \frac{\beta V_L}{2h} \qquad (1)$$

where β is the correction factor (usually taken to be 0.96) and V_L is the longitudinal wave velocity in the concrete. Gibson and Popovics demonstrated that the thickness mode frequency was at a zero-group velocity motion of the first symmetric vibration (S1-ZGV) governed by plate guided waves [4]. Moreover, a correction factor was required to compensate for the shear motion of the plate, depending on a ratio between the shear and longitudinal waves or Poisson's ratio. Many studies have utilized the concept of the S1-ZGV to detect internal defects using depth information [13–16]. Using Equation (1), the thickness of internal delamination-like defects can be estimated using information of the apparent velocity (βV_L).

However, in the case of shallow delamination, the thickness mode frequency is difficult to measure. This is because the impact source excitation frequencies have limited range up to the thickness mode frequency of shallow delamination. Instead, low frequencies from the flexural vibration of the plate are much more dominant compared to the thickness mode frequency. Oh investigated the excitability of thickness mode frequencies based on a ratio between the diameter (a) and thickness (h) of the plate under certain boundary conditions, demonstrating that the thickness mode frequency becomes dominant when the ratio (a/h) is lower than 2. Therefore, the IE testing results from shallow delamination mostly exhibit low frequencies [17]. The fundamental flexural vibration mode frequency ($f_{flexural}$) of rectangular delamination is presented as

$$f_{thickness} = \frac{k_{DL}^2 \pi}{2h^2} \sqrt{\frac{Eh^2}{12\rho(1-v^2)}} \qquad (2)$$

where E is the elastic modulus, v is the Poisson's ratio, ρ is the density, and k_{DL}^2 is the dimensionless frequency based on width to depth ratio of the delamination [18].

The excitation frequencies are directly related to the contact duration (t_c) of impactors onto the concrete surface. The shorter contact duration generates a higher frequency range of excitation. Based on contact mechanics [19], the contact duration is calculated by the sphere (impactor) and half-space (concrete) interaction, as follows:

$$t_c = 2.536 D \rho_i^{2/5} V_i^{-1/5} \left[\frac{(1-v_i^2)}{E_i} + \frac{(1-v_c^2)}{E_c} \right]^{2/5} \qquad (3)$$

where D is the diameter of the impactor, ρ_i is the density of the impactor, V_i is the impact velocity, v_i and E_i are the Poisson's ratio and elastic modulus of the impactor, respectively, and v_c and E_c are the Poisson's ratio and elastic modulus of concrete, respectively. Equation (3) expresses that the contact duration is directly related to the size of the impactor, beyond its material properties. The smaller size of the impactor reduces the contact duration, corresponding to a higher frequency range of excitation.

2.2. Air-Coupled Sensing Technology

The IE frequency components generated from concrete are measurable in air without touching the surface of the concrete. Zhu and Popovics theoretically and experimentally demonstrated that such vibration modes including the S1-ZGV mode and flexural mode could be measured by microphone [5]. In a half-space (air-concrete), the vibration modes were leaked into the air and an air-coupled sensor measured the same frequency components.

To confirm the theoretical background of the air-coupled IE method, wave propagation in the half-space was simulated using the k-wave toolbox in Matlab [20,21]. A half sine wave simulating a hammer-type impact was applied to the surface of concrete, and the corresponding stress wave field of air and concrete was solved. Forty sensors were continuously placed in air 20 mm above the surface of the concrete, 5 mm apart. The in-plane normal stress was analyzed as the output data. The details of the simulation are shown in Table 1. Figure 1a shows the wave field image of the air-coupled IE test, where the Von Mises stress is shown. From the image, the non-propagating S1-ZGV mode (thickness mode) was generated inside the concrete after the impact source was applied at the center of the concrete slab. In air, the leaky portion of the propagating waves (mainly Rayleigh waves) exhibited much greater velocity compared to the direct acoustics (semi-circle shape) generated by the impactor. Inside the semi-circle, the leaky portion of the thickness mode, measurable using air-coupled sensors, is shown. The signals obtained, as shown in Figure 1b, conveyed that the majority of frequencies were from the thickness mode vibration (S1-ZGV) while the propagating waves (including the leaky waves and direct acoustics) had a much higher amplitude, which were noise components of the air-coupled IE test. In fact, acoustic system noise has been reported during air-coupled sensing [11,12,16] so that efforts have been made to carefully measure the appropriate frequencies from the IE testing. Therefore, the impactor was required not to generate system noise itself while propagating leaky waves were quickly damped.

Table 1. Numerical simulation details.

Simulation Parameters		Value	Simulation Parameters	Value
CFL		0.1	Type	Half sine wave
Number of grids (N_x)		500	Input source Range of frequency	30 kHz
Number of grids (N_y)		1000	Location	(290, 500)
Mesh size (dx)		1 mm	Sensor Number of sensors	40
Mesh size (dy)		1 mm	Spacing	5 mm
Length of grid	Concrete	190 × 1000 mm	Concrete Elastic modulus	30 GPa
	Air	310 × 1000 mm		
Operation time		1.5 ms	Density	2400 kg/m³
dt		50 ns	Air Density	240 kg/m³

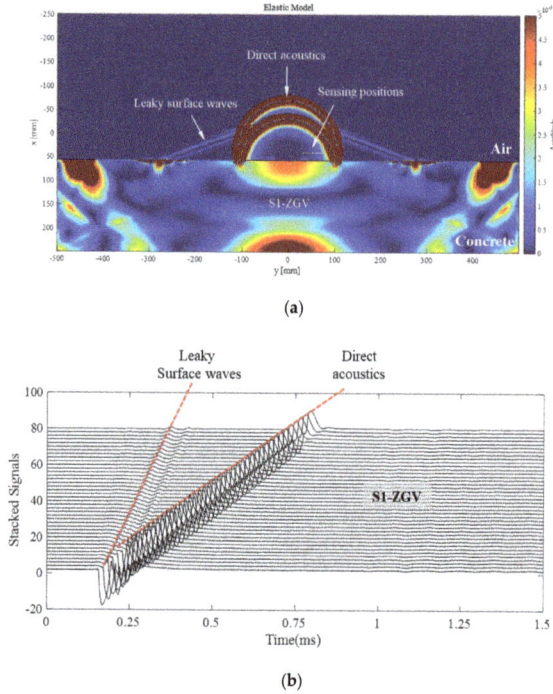

Figure 1. Illustration of air-coupled impact-echo (IE) test; (**a**) Field image of simulation, (**b**) synthetic signal data of simulation.

3. Tooth Gear Testing System

3.1. Impact Source Excitation by Tooth Gear

In this study, tooth gears were used as impactors for the air-coupled IE test. While the gear rolls and moves, continuous contact on the surface of concrete generates an impact source. Therefore, the gear has the advantage of a moving platform where the impactor itself moves and generates sources simultaneously. Moreover, the impact position is regular, based on the pitch of the gear.

Two different tooth gears were used to evaluate the excitability as impactors, including 11 and 30 teeth gears, respectively, as shown in Figure 2. The tooth width and thickness were 5.8 and 2.5 mm for the 11-tooth gear and 4.2 and 1.1 mm for the 30-tooth gear. Therefore, the potential contact area of the gears was 14.5 and 4.6 mm^2, respectively. The contact duration of the tooth gears was compared with that of a conventional steel ball impactor. The diameter of the steel ball was 12 mm. The motion of impact onto the surface of a concrete element was recorded using a high-speed camera (MEMRECAM GX-1). The videos were captured at 10,000 fps (320 × 240 resolution). Each frame was analyzed manually, and the contact duration and velocity of each impactor was estimated. The contact durations obtained were 0.1, 32, and 16 ms for the steel ball, the 11-tooth gear, and the 30-tooth gear, respectively. It is noted that the sampling rate was limited to measure the contact duration accurately; however, the different mechanism from the impactors were clearly shown. Using Equation (3), the contact duration of the steel ball was calculated as 0.08 ms (assuming the density of the steel ball to be 7850 kg/m^3, and the elastic modulus to be 200 at 30 GPa for steel ball and concrete, respectively).

As shown in Figure 3, the contact mechanism of the tooth gears is somewhat different from a hammer-type impactor. While the tooth gear continuously moves, the contact is sequentially made from one tooth to another. It is noted that the second contact was executed after waiting for the finish of the first contact. Therefore, the contact duration was longer than the contact duration from a conventional hammer-type impactor, implying low frequency excitation. The contact duration of a tooth gear depends on the pitch of the tooth (or the number of teeth) and the moving speed of the gear. The pitch of the gear (p) is defined by the radius of the gear (r) and the numbers of teeth (N), as follows:

$$p = \frac{2\pi r}{N} \tag{4}$$

Figure 2. Tooth gears used as impactors including the 11-tooth (silver) and 30-tooth (black) gears.

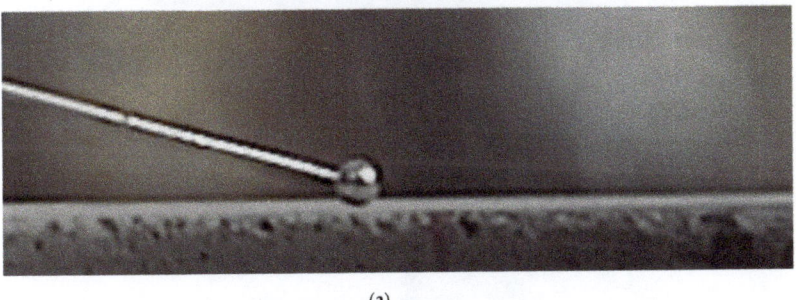

(a)

(b)

Figure 3. Photo of the contact mechanism: (**a**) 12 mm ball hammer, (**b**) 11-tooth gear.

Therefore, increasing the number of teeth reduces the pitch and the contact duration. From Equation (4), the pitches of the 11- and 30-tooth gears are 22.9 and 10.1 mm. With the same moving speed, the contact duration of the 30-tooth gear is 2.3 time shorter than that of the 11-tooth gear, implying that much a higher frequency range is excitable using the 30-tooth gear. With practical assumption of rolling speed as 250 m/s, the contact duration of 11- and 30-tooth gears are 0.09 and 0.04 ms, where corresponding excitation frequency ranges are 16.9 and 36.4 kHz.

3.2. Tooth Gear IE System Design and Data Acquisition

Based on the background information of the contact mechanism, a tooth gear IE system was designed as shown in Figure 4. For the receiver, a microelectromechanical system (MEMS, SPU0410LR5H-QB, Knowles) was used with a circuit amplified 2000 times, able to measure acoustic pressure with a sensitivity of −38 dBV/Pa. The location of the MEMS was in the same line as the impact source, 65 mm away from the tooth gear. The holder was manufactured using a three-dimensional (3-D) printer, and the system was manually controlled using a handle. The total length of the system was 200 mm so that inspectors could easily use the system in the field. While the system moves backwards and forwards, numerous signal data were measured. For data acquisition, the signals were digitized using the NI 6366, at a sampling rate of 200 kHz, and 2000 samples were collected. The total length of a signal was 10 ms. As shown in Figure 5, the testing equipment (including the laptop, data acquisition, and IE system) was placed on the concrete slab. Based on the test results, more than 100 data signals were recorded within one minute, while the system was moving.

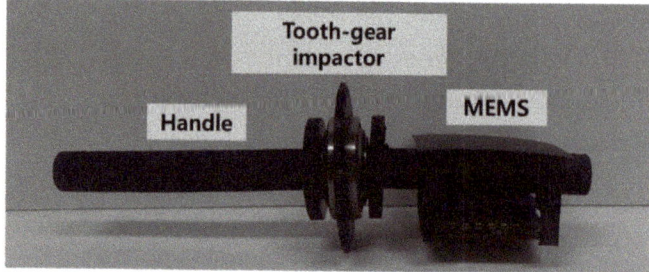

Figure 4. Designed IE system including tooth gear and microelectromechanical systems (MEMS).

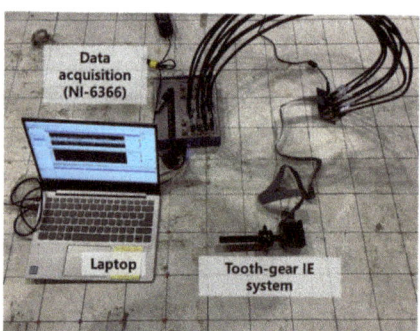

Figure 5. Data acquisition and experimental set-up for the IE system.

4. Experiment and Results

4.1. Testing Slab and the IE Results

The developed tooth gear IE system was evaluated using a reinforced concrete slab, where artificial delamination-like defects were embedded. As shown in Figure 6, the testing slab had a depth of 250 mm and the depth of the artificial delaminations were 60 (shallow) and 200 mm (deep), respectively. The artificial delamination was made from acrylic sheet (2 mm thickness), which was embedded before casting the concrete. The IE testing was performed on the testing slab using the 11 and 30-tooth gears. A 12 mm steel ball impactor was also used to perform conventional IE testing as a reference. Two locations, including the deep and shallow delaminations, were tested, where the thickness mode and the flexural mode, respectively, were obtained. Based on the Equations (1) and (2), thickness mode frequency is about 10 kHz with conventional longitudinal wave velocity of concrete (V_L) as 4250 m/s while flexural mode frequency is about 5.5 kHz with the dimensionless frequency (k_{DL}^2) as 0.2. It is noted that the width (300 mm) to depth (60 mm) ratio of shallow delamination is 0.5. In the case of the steel ball impactor, repeated impact sources were generated at the same position while the vibration was measured using an accelerometer (PCB, 352C15). Using the tooth gear IE system, a rapid scanning test was performed while the system simply moved backwards and forwards.

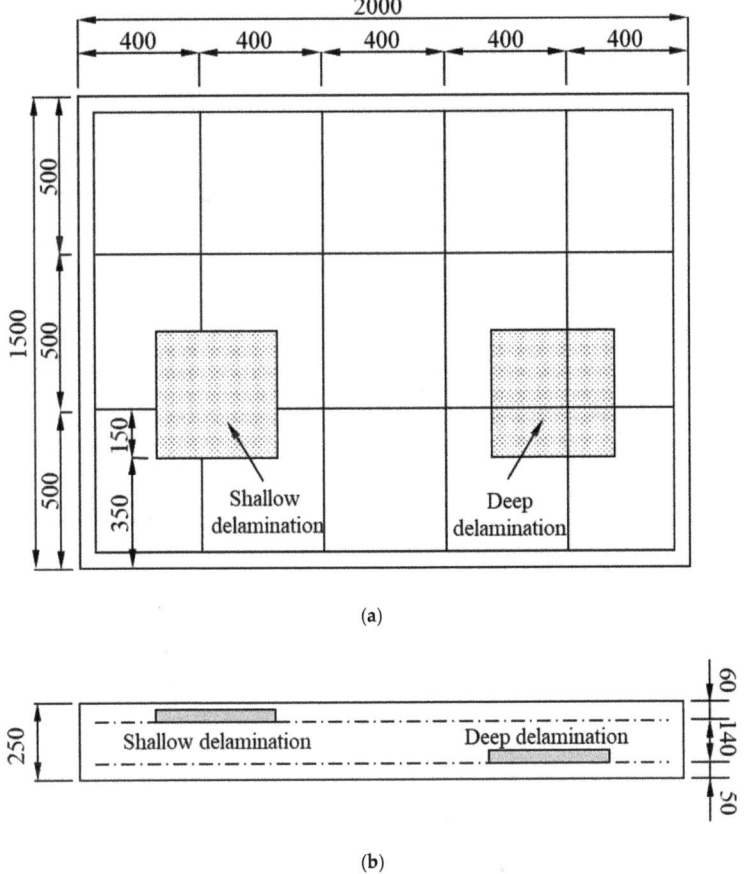

Figure 6. The details of the testing slab; (**a**) plan view (**b**) sectional view (the unit as mm).

The results of the frequencies obtained are summarized in Table 2, and examples of the signals obtained from the shallow and deep delaminations are shown in Figures 7 and 8, respectively. In the case of the shallow delamination, the measured frequency from the impactors was identified as 5.4 kHz (Figure 7). Although the waveforms from the tooth gears were noisy, the dominant frequency was clearly shown. This is because the excited frequency range of the impactors covers the low frequency range from the flexural vibration of plate, and the flexural mode is loud enough to be measured as the dominant frequency among the impactors. Based on the results from the steel ball impactor (Figure 8), the thickness mode frequency was detected as 10.2 kHz, which corresponded well with the depth of the deep delamination with an apparent velocity of 4080 m/s. In the case of the 11-tooth gear, low frequency noise was dominant. As expected from the long contact duration, the excited frequency range was clearly limited to the generation of the thickness mode. The results from the 30-tooth gear showed that reducing the pitch creates a shorter contact duration and the thickness mode becomes dominant.

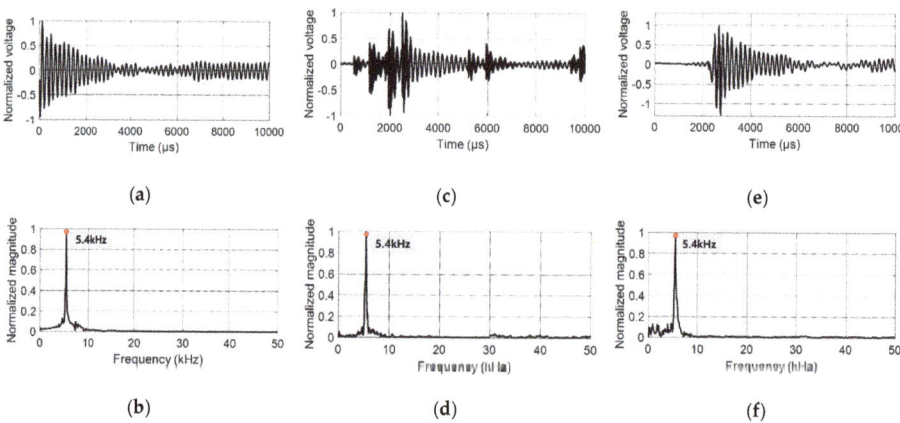

Figure 7. Obtained signals from the shallow delamination. Steel ball impactor (**a**) time signal and (**b**) frequency spectrum. 11-tooth gear (**c**) time signal and (**d**) frequency spectrum. 30-tooth gear (**e**) time signal and (**f**) frequency spectrum.

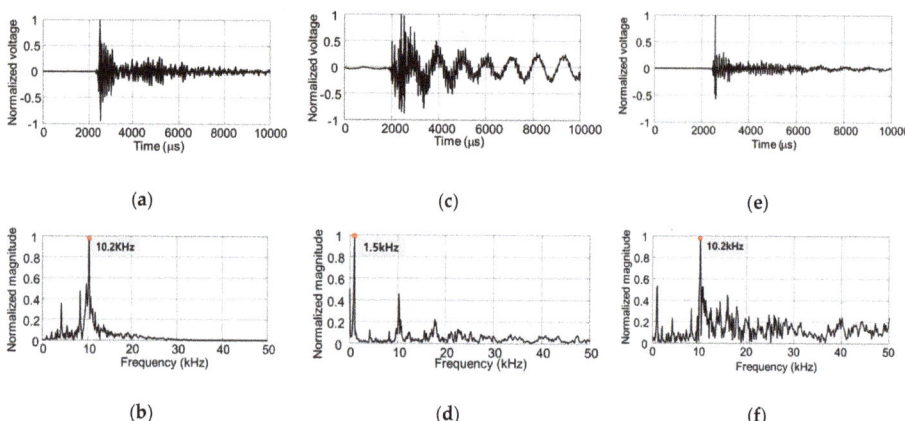

Figure 8. Obtained signals from the deep delamination. Steel ball impactor (**a**) time signal and (**b**) frequency spectrum. 11-tooth gear (**c**) time signal and (**d**) frequency spectrum. 30-tooth gear (**e**) time signal and (**f**) frequency spectrum.

Table 2. Results of obtained frequency.

Testing Locations	Depth (mm)	Measured Frequency (kHz)		
		12 mm Ball Hammer	11-Tooth Gear	30-Tooth Gear
Shallow delamination	60		5.4	
Deep delamination	200	10.2	1.5	10.2

4.2. Data Visualization

The speed of data acquisition by the tooth gear IE system was much faster than the conventional IE test. Based on numerous data, the visualized images help the operator make decisions [22,23]. The signal-processing scheme using a spectrogram was applied to visualize the obtained data from the tooth gear IE system as shown in Figure 9. After each 10 ms long signal was obtained from the tooth gear IE system, the amplitude of the signal was normalized. Next, the signals were added in the time domain—50 signals were used in this study. Spectrogram analysis was performed on the total signal set, based on a short-time Fourier transform, representing the time-frequency information with a color scheme. The dominant frequencies from the total signal set are clearly shown as a band. In addition to the data visualization, the analysis of the total signal set demonstrated the repeatability of the vibration modes during the IE testing. For a signal set of 500 ms length, the windowing samples and numbers of overlap were set to 2048 and 1900 respectively.

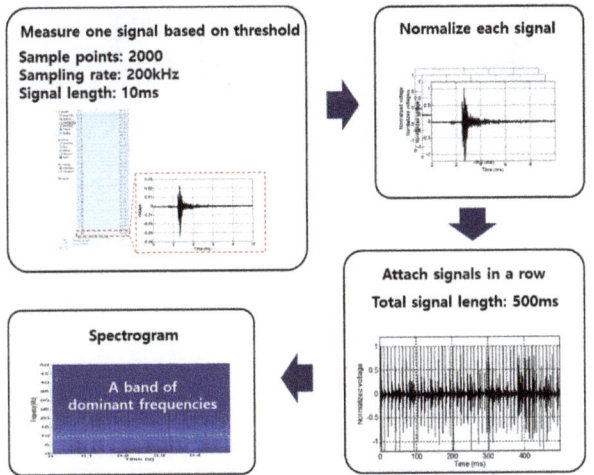

Figure 9. Applied signal processing scheme to visualize frequency information.

Figures 10 and 11 show the spectrograms from the shallow and deep delaminations between the impactors. From the shallow delamination, a band of 5.4 kHz is clearly shown from all impactors. In the case of the 11-tooth gear, a little noise content was found from the signal set. However, the band of flexural mode frequencies were clearly shown. The thickness mode frequencies from the steel ball were clearly shown in the spectrogram. In the case of the 30-tooth gear, the frequency band was detected while a portion of noise content was also captured. However, the 11-tooth gear exhibited limited results of thickness mode detection due to its long contact duration.

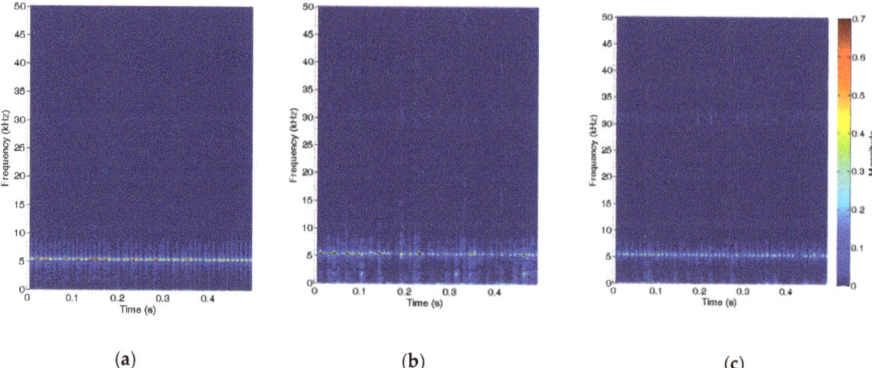

Figure 10. Results of spectrogram from the shallow delamination; (**a**) steel ball hammer, (**b**) 11-tooth gear, (**c**) 30-tooth gear.

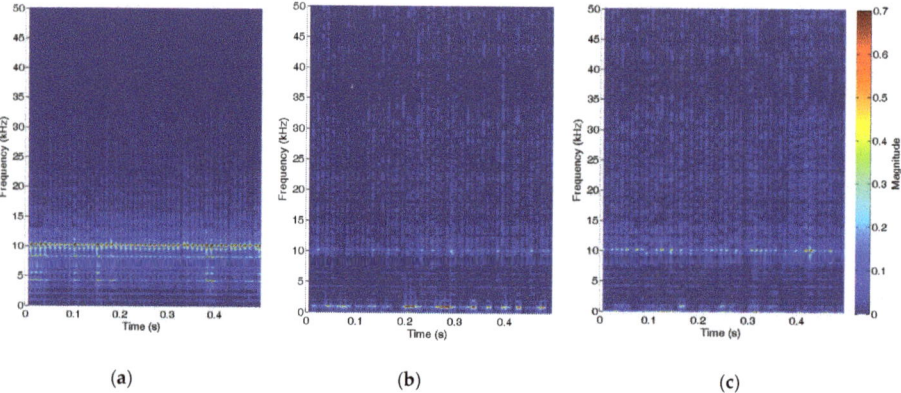

Figure 11. Results of spectrogram from the deep delamination; (**a**) steel ball hammer, (**b**) 11-tooth gear, (**c**) 30-tooth gear.

In terms of repeatability, the tooth gear showed great potential for the air-coupled IE test. Air-coupled sensing technology inherently contains noise contents from the field (traffic noise). The impacts of tooth gears are easily repeatable while the system moves and generates the source simultaneously. The large quantity of data obtained improves the decision-making process in the field. In fact, delamination in the field has arbitrary boundary conditions so that the vibration modes and frequency components can be many. Under those conditions, repeatable data with rapid data acquisition helps the analysis of the IE test.

5. Conclusions

This study demonstrated that tooth gears have great potential to produce impact sources of IE testing and generate IE vibration modes including thickness and flexural modes where the results are comparable with conventional steel ball impactors. To perform a rapid scanning test, tooth gears were used as impactors and their performance was evaluated. The contact duration of each impactor was theoretically and experimentally analyzed and a tooth gear IE system was developed. The developed system was evaluated using a testing slab, which included both shallow and deep delaminations. A further signal-processing scheme was used to visualize the signal set obtained by the rapid scanning test. The findings and conclusions of the research are summarized as follows:

- the developed tooth gear IE system is possible to apply to an impact-echo test with low acoustic noise and provides fast data collection with continuous matter. A signal set of 100 data points was measurable within one min. Using the system, the thickness mode frequencies are measurable beyond flexural mode frequencies, and measured thickness mode frequencies corresponded well with the depth of the deep delamination;
- by analyzing different contact mechanism from conventional hammer-type impact, reducing the pitch or increasing the number of teeth decreased the contact duration and provided higher frequency excitation. Based on our experimental results, the 11-tooth gear was limited at generating the thickness mode from the 200 mm depth delamination, while the 30-tooth gear was successful;
- based on the contact mechanism, the contact duration of a conventional steel ball and two tooth gears was analyzed. The theoretical contact duration corresponded well with the actual contact duration measured using a high-speed camera, implying an excited frequency range;
- data visualization using spectrogram showed a band of dominant frequencies (repeated frequency modes) from the signal set, which helps operators to make decisions based on repeatable vibration modes.

Author Contributions: H.C. conceived and designed the experiments; J.H. and H.C. performed the experiments and analyzed the data; J.H., H.C., and T.K.O. contributed device/analysis tools; J.H., H.C., and T.K.O. wrote the paper. All authors have read and agreed to the published version of the manuscript.

Funding: This research was supported by the Incheon National University Research Grant in 2018.

Acknowledgments: Authors thank Popovics for valuable discussion of the tooth-gear application on the air-coupled impact-echo.

Conflicts of Interest: The authors declare no conflict of interest.

References

1. Carino, N.J.; Sansalone, M.; Hsu, N.N. A point source-point receiver, pulse-wcho technique for flaw detection in concrete. *ACI J.* **1986**, *83*, 199–208.
2. Cheng, C.; Sansalone, M. The impact-echo response of concrete plates containing delaminations: Numerical, experimental and field studies. *Mater. Struct.* **1993**, *26*, 274–285. [CrossRef]
3. Sansalone, M.J.; Streett, W.B. *Impact-Echo: Nondestructive Evaluation of Concrete and Masonry*; Bullbrier Press: Ithica, NY, USA, 1997.
4. Gibson, A.; Popovics, J.S. Lamb wave basis for impact-echo method analysis. *J. Eng. Mech.* **2015**, *131*, 438–443. [CrossRef]
5. Zhu, J.; Popovics, J.S. Imaging concrete structures using air-coupled impact-echo. *J. Eng. Mech.* **2007**, *133*, 628–640. [CrossRef]
6. Oh, T.; Popovics, J.S.; Sim, S.-H. Analysis of vibration for regions above rectangular delamination defects in solids. *J. Sound Vib.* **2013**, *332*, 1766–1776. [CrossRef]
7. Kee, S.-H.; Gucunski, N. Interpretation of flexural vibration modes from impact-echo testing. *J. Infrastruct. Syst.* **2016**, *22*, 04160009. [CrossRef]
8. Gucunski, N.; Yan, M.; Wang, Z.; Fang, T.; Maher, A. Rapid bridge deck condition assessment using three-dimensional visualization of impact echo data. *J. Infrastruct. Syst.* **2011**, *18*, 12–24. [CrossRef]
9. Zhang, G.; Harichandran, R.S.; Ramuhalli, P. An automatic impact-based delamination detection system for concrete bridge decks. *NDT E Int.* **2012**, *45*, 120–127. [CrossRef]
10. Mazzeo, B.A.; Patil, A.N.; Hurd, R.C.; Klis, J.M.; Truscott, T.T.; Guthrie, W.S. Air-coupled impact-echo delamination detection in concrete using spheres of ice for excitation. *J. Nondestruct. Eval.* **2014**, *33*, 317–326. [CrossRef]
11. Sun, H.; Zhu, J.; Ham, S. Automated Acoustic Scanning System for Delamination Detection in Concrete Bridge Decks. *J. Bridge Eng.* **2018**, *23*, 04018027. [CrossRef]
12. Evani, S.H.; Popovics, J.S. Utility of Rolling Impactor as a Source of Excitation for Impact-Echo Tests. SMT 2016, American Society for Nondestructive Testing. August 2016. Available online: https://publish.illinois.edu/saikalyanevani/previous-projects/07112020 (accessed on 11 July 2020).

13. Clorennec, D.; Prada, C.; Royer, D. Local and noncontact measurements of bulk acoustic wave velocities in thin isotropic plates and shells using zero group velocity Lamb modes. *J. Appl. Phys.* **2007**, *101*, 034908. [CrossRef]
14. Prada, C.; Clorennec, D. Influence of the anisotropy on zero-group velocity lamb modes. *J. Acoust. Soc. Am.* **2009**, *126*, 620–625. [CrossRef] [PubMed]
15. Gomez, P.; Fernandez-Alvarez, J.P.; Ares, A.; Fernandez, E. Guided-wave approach for Spectral peaks characterization of impact-echo tests in layered systems. *J. Infrastruct. Syst.* **2017**, *1*, 0417009. [CrossRef]
16. Choi, H.; Azari, H. Guided wave analysis of air-coupled impact-echo in concrete slab. *Comput. Concr.* **2017**, *20*, 257–262.
17. Oh, T. Defect Characterization in Concrete Elements Using Vibration Analysis and Imaging. Ph.D. Dissertation, University of Illinois at Urbana-Champaign, Champaign, IL, USA, 2012.
18. Leissa, A.W. The free vibration of rectangular plates. *J. Sound Vib.* **1973**, *31*, 257–293. [CrossRef]
19. Stronge, W.J. *Impact Mechanics*; Cambridge University Press: Cambridge, UK; New York, NY, USA, 2000.
20. Firouzi, K.; Cox, B.T.; Treeby, B.E.; Saffari, N. A first-order k-space model for elastic wave propagation in heterogeneous media. *J. Acoust. Soc. Am.* **2012**, *132*, 1271–1283. [CrossRef] [PubMed]
21. Treeby, B.E.; Jaros, J.; Rohrbach, D.; Cox, B.T. Modelling Elastic Wave Propagation Using the k-Wave MATLAB Toolbox. In Proceedings of the 2014 IEEE International Ultrasonics Symposium, Chicago, IL, USA, 3–6 September 2014; pp. 146–149.
22. Matsuyama, K.; Yamada, M.; Ohtsu, M. On-site measurement of delamination and surface crack in concrete structure by visualized NDT. *Constr. Build. Mater.* **2010**, *24*, 2381–2387. [CrossRef]
23. Ohtsu, M.; Watanabe, T. Stack imaging of spectral amplitudes based on impact-echo for flaw detection. *NDT E Int.* **2002**, *35*, 189–196. [CrossRef]

© 2020 by the authors. Licensee MDPI, Basel, Switzerland. This article is an open access article distributed under the terms and conditions of the Creative Commons Attribution (CC BY) license (http://creativecommons.org/licenses/by/4.0/).

Review

Electrical Resistivity and Electrical Impedance Measurement in Mortar and Concrete Elements: A Systematic Review

Gloria Cosoli [1,*], Alessandra Mobili [2,*], Francesca Tittarelli [2], Gian Marco Revel [1] and Paolo Chiariotti [3]

1. Department of Industrial Engineering and Mathematical Sciences, Università Politecnica delle Marche, 60131 Ancona, Italy; gm.revel@staff.univpm.it
2. Department of Materials, Environmental Sciences and Urban Planning, Università Politecnica delle Marche, INSTM Research Unit, 60131 Ancona, Italy; f.tittarelli@univpm.it
3. Department of Mechanical Engineering, Politecnico di Milano, 20156 Milan, Italy; paolo.chiariotti@polimi.it
* Correspondence: g.cosoli@staff.univpm.it (G.C.); a.mobili@staff.univpm.it (A.M.)

Received: 6 December 2020; Accepted: 18 December 2020; Published: 21 December 2020

Featured Application: Systematic review of electrical resistivity and impedance measurements methods and applications on cement-based materials.

Abstract: This paper aims at analyzing the state-of-the-art techniques to measure electrical impedance (and, consequently, electrical resistivity) of mortar/concrete elements. Despite the validity of the concept being widely proven in the literature, a clear standard for this measurement is still missing. Different methods are described and discussed, highlighting pros and cons with respect to their performance, reliability, and degree of maturity. Both monitoring and inspection approaches are possible by using electrical resistivity measurements; since electrical resistivity is an important indicator of the health status of mortar/concrete, as it changes whenever phenomena modifying the conductivity of mortar/concrete (e.g., degradation or attacks by external agents) occur, this review aims to serve as a guide for those interested in this type of measurements.

Keywords: electrical resistivity; electrical impedance; concrete; mortar; measurement accuracy

---

1. Introduction

An effective maintenance strategy of concrete structures asks for a strong involvement of Structural Health Monitoring (SHM) because this makes it possible to minimize the required interventions and to optimize their cost-effectiveness. However, despite, in developed countries, the maintenance and repairing costs of concrete structures being estimated at around 3–4% of gross national product [1], at present, no regular monitoring techniques are widely applied. Inspections are preferred, even if their efficiency is more limited than the one provided by monitoring, which continuously provides data and information on the health status of the target structure. Furthermore, it is worth noting that the service life of a non-temporary concrete structure is quite long: for a common structure, it accounts to 50 years, for a strategic structure (e.g., a monument, a bridge, or other civil engineering structures), at least 100 years [2].

Since non-destructive techniques (NDTs) can provide data on the structure without the necessity of sampling the structure (therefore exposing the structure itself to higher risks of damage), they are preferred for SHM. Among NDTs, the measurement of electrical resistivity of concrete is gaining wider and wider consensus among the scientific and technical community. Indeed, the technique is easy to adopt, tests can be performed rapidly, and several factors related to concrete durability [3–5] can modify the resistivity of concrete, like water and chlorides penetration, carbonation, or presence of cracks.

Electrical resistivity (ρ [$\Omega \cdot$m]) is the ability of a material to oppose the flow of electric current. The electrical resistivity of concrete varies from 10^1 to 10^5 $\Omega \cdot$m, mainly in relation to its moisture content and composition [6–10]. The composition includes cement type (fineness, composition, and soluble salts content), water to cement (w/c) ratio, type and size of aggregates, and presence of additions, either conductive or not. Indeed, as aggregates are concerned, they act as an obstacle in the electric current path, hence increasing the electrical resistivity of concrete [11]. Electrical resistivity depends on cement paste microstructure, characterized by the volume, the size, and the interconnection of pores, besides the pore solution, where different salts are dissolved, releasing ions (mainly K^+, Na^+, OH^-, and, in smaller extent, also Ca^{2+} and SO^{2-}) that enable conduction [12]. Porosity is governed by the w/c ratio: the higher the w/c ratio, the higher the total porosity [13] and, consequently, the lower the electrical resistivity [14,15]. However, in highly saturated concrete, the effect of w/c ratio on electrical resistivity is less significant [6,16] because moisture (thus water) plays a major role in defining concrete electrical properties. Supplementary cementitious materials (e.g., fly ash, silica fume, and ground granulated blast furnace slag (GGBFS), just to cite those mostly used) reduce both pores size and volume. A refinement of the pore structure increases ρ. Indeed, these mineral additions (e.g., blast-furnace slag [17], fly ash [18], and silica fume [19]), are also generally added to release more durable concretes [20]. On the other hand, if conductive fillers or fibers are added to the mix-design, electrical resistivity decreases, and the material conduction partially moves from electrolytic to electric [21–23]. The electrolytic conductivity is related to mobile ionic species that form when water and binder are mixed together; concrete can be thought of as an electrolytic conductor, since the current mainly passes through the pore solution [24]. On the other hand, the electronic conductivity is related to the transportation of free electrons through the conductive phase (carbon or metal fillers/fibers). Contrarily, non-conductive polymeric fibers seem not to have a significant effect on the matrix electrical resistivity [22]. The literature reports some reference values of electrical resistivity related to cement type and exposure conditions [25,26]; however, it is worth considering that each mix-design would provide different electrical resistivity values.

In reinforced concrete structures (RCS), another parameter affecting electrical resistivity is the short circuit effect related to the embedded steel reinforcement [27]; in fact, reinforcement, being metallic, attracts the electric current [28], distorting the measurement result. Besides the intrinsic characteristics of concrete, also the environmental conditions, particularly temperature (T) and relative humidity (RH) [29], impact concrete electrical resistivity. Concerning T, a higher value increases the ions mobility, making ρ decrease. Concerning RH, since water has a good electrical conductivity, the material electrical conductivity increases with its saturation degree. In this regard, also aging of concrete influences the electrical resistivity, since it gives a reduction in the moisture content due to the continuous hydration of the cement paste, which modifies the paste microstructure.

Electrical resistivity also correlates with concrete degradation status, represented by different durability indicators as saturation degree, porosity, and chloride penetration [30]; electrical resistivity of concrete makes it possible to detect the penetration of aggressive agents prior to the onset of the degradation process [31]. If porosity is high, permeability increases and, consequently, so does the ingress of aggressive substances. This ingress mines the integrity of concrete and reinforcements, especially if the pore interconnection is high. In addition, thanks to the piezoresistive behavior of concrete (piezo-resistivity is a physical characteristic of electrically-conductive materials, whose electrical resistivity changes with deformations [32]), electrical resistivity measurements make it possible to monitor the stress/strain behavior of concrete [33].

The electrical resistivity of concrete can be measured by direct current (DC) or alternating current (AC). The type of current used (DC or AC) and the electrodes configuration (e.g., electrode spacing [34,35]) adopted influence the measurement result, especially in terms of accuracy. A minimum of two electrodes is needed to excite the material with an electric current (or potential) and to measure the corresponding electrical potential difference (or current). Another configuration can adopt 4-electrode measurement points, separating excitation and measurement electrodes in order to avoid sharing electric current and potential. Also, the electrodes positioning, which the current path depends from [36], heavily affects the measurement results. Different types of electrodes are used in the literature, such as wires, plates, meshes, bars, tapes, and conductive paints.

However, one of the major problems regarding electrical resistivity measurement in concrete is that no widely accepted standards are currently available, and many different methodologies are reported in the literature. Therefore, this review paper aims at providing the reader with an overview of the different approaches suggested by the scientific community when addressing electrical resistivity measurements, as well as highlighting some issues that may be tackled by the experimenter as well as the real missing item related to the topic, i.e., measurement uncertainty. Also, several target applications of this measurement method are discussed, with a focus on inspection and monitoring of cement-based structures. Finally, suggestions on how to perform electrical resistivity/impedance measurements, either at lab-scale or in-field, are provided.

2. Electrical Resistivity/Impedance Measurement: Is It Really that Simple?

2.1. DC Measurement of Electrical Resistivity

2.1.1. 2-Electrode Configuration

The two-point probe method, also known as uniaxial method, enables the measurement of bulk electrical resistivity. The specimen is placed between two electrodes (commonly parallel metal plates) with moist sponge contacts, then a DC electric current is injected and the consequent potential drop is measured (Figure 1).

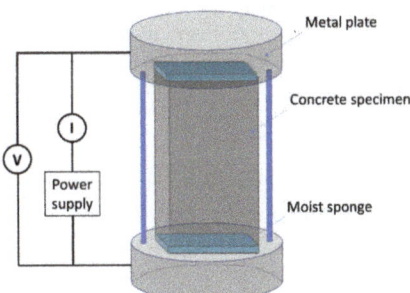

Figure 1. Uniaxial method (please note that electric current is measured in series to the power supply, whereas the corresponding potential drop is measured in parallel).

The standard to perform electrical resistivity measurements with the uniaxial method is the ASTM C1760 "Standard test method for bulk electrical conductivity of hardened concrete" [37]. The standard foresees the use of saturated cylindrical (100 mm diameter, 200 mm length) hardened concrete specimens, with a standard 28-day curing (56-day curing in the presence of supplementary cementitious materials, such as GGBFS or fly ash, since they prolong the time to dry). A 60 V DC potential difference is applied and the electric current is measured after 1 min; therefore, the electrical conductivity is obtained according to Equation (1):

$$\sigma = k \cdot \frac{I_1}{V} \cdot \frac{L}{D^2} \ [\text{mS/m}] \tag{1}$$

where:

- I_1 is the electric current at minute 1 (±5 s), in mA

- V is the applied potential difference, in V
- L is the average length of the sample, in mm
- D is the average diameter of the sample, in mm
- K is the conversion factor, equal to 1273.2

Literature reports the use of different electrode configurations as well as different geometrical configurations of the specimen with respect to the standard (see Table 1). Indeed, the standard is rarely followed: in the last 20 years, no study has been carried out in compliance with ASTM C1760, probably because authors prefer to use their own instrumentations and to prepare specimens with molds available in their laboratories, adopting the measurement configuration that better fits their study objectives.

Stainless steel meshes are often used as electrodes. D'Alessandro et al. [38] used a wire net as electrode in concrete cubic specimens. These specimens would have been embedded in concrete structures as distributed sensors for SHM purposes. Zhang et al. [39] used the same configuration but in concrete prisms manufactured with carbon nanotubes and nanocarbon black filler. Ding et al. [40] realized stainless steel gauzes to be embedded in cementitious prisms, whereas Dong et al. [41] inserted these electrodes in concrete prisms reinforced with super-fine stainless wires.

Also, copper is used for the electrodes; Melugiri-Shankaramurthy et al. [42] used copper mesh electrodes embedded in cubic cement paste specimens. Huang et al. [43] used copper tapes, glued with silver conductive epoxy around the surface of dog-bone paste specimens manufactured with fly ash and carbon black. Wen and Chung [44] used silver paint and copper wires as electrodes in prismatic specimens. Brass-made wires were used as electrodes by Chen et al. [45] in both prismatic and cubic specimens.

Plates are another type of electrodes used in the literature. Lim et al. [46] embedded stainless steel plate electrodes in prisms. Dehghanpour and Yilmaz [47] guaranteed a good electric and mechanical contact between metal plate electrodes and specimens surface by means of a cloth moistened with NaCl solution and a clamp, testing concrete specimens manufactured with conductive carbon fibers different in terms of shape and dimensions.

2.1.2. 4-Electrode Configuration

In the 4-electrode configuration, excitation and measurement electrodes are distinct. In particular, the two external electrodes (namely Working Electrode, WE, and Counter Electrode, CE) are used to inject the excitation electric signal (current or potential difference), whereas the two internal ones (namely Sensing Electrode, SE; and Reference Electrode, RE) are used to measure the electrical response (potential difference or current). Different types of electrodes and specimens are used in the literature (see Table 2).

Mesh electrodes were often used in the literature, manufactured with different materials: stainless steel [48,49], titanium [22], and titanium platinum-coated [50]. Also, copper is widely used in different forms: wire mesh [33,51,52], tapes [53,54] and loops [55]; silver paint is employed to realize electrodes on specimens surface [56–64], e.g., in the form of silver paint stripes perpendicular to the length of the specimen or combined with copper wires. Other types of electrodes are brass wires (with carbon powder interposed between electrodes and paste, to ensure a good electric contact) [65,66] and small pieces of reinforcing rebars plus Ag/AgCl reference electrodes (with wet sponges guaranteeing a good electric contact) [27].

Table 1. DC measurements with 2-point probe method.

Reference	Specimens	Electrodes	Type of Material	Additions
D'Alessandro et al. [38]	Cubes: 51 mm (side)	Stainless steel nets (1 mm diameter wires at 0.6 mm from each other)	Concrete	CNTs
Zhang et al. [39]	Prisms: 40 × 40 × 80 mm	Stainless steel gauzes (20 mm spacing)	Concrete	CNTs, nanocarbon black filler
Ding et al. [40]	Prisms: 20 × 20 × 40 mm	Stainless steel gauzes (0.5 mm wire diameter, 2 × 2 mm mesh size)	Paste	-
Dong et al. [41]	Beams: 40 × 40 × 160 mm	Stainless steel gauzes	Concrete	-
Melugiri-Shankaramurthy et al. [42]	Cubes: 50 mm (side)	Copper meshes	Paste	-
Huang et al. [43]	Dog-bones	Copper tapes (4 mm width, 20 mm length)	Paste	Fly ash, carbon black
Wen and Chung [44]	Prisms: 150 × 12 × 11 mm	Silver paint and copper electrodes	Paste	-
Chen et al. [45]	Prisms: 56 × 56 × 280 mm	Brass wires	Concrete	-
Lim et al. [46]	Cubes: 100 mm (side)	Stainless steel plates (10 × 10 × 2 mm)	Paste	-
Dehghanpour and Yilmaz [47]	Beams: 100 × 100 × 400 mm Cylinders: 75/100/150 mm (diameter) × 150/200/300 mm (length) Prisms: 75/100/150 × 75/100/150 × 150/200/300 mm Cubes: 75/100/150 mm (side)	Metal plates	Concrete	-

Table 2. DC measurements with 4-point probe method.

Reference	Specimens	Electrodes	Material	Additions
Belli et al. [48]	Prisms: 40 × 40 × 160 mm	Steel meshes (30 × 30 mm immersed, 40 mm spacing)	Mortar	Graphene nanoplatelet filler and/or carbon fibers (virgin or recycled)
Liu and Wu [49]	Cylinders: 150 mm (diameter) × 130 mm (height)	Stainless steel nets	Concrete	Limestone powder, graphite powder, and carbon fibers
Berrocal et al. [22]	Prisms: 40 × 40 × 160 mm	Titanium meshes (40 × 60 × 0.7 mm) and copper wires (at 30 mm from meshes)	Mortar	-
Boulay et al. [50]	Cylinders: 110 mm (diameter) × 220 mm (height)	Titanium platinum-coated grids	Concrete	-
Han and Ou [67]	Prisms	Gauzes (2.03 × 2.03 mm opening)	Paste	Carbon fibers and carbon black
Teomete and Kocyigit [51]	Cubes: 50 mm (side)	Copper wire meshes	Paste	Steel fibers
Lee et al. [33]	Cubes: 50 mm (side)	Copper electrodes (20 × 75 × 0.3 mm, 10 mm spacing)	Paste	Silica fume, carbon fibers, and multi-walled carbon nanotubes
Wen and Chung [56,58,59]	Dog bones: 30 × 20 mm (narrow part) Cubes: 51 mm (side)	Silver paint and copper wires	Paste	-
Cao and Chung [60]	Cylinders: 125 mm (diameter) × 250 mm (height) Prisms: 40 × 40 × 160 mm	Silver paint and copper electrodes	Concrete	-
Chen and Liu [61]	Prisms: 150 × 12 × 11 mm	Copper wires and silver paint	Concrete	-
Cao and Chung [62,63]	Prisms: 160 × 40 × 40 mm Cubes: 51 mm (side)	Silver paint	Paste	Fly ash or silica fume; coke powder
Cao and Chung [64]	Prisms: 150 × 12 × 11 mm Cubes: 51 mm (side)	Silver paint	Paste	Carbon fiber or carbon black
Wen and Chung [57]	Prisms: 25 × 25 × 70 mm	Silver paint and copper wires	Paste	Carbon fibers
Wang et al. [65,66]	Mortar strips (90.0 × 14.0 × 13.3 mm and 95.0 × 14.2 × 13.9 mm) overlapped at 90°	Brass wires and carbon powder	Paste	Ground granulated blast-furnace slag
Luo and Chung [52]	Bell-shaped (25 × 50 mm rectangular section)	Copper wires fixed with silver paint	Mortar	-
Nguyen et al. [53,54]	Prisms: 20 × 20 × 40 mm	Copper tapes and silver paste	Concrete	-
Han et al. [55]	Cylinders: 150 mm (diameter) × 300 mm (length)	Copper loops	Paste	-
Garzon et al. [27]	Prisms: 40 × 40 × 160 mm	Reinforcing steel (external electrodes) and Ag/AgCl reference (internal) electrodes	Concrete	-

2.2. AC Measurement of Electrical Resistivity

With respect to DC measurements, AC measurements of electrical resistivity give electrical impedance as result value, which includes a real part (R, electrical resistance) and an imaginary part (X, electrical reactance). It is worth noting that only electrical resistance represents the ionic movement in the pore network and is related to concrete durability [22]. Recent studies [68] reported that the real part of impedance can be correlated to degradation phenomena in cement-based materials.

2.2.1. 2-Electrode Configuration

The measurement setup defined by the uniaxial method of Section 2.1.1 can be also adopted with AC excitation (see Table 3). However, there are no general statements on optimal frequency to be used and both sine and square waves are deployed. In fact, it is worth noting that materials electrical properties are frequency-dependent [69] and, consequently, the measurement frequency deeply affects the results.

AASHTO TP 119 [70] suggests the use of an AC 60 V voltage, a testing period of 6 h, and a 100 × 50 mm (or 150 × 75 mm) cylindrical specimen, where two stainless steel plates connected to the Wenner probe tips (short-circuiting WE and SE electrodes and RE and CE electrodes, respectively) are applied as electrodes. As for uniaxial test of ASTM C1760 [37], this procedure is not widely adopted, and everyone prefers to set up its measurement bench configuration.

Electrodes in pin/bar forms are often used for this configuration: stainless steel pins [71], also covered with heat-shrink tubes apart from the 10 mm tip [72], brass electrodes [73] or screws can be used to improve the contact between electrodes and material, at different depths to obtain information on different layers [74]. Also, gauzes can be used, in different materials, such as copper (with water-saturated paper towel strips to ensure a good electric contact between electrodes and specimen surface) [75], titanium [22], stainless steel [21], and iron [76]. Plate electrodes were employed [77], maintaining the contact through different methods: clamps [78], wet sponges [11], absorbent cloth with a soap diluted solution by placing 5 kg on the top face of the specimen [79], sponges saturated with $Ca(OH)_2$ solution [80], or saturated with 20 wt.% NaCl solution [81], applying an external pressure of 6 kPa on the top electrode to improve the electrode/specimen contact [82]. Chiarello and Zinno [23] used G-clamps or wood cubes to ensure a good contact between electrodes and specimens; furthermore, they spread a paste of ethyl alcohol and carbon dust on the specimen ends to apply the excitation signal. Other electrodes used in the literature are stainless steel circular electrodes [83,84], red-copper electrodes with thin absorbent sponges to guarantee the contact [85], and graphite electrodes [86].

Commercial systems are also used to measure electrical resistivity. For example, Carsana et al. [87] used a commercial conductivity meter. As electrodes, parallel wires of mixed metal oxide activated titanium were embedded in the concrete specimens. Ghosh and Tran [88] employed the commercial Merlin meter (working at 325 Hz) to perform electrical resistivity measurement in cylindrical concrete specimens (100 mm diameter, 200 mm height). The electric contact between the clamp electrodes and the specimen faces was guaranteed by means of interposed sponges. Finally, Velay-Lizancos et al. [89] used an analogue sensor (65 × 8 × 3 mm) embedded in prismatic concrete specimens (300 × 200 × 110 mm) to measure temperature and electrical resistivity at 100 kHz.

Table 3. AC measurements with 2-point probe method.

Reference	Specimens	Electrodes	Material	Additions	Frequency
Kim et al. [72]	Cubes: 50 mm (side)	Stainless steel pins (2.4 mm diameter, 10 mm electrode spacing), embedded for 30 mm	Mortar	-	20 Hz–1 MHz
Mason et al. [71]	Prisms: 25 × 25 × 100 mm	Stainless steel electrodes (20 × 30 × 0.5 mm)	Paste	Steel/carbon fibers	5 mHz–10 MHz
Osterminski et al. [73]	Cubes: 150 mm (side)	Brass electrodes (6 mm diameter), at a center-to-center distance of 50 mm	Concrete	SCMs	107, 108, and 120 Hz
Polder and Peelen [74]	Prisms: 100 × 100 × 300 mm	Stainless steel screws, at depths of 10 mm and 50 mm	Concrete	-	120 Hz
Peled et al. [75]	Prisms: 25 × 180 × 4(8) mm	Copper gauze strips (18 × 18 mm), with 63 mm spacing	Paste	-	0.5 Hz–10 MHz
Berrocal et al. [22]	Prisms: 40 × 40 × 160 mm	Titanium mesh	Mortar	Steel fibers	100 mHz–1 MHz
Donnini et al. [21]	Prisms: 40 × 40 × 160 mm	Stainless steel nets	Mortar	-	10 Hz–1 MHz
Zeng et al. [76]	-	Iron nets	Concrete	-	-
Rovnaník et al. [78]	Prisms: 40 × 40 × 160 mm	Brass plates (30 × 100 mm) Gauzes (2.5 mm mesh size)	Concrete	-	40 Hz–1 MHz
Hou et al. [11]	Cubes: 100 mm (side) Cylinders: 100 mm (diameter) × 200 mm (height)	Stainless steel plates	Concrete	-	1 kHz, according to ASTM C1760 [37]
McCarter et al. [77]	Cubes: 50 mm (side)	Stainless steel plates (50 × 50 mm)	Paste	-	1–100 kHz
McCarter et al. [80]	Cubes: 150 mm (side)	Stainless steel plates (with a polymethyl methacrylate backing)	Concrete	-	1 kHz
Wang et al. [90]	Cylinders: 50 mm (diameter) × 100 mm (height)	Stainless steel plates	Mortar	-	1 kHz
Van Noort et al. [79]	Cubes: 150 mm (side)	Stainless steel plates	Concrete	-	1 kHz

Table 3. Cont.

Reference	Specimens	Electrodes	Material	Additions	Frequency
Sengul et al. [5]	Cylinders: 100 mm (diameter) × 200 mm (height)	Steel plates	Concrete	Fly ash, GGBFS, silica fume, ad natural pozzolan	1 kHz (correlation coefficient R = 0.999 when comparing the results with those obtained in compliance with ASTM C1760 [37])
Chiarello and Zinno [23]	Prisms: 40 × 40 × 160 mm	Copper sheets (0.3 mm thickness, 40 × 18 mm transverse dimensions, 8 mm set in the composite); lead plates (1 mm thickness, with 40 × 50 mm transverse dimensions)	Concrete	-	100 kHz
Zhu et al. [91]	Prisms: 40 × 40 × 160 mm	Special metal electrodes resistant to high alkaline environments	Mortar and engineered cementitious composite (ECC)	Fly ash and polyvinyl alcohol (PVA) fibers	0.01 Hz–10 kHz
Sato and Beaudoin [83]	Cylinders: 9.64 mm (diameter) × 19.05 mm (length)	Stainless steel discs with soldered wires	Paste	-	100 Hz–5 MHz
Perron and Beaudoin [84]	Cylinders:8 mm (diameter) × 16 mm (height)	Stainless steel electrodes with soldered wires	Paste	-	-
Zhou et al. [85]	Cylinders: 150 mm (diameter) × 400 mm (height)	Red copper	Concrete	-	500 Hz
Díaz et al. [86]	Cylinders: 200 mm (diameter) × 90 mm (length)	Graphite electrodes	Mortar	-	100 Hz–40 MHz
Carsana et al. [87]	-	Parallel wires of mixed oxide activated titanium	Concrete	-	-
Ghosh and Tran [88]	Cylinders: 100 mm (diameter) × 200 mm (height)	External clamp	Concrete	-	325 Hz
Velay-Lizancos et al. [89]	Prisms: 300 × 200 × 110 mm	Commercial analogue sensor (65 × 8 × 3 mm)	Concrete	-	100 kHz

2.2.2. 4-Electrode Configuration

The AC 4-point probe method, also known as Wenner array method, derived from soil mechanics to evaluate the possible corrosion of buried structures, is one of the most applied methodologies for field concrete structures [35,92]. Four electrodes equally spaced are used, guaranteeing their electric contact with the surface e.g., by water pre-wetting, contact graphite gel or pastes; the two internal electrodes measure the voltage created by the application of an electric current between the two external electrodes (Figure 2).

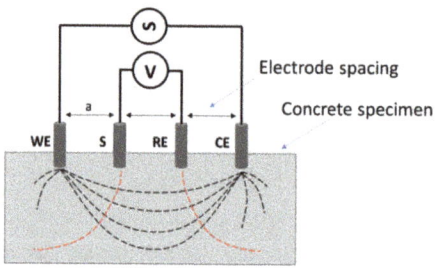

Figure 2. Wenner array method.

The frequency range commonly used is wide (Table 4). The Wenner-array method appears a suitable and reliable test method for in-field measurements and performance-based quality control of concrete durability [4]; the electrical resistivity measured by this approach is the average electrical resistivity of a hemisphere of radius equal to the electrode spacing (a), as reported in Equation (2):

$$\rho = 2\pi a \frac{V}{I} \qquad (2)$$

This relationship is valid with the hypothesis of homogeneous medium and semi-infinite dimensions. In real conditions and with the use of surface electrodes, the apparent resistivity (ρ_a) is measured; in fact, because of non-homogeneity and finite geometric dimensions of the specimen under test, a cell constant correction factor (K) needs to be applied. The correction factor K can be obtained in two different ways: from theoretical considerations or from calibration by means of standard concrete samples or electrolytes of known resistivity. The relationship between the electrical resistivity and its apparent value is reported in Equation (3):

$$\rho = \frac{\rho_a}{K} \qquad (3)$$

There are no widely accepted standards for the measurement of electrical resistivity using a Wenner array (ASTM G57-06 "Standard Test Method for Field Measurement of Soil Resistivity Using the Wenner Four-Electrode Method" [93] is used in soil mechanics). A Spanish standard (UNE 83988-2 "Concrete durability. Test methods. Determination of the electrical resistivity. Four points or Wenner method") suggests the use of cylindrical samples (100 mm diameter, 200 mm height); measurements should be carried out in saturated surface dry (s.s.d.) conditions, in order to consider the means homogeneous (since the inter-particle voids are water-saturated) [94]. Another provisional standard, namely AASHTO T 358 [95], was made in 2015 by the Department of Transportation (DoT) community of the American Association of State Highway and Transportation Officials (AASHTO); this standard suggests to apply current (flat-topped trapezoidal wave at approximately 13 Hz and a nominal peak-to-peak voltage of 25 V) by a Wenner probe located on the specimen surface (electrode spacing of 38 mm) and measure the voltage drop to calculate the impedance of the specimen. Tibbets et al. [96] applied this standard to cylindrical specimens (102 mm diameter, 205 mm height) containing different SCMs. They found a very strong linear correlation with the bulk resistivity measured according to the AASHTO TP119 "Standard Method of Test for Electrical Resistivity of a Concrete Cylinder Tested in a Uniaxial Resistance Test" [70] (y = 1.93x, with R^2 = 0.99). Also Spragg et al. [81] performed measurement according to AASHTO T 358, using stainless steel plate electrodes, ensuring a good electric contact with the specimen by lime-water saturated sponges. They compared their results with the surface resistivity method described in FM 5-578 [97], obtaining very good correlation (R^2 = 0.9986); moreover, they observed variation coefficients of approximately 4% and 13% for intra- and inter-operator measurements, respectively.

A resistivity profile can be obtained by moving the electrodes along a horizontal section at constant depth, whereas a vertical electric sounding can be achieved by progressively widening the electrode spacing and keeping the quadrupole center fixed. By realizing more resistivity profiles at increasing depths, a pseudo-section of the apparent resistivity can be obtained through inversion algorithms: this is known as electrical resistivity tomography [98]. A system suitable for this aim was employed by Alhajj et al. [99], using a multi-electrode probe consisting of 14 electrodes equally spaced at intervals of 20 mm; four different electrode spacing are possible, namely 20, 40, 60, and 80 mm. It is worth underlining that the measurement results are influenced by cracks, reinforcing rebars, humidity, and voids.

Commercial systems are also available: Simon and Vass [12] used a commercial resistivity meter, RESI (whose current version is Resipod by Proceq), intended for the assessment of the risk of corrosion. This meter exploits the Wenner method (with a measurement frequency of 40 Hz) and gives an estimation of corrosion rate, besides evaluating concrete homogeneity and curing conditions. Ramezanianpour et al. [100] used a surface resistivity meter [97] with four equally spaced electrodes and an excitation signal at 13 Hz with a peak-to-peak amplitude of 25 V (trapezoidal wave), as indicated in [101]. They took measurements at 4 quaternary longitudinal locations and considered the average value. They stated that the four-point Wenner array probe is one of the best methods to measure electrical resistivity, being non-destructive, easy to perform, and fast. Also, Ferreira and Jalali [16] used a commercial resistivity meter based on 4-point probe method, using an electrode spacing of 30 mm and a low-frequency (13 Hz) AC excitation current. They considered both cubic (150 mm side) and cylindrical (100 mm diameter, 200 mm height) specimens; for the former, two measurements on three perpendicular surfaces were performed, whereas for the latter two opposite headings (180°) measurements parallel to the length of the cylinders, 120° apart from each other. Presuel-Moreno et al. [102] applied a commercial surface resistivity meter on cylindrical specimens; Medeiros-Junior and Lima [103] used a commercial system based on the four-point Wenner probe (50 mm spacing) on cubic specimens of 250 mm per side (cast with three different w/c ratios of 0.4, 0.5 and 0.6) after 28 days of curing in a moist chamber and then in unsaturated condition in a laboratory environment (air-dried, T = 22 ± 3 °C, RH ± 65%) for two years. CNS RM MKII commercial resistivity meter was used by Kessler et al. [104], who considered three different sets of concrete specimens; Nilsson model 400 soil resistivity meter was used also in [105], considering cylindrical test specimens, with 4 rebars segments. Chen et al. [35] used a resistivity meter to investigate the effect of the probe spacing on resistivity by testing cylindrical and prismatic specimens. Ghosh and Tran [88] used a commercial surface resistivity meter (50 mm electrode spacing) on cylindrical concrete specimens.

Surface electrodes are widely used in 4-point probe AC configuration. Gowers and Millard [106] considered prisms, as well as Lubeck et al. [31]. Mendes et al. [94] used the four-point Wenner probe (50 mm spacing) according to the standard UNE 83988-2 "Concrete durability. Test methods. Determination of the electrical resistivity. Part 2: Four points or Wenner method." [107], averaging the results on three measurements spaced at 120°. Cylindrical specimens (manufactured with the addition of fly ash, high-volume fly ash (HVFA), and fine limestone powder) were considered also by Tanesi et al. [101] and the same method is used by Wiwattanachang and Giao [98], adopting an increasing spacing of the electrodes in order to perform a vertical sounding of the tested prisms. Sengul et al. [4,108] compared the results obtained with the Wenner method with those obtained with 2-point probe method. Millard and Sadowski [109] used the Wenner technique with a modified electrode array, with two copper-copper sulphate reference electrodes to give a stable surface potential for the measurement of electrical potential difference caused by the excitation given through the outer electrodes; the same configuration was used in [110–112]. Moreover, Sadowski [113] used the four-point Wenner array in the frequency range of 50–1000 Hz on slab specimens; the area where the measurement was made had been prior wetted with tap water, to optimize the electrical contact. A grid of measurement points was defined on the specimen surface, to obtain a resistivity map. Nguyen et al. [28] used surface electrodes to perform electrical resistivity measurements on concrete prisms; they guaranteed a good electric contact by means of conductive gel based on potassium hydroxide. Zhang et al. [114–116] used a four-electrode array on prisms and performed the measurement in the 0.01–1000 Hz band. Also, Nguyen et al. [117] used the 4-point probe method, but they did not consider a constant spacing among electrodes.

Goueygou et al. [118] used a specific square probe (50 mm spacing) on prisms reinforced with rebars (10 mm diameter), repeating the measurement twice for each tested location; the same probe is used by Lataste et al. [119]. A square array is used also by Yim et al. [120], employing an excitation signal of 1 V (peak amplitude) at 100 Hz to scan a series of 12 electrodes placed around a cylindrical concrete specimen, in order to average the results on 12 measurements. They used brass electrodes, inserted for 10 mm in the specimen. The measurement was performed 30 min after the cast. The square array was reported to be sensitive to the specimen anisotropy.

Also attached or embedded electrodes can be used for 4-electrode configuration. Azhari and Banthia [121] used copper electrodes attached with silver paste on cylindrical specimens, as well as Lee et al. [122]. Fan et al. [123] used copper tapes as electrodes, attached with conductive silver colloidal paste. Loche et al. [124] used two platinum electrodes and two reference electrodes, controlling the field stability in the specimen. Yim et al. [125] used copper electrodes, with insulating coating at the extremities.

Chiarello and Zinno [23] tested 4-electrode configuration with different materials, both embedded or coupled externally to the specimens. In particular, they used copper electrodes, lead plates, and copper wires. They tested three different electrodes configurations: 4 copper electrodes (internal electrode distance of 80 mm, external electrodes distance of 120 mm), 2 copper electrodes (internal, at a distance of 120 mm) and 2 lead plates (external, clamped on the specimens), and 4 copper wires (internal electrode distance of 80 mm, external electrodes distance of 120 mm). They reported that the most reliable system is the one composed by 4 copper sheets. Two different types of electrodes were tested by Yu et al. [126]: square sheets of conductive film and cylindrical urethane sponges penetrated by a gold-coated needle along the longitudinal axis and containing water; the electrode spacing was equal to 75 mm. Also, McCarter et al. [80] tested two different electrode configurations: four stainless steel rod-electrodes with heat-shrink sleeving (keeping uncovered a 10-mm tip), and a spacing of 30 mm; two internal rod-electrodes and two external stainless steel plate-electrodes with a polymethyl methacrylate backing. They performed the measurements at 1 kHz on cubic concrete specimens (150 mm side), noticing that the latter configuration is not suitable at higher resistivity values, when the boundary effects at the concrete/air interface become significant, overestimating the measured resistivity. The same measurement frequency was adopted by Hope and Ip [34], making the two external electrodes adhere to the external opposite prismatic specimen faces by means of gel, whereas the internal electrodes were brass or stainless knurled steel bars cast in concrete. Finally, Lübeck et al. [31] performed electrical resistivity measurements on prismatic specimens containing blast furnace slag.

The Wenner probe method, using a probe with 50 mm spacing and an AC excitation at 40 Hz, was exploited also within the Long-Term Bridge Performance (LTPB) Program of the Federal HighWay Administration (FHWA) [127], where the RABIT-CETM multifunctional Non-Destructive Evaluation (NDE) platform has been employed to enhance the assessment of bridge decks. The electrodes for the measurement of electrical resistivity were equipped with sponges, which were saturated with soapy water to assure electrical coupling between electrodes and concrete. The analysis of the data measured (which should be repeated five times at each location) enables to derive contour maps reporting the spatial distribution of the electrical resistivity values: areas with low resistivity describe highly corrosive environment, facilitating high corrosion rates. It is worth noting that, contrarily to AC 2-electrode measurement (Section 2.2.1), literature (Table 3) reports more documents involving concrete than paste or mortar specimens when addressing the 4-electrode configurations. To the best of the authors' knowledge, there are no specific reasons for this; a hypothesis could be that this is a simple costs-benefits issue, being paste/mortar specimens easier to be manufactured (and, often, smaller specimens are realized with respect to concrete ones, so that 2-electrode configuration appears more practical to be applied), being the cost of devices involved lower and being the overall test set-up easier to be prepared. Of course, the lower precision and accuracy of a 2-electrode test with respect to a 4-electrodes test still holds.

2.2.3. Other Methods Exploiting AC Excitation

Also other methods are reported in the literature. Li et al. [128] invented and patented [129] a non-contact device for monitoring electrical resistivity during the hydration process of cement paste. This system is based on the transformer principle to avoid the issues related to the electrode-specimen interface. Tang et al. [130,131] used the same technique with a 1–100 kHz sine wave in frequency domain, inducing the signal on a ring specimen of 1.4 L volume identified as the secondary coil of the transformer. The transformer-based method [132] for the measurement of electrical resistivity was used also by Wei and Xiao [133] and by Wei et al. [134,135] on ring-shaped specimens (1.67 L volume); the sampling interval was 1 min and the test was performed for 24 h. Similarly, Dong et al. [14] performed measurements on cement paste ring-shaped specimens for 1 day every 5 s. Lianzhen et al. [136] adopted the same test setup, prolonging the measurement up to 72 h. The same method was used by Shao et al. [137]. A specimen volume of 1.4 L was required and the test was carried out in a thermo-chamber with T = 20 °C, equal to the curing room. An electrical induced potential of 0.1 V at 1 kHz was used, and the measured data were recorded. The test lasted 3 days, starting when cement is mixed with water. Finally, Xiao and Li [138] used the non-contact method in specimens with a trapezoidal section (1.67 L in volume), carrying out the measurements for 48 h with 1-min sampling intervals.

Guthrie et al. [139] used a rolling probe, including a guard ring and a center electrode, for testing concrete bridge decks. They proved that this method is significantly faster (190 m^2/h) than static probes (30 m^2/h [140]), with a spatial resolution (0.1 m) 12 times higher than the latter. An excitation 3.3 V peak-to-peak sinusoidal signal at 190 Hz is used [141]. Pre-wetting of concrete surface (using a water and liquid soap solution) was necessary to guarantee a proper electric contact with the sensing electrodes [142]; the alterations caused by water were considered negligible. This method can provide an impedance map with valuable information on the reinforcement steel status related to protection against corrosion.

Finally, Melara et al. [143] used a three-electrode electrochemical cell, with an excitation signal of low-amplitude (25 mV) in the frequency range of 0.01 Hz–100 kHz, considering prismatic specimens (40 × 90 × 100 mm). They used the reinforcement (carbon steel rebar, 6.3 mm diameter) with a copper wire at its end and a graphite bar as electrodes. They performed the measurement according to ASTM G106 [144].

Table 4. AC measurements with 4-point probe method.

Reference	Specimens	Electrodes	Material	Additions	Frequency
Tibbetts et al. [96]	Cylinders: 102 mm (diameter) × 205 (height)	Surface electrodes	Concrete	Fly ash, silica fume, metakaolin, ground glass, and sugarcane bagasse ashes	13 Hz
Spragg et al. [81]	Cylinders: 102 mm (diameter) × 205 (height)	Surface electrodes	Concrete	-	13 Hz
Alhaj et al. [99]	Slabs: 900 × 700 × 150 mm	Multi-electrode array (14 electrodes, equally spaced at 2C mm)	Concrete	-	2 Hz (square wave)
Simon and Vass [12]	-	Surface electrodes (RESI system)	Concrete	-	40 Hz
Ramezanianpour et al. [100]	Cylinders: 100 mm (diameter) × 200 mm (height)	Surface electrodes (Florida method resistivity meter)	Concrete	-	13 Hz
Ferreira and Jalali [16]	Cubes: 150 mm side Cylinders: 100 mm (diameter) × 200 mm (height)	Surface electrodes (commercial resistivity meter)	Concrete	-	13 Hz
Presuel-Moreno et al. [102]	Cylinders: 101.6 mm (diameter) × 203.2 (height)	Surface electrodes (commercial resistivity meter)	Concrete	-	-
Medeiros-Junior and Lima [103]	Cubes: 250 mm (side)	Surface electrodes (commercial equipment)	Concrete	-	-
Kessler et al. [104]	Cylinders: 101.6 mm (diameter) × 203.2 mm (height)	Surface electrodes (CNS FM MKII resistivity meter)	Concrete	-	13 Hz
Morris et al. [105]	Cylinders: 150 mm (diameter) × 200 mm (height) Cylinders: 100(150) mm (diameter) × 200(300) mm (height)	Reinforcement rebars	Concrete	-	97 Hz (square wave)
Chen et al. [35]	Prisms: 200 × 200 × 175 mm; 160 × 160 × 140 mm; 120 × 120 × 110 mm	Surface electrodes	Concrete	-	-

Table 4. Cont.

Reference	Specimens	Electrodes	Material	Additions	Frequency
Ghosh and Tran [88]	Cylinders: 100/150 mm (diameter) × 200/300 mm (height)	Surface electrodes	Concrete	-	-
Lubeck et al. [31]	Prisms: 100 × 100 × 170 mm	-	Concrete	Slag (50% and 70% by mass)	-
Mendes et al. [94]	Cylinders: 100 mm (diameter) × 200 mm (height)	Surface electrodes	Concrete	Fly ash, HVFA, fine limestone powder	-
Tanesi et al. [101]	Cylinders	Surface electrodes	Concrete	HVFA and fine limestone powder	-
Wiwattanachang and Giao [98]	Beams: 100 × 200 × 1500 mm	Surface electrodes	Concrete	Fibers	-
Sengul et al. [4,108]	-	Surface electrodes	Concrete	-	-
Millard and Sadowski [109]	Slabs: 400 × 300 × 100 mm	Surface electrodes	Concrete	-	-
Sadowski [110–112]	Slabs: 400 × 300 × 100 mm	Surface electrodes	Concrete	-	-
Sadowski [113]	Slabs: 200 × 750 × 750 mm	Surface electrodes	Concrete	-	50–1000 Hz
Nguyen et al. [28]	Slabs: 250 × 250 × 500 mm	Surface electrodes	Concrete	-	-
Zhang et al. [114–116]	Blocks: 89 × 114 × 406 mm	Surface electrodes	Concrete	-	0.01–1000 Hz
Nguyen et al. [117]	Slabs: 500 × 250 × 120 mm	Surface electrodes	Concrete	-	5 Hz
Goueygou et al. [118]	Slabs: 600 × 600 × 120 mm	Specific square probe	Concrete	-	-
Lataste et al. [119]	Slab	Specific square probe	Concrete	-	-
Yim et al. [120]	Cylinder: 150 mm (diameter) × 540 mm (height)	Brass electrodes (5 mm diameter)	Concrete	-	100 Hz
Azhari and Banthia [121]	Cylinders: 50.8 mm (diameter) × 100 mm (height)	Copper electrodes attached with silver paste	Paste	Carbon fibers and CNTs	100 kHz
Lee et al. [122]	Cylinders: 100 mm (diameter) × 200 mm (height)	Copper wires and silver paste	Concrete	-	100 kHz
Fan et al. [123]	Prisms: 152 × 50 × 12.5 mm	Copper tape (spacing of 100 mm and 60 mm for external and internal electrodes, respectively) attached with silver paste	Mortar	-	0.1 Hz–1 MHz

89

Table 4. Cont.

Reference	Specimens	Electrodes	Material	Additions	Frequency
Loche et al. [124]	Prisms: 150 × 150 × 180 mm	Two platinum electrodes (100 mm diameter) and two reference electrodes	Mortar	-	50 mHz– 10 MHz
Yim et al. [125]	Prisms	Copper electrodes (1.78 mm diameter)	Mortar	-	500 kHz
Chiarello and Zinno [23]	Slabs: 40 × 40 × 160 mm	4 copper electrodes (0.3 mm thickness, 40 × 18 mm transverse dimensions, 8 mm set in the composite), 2 copper electrodes and 2 lead plates (1 mm thickness, with 40 × 50 mm transverse dimensions, G-clamped on the specimen); 4 copper wires (0.5 mm diameter, fixed by a conductive silver paint) attached with silver paint.	Concrete	-	100 kHz
Yu et al. [126]	Prisms: 300 × 300 × 100 mm	Square sheets of conductive films (15 mm side); cylindrical urethane sponges (10 mm diameter)	Concrete	-	0.01 Hz– 1 MHz
McCarter et al. [80]	Cubes: 150 mm (side)	Stainless steel rod-electrodes (6 mm diameter), also in combination with external stainless-steel plate-electrodes (150 × 150 mm)	Concrete	-	1 kHz
Hope and Ip [34]	Prisms: 25 × 25 × 100 mm	External plates; embedded brass or stainless knurled steel bars	Concrete	-	1 kHz
Lübeck et al. [31]	Prisms: 100 × 100 × 170 mm	-	Concrete	Blast furnace slag	-

2.3. Effect of Reinforcements in Electrical Resistivity Measurements

In case of **embedded steel**, the short circuit effect can be minimized by making the excitation current flow orthogonally with respect to the rebars (e.g., placing the sensing probe perpendicular to the rebars, in case of Wenner's method [102]). Garzon et al. [27] examined this effect both through experimental and numerical tests highlighting its dependence on both the specimen geometry and the electrode spacing; they introduces a rebar presence factor quantifying the effect on the measured resistivity. Also, Alhajj et al. [99] used numerical simulations to evaluate the effect of steel rebars on electrical resistivity measurements, carrying out parametric studies to evaluate the effect of different geometries and sensor positioning. As a rule of thumb, the distance between the Wenner's array and the reinforcement rebar should be at least twice the electrode spacing. Mc Carter et al. [80] have found that also the sponge used for electric contact can contribute to the measured result, increasing the electrical resistivity; therefore, particular attention should be paid to each component of the measurement chain. Millard et al. reported numerous factors causing deviations from the real resistivity value when using the Wenner array [145], such as size and geometry of specimens, rebars positioning, and presence of surface layers with different conductivity. Sengul et al. [4,108] compared the results obtained with the Wenner method with those obtained with 2-point probe method; they highlighted the necessity of performing the measurement as far as possible from the metallic reinforcement of concrete structures or adopting a sufficiently small electrode spacing (in relation to the cover depth), concluding that the Wenner method is suitable for a reliable quality control of concrete durability also during the construction process.

2.4. Uncertainty and Calibration in Electric Resistivity/Impedance Measurements: The Missing Items in Literature

Uncertainty [146] is a key aspect in every measurement system. Yet, it is also one the most missing item even in scientific documents targeting applications that involve measurement systems. This is also the case for electrical resistivity measurements. Indeed, to the best of the authors' knowledge, the uncertainty associated with an electrical resistivity measurement has been rarely tackled by the scientific community dealing with this approach in concrete. Actually, very few studies face this issue. Bourreau et al. analyzed the uncertainty associated with resistivity measurement for a coastal bridge [147]. The uncertainty was assessed based on measurements repeatability (1800 measurements) and considering local material anisotropy. The Wenner 4-probe approach was adopted to perform the measurements. The authors also tried to correlate the results of the uncertainty analysis with the corrosion risk from RILEM TC154 recommendation [148]. This way, they suggested a parameter, namely probability of wrong assessment (PWA), that was adopted, together with uncertainty, to help the bridge owner in the maintenance of the infrastructure.

Spragg et al. [81] focused on the variability of bulk electrical resistivity measurements (uniaxial method) on concrete cylinders. They considered both within and multi-laboratory variability exploiting the coefficient of variation (COV) as parameter addressing the variability. They claimed that, in both conditions (within and multi-laboratory), variability increases over time. Operator variability, variability of specimens, and the inherent variability in the mixture are all associated together into the within-laboratory variability, which does not exceed 4.36%. As the multi-laboratory variability is concerned, this was evaluated from ten laboratories and twelve differing mixtures, resulting in a variability of 13.22%. An important aspect underlined by the authors is that specimen geometry can greatly influence the results of an electrical test. Hence, they suggest using a geometry correction factor (e.g., in cylinders the ratio of sample area to sample length). As for the electrode resistance, which was addressed using a series model, they showed that, for the materials used in their evaluation, the needed correction is quite small.

Indeed, the variabilities reported by Spragg et al. can become even higher (10–40%), as discussed in [149,150], due to both concrete heterogeneity or measurement system metrological characteristics [151], just to cite some.

The concept of uncertainty associated with resistivity/impedance measurements becomes even more relevant when in-field testing is considered. In fact, in in-field applications, the whole measurement chain (instrument to inject/measure electrical quantities, wires, electrodes, concrete) related to the resistivity/impedance measurement becomes difficult to be controlled, mainly because it is highly affected by the conditions of concrete (it is indeed non-stationary) and the interaction of the whole chain with the environment (e.g., temperature, relative humidity, etc.).

Su et al. [152] discussed the decrease of electrical resistivity with the increase of temperature and RH. As RH is concerned, Su et al. identified a linear relationship between concrete resistivity and RH (correlation coefficient > 0.83). The higher the RH level, the higher the moisture content in concrete. Hence, the easier the flow of current within concrete. The results provided by Su et al. well match with the findings previously discussed in their paper of 2002 [153], in which they investigated the effect of moisture content on concrete resistivity on both air-dried specimens and oven-dried specimens. They demonstrated that resistivity decreases with an increase in w/c ratio, but for the saturated concrete specimens with high w/c ratios (w/c > 0.55) the resistivity difference is small. Mostly important, they showed that with consistent water loss ratio (>3%) concrete resistivity increases no matter the mix adopted. This is indeed justified by the fact that current is carried by ions [152], and ions constitute the conductive phase of concrete as pore solution, whose electrical resistivity is linked to concrete resistivity through the empirical law of Archie [154]. As for temperature, electrical resistivity and temperature are linked through the Arrhenius equation, involving the empirical computation of activation energy [155], which in turn is linked to moisture content itself [154]; when temperature rises, the movement of the charged ions dissolved into the pore solution increases according to a linear function [152].

Given this high-variability of concrete resistivity/impedance, it becomes clear that in-situ calibration becomes relevant when addressing in-field applications. Despite this is a recognized need, it is not widely addressed in the literature. Corva et al. [156] proposed to use LCR meters to derive proper correction factors to be applied to the results to improve their accuracy. Priou et al. [155] combined numerical models with laboratory tests to evaluate the uncertainty associated with the measurement and using water (instead of concrete) as a reference for calibrating the sensor to be used in-field. Recommendations on calibration are available mostly for in-lab tests, given that in-field measurements are much more complicated and often just connections and contacts are verified [25]. Indeed, calibration of the measurement system is carried out in-situ more as a performance verification. "Dummy cells" (electric circuits with high precision components of known value) are typically used for this purpose. However, the characteristics of these reference devices may differ quite a lot from those of the target structure.

In this sense, long-term monitoring applications can be considered more significant than inspections, because they make it possible to evaluate trends over time on a more solid statistical basis than the one provided by single inspections (even though multiple measurements are performed).

3. Electrical Resistivity/Impedance Measurement: Measuring One Variable Aiming at Several Target Applications

3.1. Curing and Hydration

The hydration process of cement-based materials varies with chemical composition, eventual additions and admixtures, mix proportions and curing conditions, such as T and RH. While time passes, porosity decreases because of the formation of more hydration products [157] (the decrease of porosity is associated with the difference between hydration products and hydrated cement in terms of volume [138]). In particular, in the early stages, the pore structure densifies and the formed hydration products block ions conductive paths, hence electrical resistivity increases. When higher strength cement is used, porosity decreases more rapidly (while the solid phase increases) and the electrical resistivity increase is steeper [135]. Indeed, electrical resistivity is highly correlated to porosity and pore solution characteristics (conductivity reflects the volume fraction and the interconnectivity of pores, besides the conductivity of pore solution and the saturation degree [81]), hence it can be related also to hydration. This measurement can be performed in situ in real-time [158], also when concrete is in a weak and plastic state [159], avoiding the influence on the development of the microstructure [130]. Non Destructive Techniques (NDT) and in particular electrical resistivity are used to monitor concrete porosity during its early age evolution [160] and, generally, the microstructure evolution in cement-based materials during hydration processes [77]. According to Wiwattanachang and Giao [98], during the first week of curing, electrical resistivity rapidly increase up to 60% of the final value, according to a logarithmic trend. Approximately 60 days of curing are necessary to reach an almost stationary value in mortar with carbon fibers [21]. McCarter et al. [77] underlined how electrical conductivity reduces with the hardening process considering 1, 10, and 100 kHz measurements.

Also non-contact devices based on transformer principle [128] can be used for measuring electrical resistivity during early hydration. This makes it possible to identify five distinct phases, namely dissolution, induction, acceleration, deceleration, and diffusion-controlled periods [161]. The ionic concentration of pore solution gives a significant contribution to electrical resistivity [160]. Tang et al. [130] used non-contact electrical impedance measurements [129] for the characterization of pore structure in cement pastes for three days from casting time. Through the use of pore fractal theory (combining fractal electrical network and pore structure network [162]), they obtained a relationship between the cumulative pore volume (important parameter to understand the contribution of different size pores in hydration kinetic and transportation [163]) and electrical impedance, which proved to be suitable for the assessment of pores ranging from μm to mm. Xiao and Li [138] measured the electrical resistivity of concrete during the first 48 h of hydration; they noticed that the characteristic points on the resistivity curve over time mirror the transition processes of hydration, namely the supersaturation point (lowest point) and the final setting of concrete (first peak point of the derivative curve). Dong et al. [14] evidenced that the electrical resistivity of concrete decreases with the gradual increase of ion concentration in the hydration process; moreover, there is a linear relationship between electrical resistivity and compressive strength at an early age, which means that standard compressive strength could be predicted by ρ. Carsana et al. [87] evidenced that in the first hours of curing there is a conductivity peak, linked to the ions released from the binder to the solution; then, hydration products form, and the pore structure develops, making electrical resistivity increase. The increase of electrical resistivity over time is confirmed by Polder and Peelen [74], who attributed this rise to drying out and hydration of concrete. In particular, after 3 weeks in air (T = 20 °C, RH = 80%) ρ increases by a factor 3-5, stronger in outer layers (due to pore water evaporation); at approximately 30 weeks, ρ increases by a factor 6–20 in the outer layers (10 mm depth) and by a factor 10–30 in the inner parts (50 mm depth) compared to 1-week values.

Tibbetts et al. [96] highlighted that, since pozzolanic reactions occur later, after the first 28 days in concrete with SCMs the electrical resistivity increases more than in plain concrete [164]. In fact, for concrete mixes with the addition of SCMs a curing period of 56 days is recommended [165]. Presuel-Moreno et al. [102] evaluated the effects of different curing regimens (immersion in tap water or 3.5% NaCl solution, exposure to fog room, high humidity, and laboratory humidity) on the electrical resistivity of concrete. In general, the exposure to fog room and high humidity levels cause lower resistivity values with respect to curing at room conditions. Tomlinson et al. [166] evaluated the effect of thermal cycling (between −24 °C and +24 °C, at a rate of 1 °C/h) on the electrical resistivity of concrete at early ages. Resistivity increases when temperature decreases and when the pore solution freezes (phase transition temperature, typically between −5 °C and 0 °C); in fact, ice is an insulator and limits ion mobility [167]. It is worth noting that phase transition temperature varies with concrete composition (e.g., GGBFS decreases it, because of changes in the ionic combination of the pore solution) and decreases with the ageing of concrete (because of different water content and, consequently, different ionic concentrations influencing the ice formation). Moreover, the phase transition temperature is higher during thawing cycles than freezing ones, meaning that in the former, there is more ice than in the latter, at the same temperature.

Also, the setting time of concrete has been evaluated by means of electrical resistivity measurements. Dong et al. [14] explored the relationship between setting time and electrical resistivity in the early hydration of cement-based materials; in particular, they identified linear relationship between the minimum of electrical resistivity curve and initial setting time (with $R^2 > 0.956$) and between the first peak of electrical resistivity curve and final setting time (with $R^2 > 0.986$). Also, Yim et al. [125] evaluated the setting time of mortar specimens by means of electrical resistivity, noticing that electrical resistivity increases with hydration depending on mix and chemical admixtures. In particular, they considered the rising time of electrical resistivity (indicating the increase onset) and the increasing ratio over time, which reflects the evolution of the cement-based materials microstructure.

3.2. Compressive Strength and Elastic Modulus

The prediction of concrete strength in a simple and effective way is fundamental in order to optimize the construction times. There are some applications of electrical resistivity measurement related to the prediction of **compressive strength** (commonly controlled at 28 days) and **elastic modulus**. The increase of hydration products goes together with the microstructure formation; when hydration products start blocking the pores conduction path and tortuosity increases, electrical resistivity sharply increases, as well as a compressive strength gain [35]. Therefore, the electrical resistivity can be considered as a predictor of the compressive strength of cementitious materials; since compressive strength value is used for quality control, also electrical resistivity measurement can be employed with that aim.

Many authors identified **linear relationships between compressive strength and electrical resistivity** of concrete [31,168], since both ρ and compressive strength depends on the matrix porosity and structure compactness. Hope and Ip [34] stated that electrical resistivity of concrete increases with increasing age—the more the lower w/c ratio is— cement, and aggregates used [137]. Wei et al. [135] evidenced a linear relationship (y = 8.7648x + 20.406, with $R^2 = 0.9634$) considering compressive strength at 28 days and electrical resistivity at 24 h. Ferreira and Jalali [16] found relationships between 28-day compressive strength and 7-day electrical resistivity values. They report errors < 10% in the compressive strength estimation when a model based on a theoretical equation related to concrete hydration process is used, considering also a correction factor for the surface temperature of concrete. Shao et al. [137] observed a linear relationship between electrical resistivity and concrete strength, reporting a strong correlation ($R^2 = 0.97$). This was justified by the presence of dissolved ions. Similarly, Lianzhen et al. [136], after temperature correction, established a linear relationship between compressive strength and concrete electrical resistivity, independent of curing temperature and w/c ratio (y = 4.51x − 3.80, with $R^2 = 0.98$). Dong et al. [14] established a linear relationship between 1-day compressive strength and 24-h electrical resistivity of cement paste (y = 0.1725x + 36.106, with $R^2 = 0.9371$). Lübeck et al. [31] found a linear relationships between electrical resistivity and compressive strength at 28 and 91 days of curing also in the presence of GGBFS. Wei et al. [134] established a power relationship between 1-day electrical resistivity and w/c ratio similar to that between compressive strength and w/c ratio [169]. On the other hand, they established a linear relationship between 2-day compressive strength and 1-day resistivity (y = 1.609 + 3.8497x, with $R^2 = 0.993$) and a logarithmic relationship between 7-day (or 28-day) compressive strength and 1-day resistivity (y = 5.4302 + 19.179lnx with $R^2 = 0.987$ and y = 15.427 + 21.835lnx with $R^2 = 0.976$ for 7-day and 28-day compressive strength, respectively).

Finally, Wei and Xiao [133] established a logarithmic relationship between 7-day or 28-day elastic modulus and 1-day electrical resistivity ($R^2 > 0.90$). Moreover, knowing the aggregate content, the 7-day (or 28-day) compressive strength can be derived from the 1-day electrical resistivity, according to a linear relationship ($R^2 > 0.95$). Aggregates largely increase the concrete bulk electrical resistivity with an exponential correlation [120]. On the other hand, Ramezanianpour et al. [100] do not recommend to use ρ as a compressive strength indicator since the pore solution greatly affects electrical resistivity, but has no influence on compressive strength.

Also the **elastic modulus can be related to electrical resistivity**; Shao et al. [137] found a non-linear regression during the early age hydration process of concrete [128]. As compressive strength, also elastic modulus increases as hydration proceeds. Using electrical resistivity for the evaluation of the modulus of elasticity is a meaningful, cheap, sensitive, and non-destructive method, as opposed to the analysis of the load-deformation curve of concrete specimens, which turns to be quite complicated and time consuming, requiring the cast of several dedicated specimens. Moreover, hydration products formation determines the electrical resistivity value, as the propagation velocity of ultrasound waves does. Since the ultrasonic pulse velocity measurement method can be used to investigate the elastic modulus, the combination of electrical resistivity and the elastic modulus obtained with ultrasound techniques enables a more accurate measurement with a correlation coefficient of $0.92 \leq R^2 \leq 0.98$.

3.3. Water Penetration

Water penetration can be evaluated according to EN-12390-8 "Testing hardened concrete—Part 8: Depth of penetration of water under pressure" [170]. Water penetration test is used to evaluate the concrete permeability and, consequently, its performance and durability [171]. Water is applied on one face of a cubic specimen under a pressure of 0.5 MPa, kept constant for 72 h; then, the specimen is split into two halves and the maximum penetration depth is considered as an indicator of water penetration. When water penetrates, some soluble salts (e.g., chloride ions) can go through concrete, easing corrosion of embedded reinforcements and undermining its durability. Concrete electrical resistivity has proved to be correlated with concrete permeability and water penetration [100].

Ramezanianpour et al. [100] highlighted a strong power relationship between concrete electrical resistivity (measured as surface resistivity—SR—with four-point Wenner array probe) and water penetration (measured in compliance to EN 12390-8) when the same type of cementitious materials are considered (y = 107.88x − 0.777, with $R^2 \approx 0.87$). On the other hand, when different cementitious materials are used, the correlation is slightly reduced ($R^2 \approx 0.83$), since the surface resistivity test depends on microstructure and pore solution, whereas water penetration only on pore solution. Tibbets et al. [96] investigated the correlation between water permeability and electrical resistivity; they used a pressurized, uniaxial, steady-state flow permeameter [172] and the procedure reported in AASHTO T358 [95] (surface resistivity) and AASHTO TP119 [70] (bulk resistivity). They proved that electrical resistivity increases as water permeability decreases; different trends are observable with different designs and SCMs; in particular, electrical resistivity measurements in specimens containing sugarcane bagasse ashes and ground glass tend to overestimate the concrete penetrability. Tang et al. [131] used non-contact electrical impedance measurements [129] for the development of a fractal permeability model to evaluate the permeability of young cement pastes (ring-shaped specimens); pores with a diameter < 6.2 nm were not considered, since the ions movement is constrained by pore walls [163]. Finally, Nguyen et al. [28] found a power relationship between electrical resistivity and saturation degree (y = 24.848$x^{-3.262}$, with $R^2 = 0.9702$), since resistivity increases while water content decreases.

3.4. Chloride Penetration

Chloride penetration is the main degradation cause of reinforced concrete structure. When a critical threshold is overcome, depassivation induces the corrosion of reinforcements. Therefore, a proper service life design of concrete infrastructures, verifying the durability requirements for concrete exposed to aggressive environments due to seawater or de-icing salts [173,174], is required for minimizing repair and maintenance costs.

Good correlations has been found [5] between concrete **electrical resistivity** [74] and **chloride ingress** [175] and this relationship can be used as a durability indicator, also for quality control and concrete classification [176]. In general, the chloride diffusion coefficient is inversely proportional to the electrical resistivity of concrete [177]. Similarly, Layssi et al. [178] evidenced a linear correlation ($R^2 = 0.93$) between the diffusion coefficient and the electrical conductivity of concrete. Electrical resistivity reflects the ions mobility in pore solution; therefore, the relationship with chloride penetration is reasonable [100]. Indeed, electrical resistivity can be used as a NDT for the evaluation of the service life [92], since it reflects charge flow and ions mobility, easing the corrosion process [103]. Chloride ions modify the electrical impedance response of cementitious materials; in particular, as chloride migration proceeds, first the bulk electrical resistance increases, then decreases [124]. McCarter et al. [77] monitored concrete blocks (300 × 200 × 200 mm) exposed in a marine site in three different locations: just above the low-water mark, just below the high-water mark, and well above it. They employed multielectrode arrays to evaluate the advancement of NaCl in the concrete cover, given that the increasing ionic concentration in the pore fluid, due to seawater penetration, decreases the concrete electrical resistivity.

Several works in the literature aim at correlating electrical resistivity and chloride penetration [7]; however, the relationships found are very different from each other, and this can lead to a difficult interpretation of measurement results, also considering that in-field measurements cannot benefit from controlled conditions, contrary to laboratory setups [92]. In fact, electrical resistivity mainly depends on the chemical composition of the pore solution and the presence of ions different from chlorides could be misleading in the evaluation of chloride permeability through electrical resistivity assessment. As a matter of fact, Archie's law provides the estimation of chlorides through multilinear regression considering the relationship between concrete electrical resistance and porosity, saturation degree, and the interstitial fluid resistance [30].

The **Surface Resistivity (SR) Test** has been judged as a promising alternative to Rapid Chloride Permeability Test (RCPT), since it is easier, more rapid, does not require sample preparation, and with results affected by lower variability [101]. However, SR Tests needs to be accurately set-up, as electrode configuration (Figure 2), contact force between electrodes, and specimen might severely affect the results of test [179]. In fact, the electrode-specimen contact area influences the measurement, with a maximum error in electrical resistivity estimated at 6% when passing from 1 mm to 40 mm electrode diameter (it is worth underlining that this effect is much more problematic in 2 electrode configuration, also limiting the suitable measurement frequency range) [155]. RCPT method, specified by ASTM C1202, is really a measure of electric conductivity, since it monitors the amount of electric current passing through a cylindrical specimen when a 60 V DC electrical potential difference is applied for 6 h [165]. The total charge is obtained by integrating the measured current over time, and corresponding levels of penetrability to chlorides are given (high, moderate, low, very low, or negligible) [180]. Even if this test is relatively simple and rapid (despite requiring an extensive specimen preparation), some issues have been raised, such as absence of steady-state conditions, heat evolution (its increase enhances the total passing charges, besides increasing the microstructure damages and possibly changing the pore solution chemical composition; especially in younger concretes or with high w/c ratio [181]), and pore solution alteration in the presence of pozzolanic materials [182]. Tanesi et al. [101] reported a decreasing exponential relationship between RCPT and SR test results, with a high correlation ($R^2 = 0.92$) and a quite low variability (5.3%). Balestra et al. [92] found a power relationship between chloride concentration and SE ($y = 25.724 \cdot x^{-0.5}$, with $R^2 = 0.9568$). Lower electrical resistivity values can indicate an easier chloride penetration through pores; in particular, they state that for $\rho > 17$ kΩ·cm the chloride penetration is lower, whereas it becomes very low for $\rho > 41$ kΩ·cm and negligible for $\rho > 220$ kΩ·cm. on the other hand, there is a very high chloride penetration for $\rho < 5$ kΩ·cm.

Ramezanianpour et al. [100] found a strong power relationship ($y = 67998x^{-1.028}$, with $R^2 \approx 0.90$) between RCPT and SR for a wide range of concrete specimens also containing SCMs, such as natural pozzolans (pumice and tuff), rice husk ash, silica fume, and metakaolin. The relationship between electrical resistivity and total passing charge would be linear if the specimen temperature was kept constant during the test ($R^2 \approx 0.99$) [181]. Similarly, good correlations were found by Kessler et al. [104] ($y = 5801.2x^{-0.819}$, with $R^2 \approx 0.95$) and Tibbets et al. [96] ($y = 7.535x^{-0.88}$ with $R^2 = 0.95$ and $y = 15.712x^{-0.90}$ with $R^2 = 0.98$, for bulk and surface resistivities, respectively), measuring surface/bulk electrical resistivity according to AASHTO T358 and AASHTO TP119, respectively. Therefore, SR can be considered an indicator of permeability and represents a valuable alternative (faster and more precise) to RCPT test, changing the type of measured data (electrical resistivity instead of total passing charge) but remaining a measurement linked to concrete electrical resistance. Surface resistivity can be used as a quality control predictor of the chloride penetration resistance, but not of **diffusion behavior**, requiring dedicated long-term diffusion tests.

However, according to Shi et al. [183] the **AC impedance measurement** is a simple, reproducible and therefore valuable technique for determining the diffusion coefficient in concrete. Similarly, Sengul [5] stated that electrical resistivity can be used as an indirect control of chloride diffusivity thanks to the Nernst-Einstein's equation, correlating electrical resistivity and ion diffusivity ($y = 49.13x^{-0.91}$, $R^2 = 0.97$). Connectivity and pores play a fundamental role: fewer pores with lower connectivity, higher electrical resistivity and lower diffusivity are obtained.

When chlorides penetrate, the material microstructure evolves, changing porosity and tortuosity. Electrical resistivity mirrors these variations, decreasing together with the ionic concentration [86]. Electrochemical Impedance Spectroscopy (**EIS**) can be used to monitor reinforced specimens during chloride ingress, allowing the detection of chloride arrival on the reinforcement as well as the steel depassivation, the corrosion onset, and the chloride exit [143]. Finally, Van Noort et al. [79] found a linear correlation between conductivity and **chloride migration coefficient**, ($y = 1195x$, with $R^2 = 0.88$); therefore, electrical resistivity measurement can be considered as a quicker and cheaper alternative method to the standard Rapid Chloride Migration (RCM) test [184].

3.5. Corrosion Risk of Reinforcements

The corrosion of steel rebars is the principal cause of structural deterioration of concrete [31], heavily impacting the service life of a structure. The corrosion rate depends on both the oxygen availability and the electrical resistivity of concrete, which mirrors the ion mobility in concrete and, consequently, the corrosion speed [7]. A low resistivity is generally linked to the concrete susceptibility to corrosion, since it is related to more rapid chloride penetration and corrosion rate [25]. Indeed, during corrosion, local electrical cells continuously generate a very little current, which can cause great harm, especially if the electrical resistivity of concrete is low. The electrical resistivity of concrete, being related to the ion mobility in the solution and to the microstructure itself, can act as an indicator for the ingress of CO_2 and chlorides [185]. This holding, electrical resistivity is directly linked to structural durability [31]. The literature proposes several threshold values; however, these thresholds are very variable. For instance, Morris et al. [105] indicated active corrosion when $\rho < 10$ kΩ·cm, whereas low corrosion probability when $\rho > 30$ kΩcm. Sengul [5] highlighted the importance of controlling moisture and temperature on specimens to obtain reliable and comparable data. Despite such different ranges, concrete electrical resistivity is considered as a valuable parameter for the evaluation of **corrosion risk**, provided that the concrete composition influencing the "baseline value" of electrical resistivity is considered. The quantification of corrosion risk is decisive for maintenance, protection, and repair decision-making [186,187]. Moreover, corrosion can occur only in a small range of temperature, w/c ratio, and relative humidity; Yu et al. [188], developed a probabilistic evaluation method for the estimation of corrosion risk by means of electrical resistivity measurements, taking into account the corrosion rate and identifying different corrosion risk levels (negligible, low, moderate, and high). Sadowski [113] proposed a methodology to evaluate the corrosion probability in concrete slabs by using electrical resistivity measurements (according to the four-point Wenner array method) in combination with half-cell potential method (defined in ASTM C876 [189]). The latter requires direct access to the steel reinforcement for the connection of a high impedance voltmeter and a reference half-cell, to obtain a map of the potentials that could be associated with areas of active corrosion. Among the other factors, corrosion probability depends on the ionic conductivity. Ionic conductivity can be estimated by measuring the electrical resistivity of concrete [190], therefore the two quantities may be correlated. The combined evaluation of electrical resistivity and corrosion potential may improve the quantification of the probability of corrosion, which seems higher when $\rho < 5$ kΩ·cm. In fact, the corrosion potential sharply increases for ρ < 4 kΩ·cm. On the other hand, Hope and Ip [34] considered 10 kΩ·cm as the upper limit for the corrosion probability in concrete. However, both the methods have their own disadvantages and provide better results when used in combination, even if they do not directly assess the instantaneous corrosion rate [110]. Polder [191] found a reasonable linear relationship between corrosion probability and concrete electrical resistivity (y = −0.11x + 118, with R^2 = 0.89), indicated by steel potential after 20 weeks of salt/drying cycles. However, this does not imply a direct relationship between electrical resistivity and critical chloride content (i.e., the chloride quantity that is believed tolerable before corrosion starts); electrical conductivity is strongly related to chloride penetration, as previously demonstrated [192]. The same author in [74] highlighted that electrical resistivity reflects the concrete properties related to chloride penetration, corrosion initiation (i.e., corrosion probability), and propagation (i.e., corrosion rate).

Since electrical resistivity of concrete reflects the transport of ions (including chloride), it is quite immediate to infer a possible correlation between resistivity and corrosion initiation. Poupard et al. [193] quantified the chloride concentration threshold leading to corrosion initiation by means of low-frequency impedance response (in the frequency range of 10 mHz–10 Hz). This threshold corresponds to the polarization resistance drop, which increases with low w/c ratio. The microstructure properties are reflected by polarization resistance, which consequently is able to characterize the activation of the corrosion process. However, this methodology can be used only as a comparative test because the external field used for the measurement accelerates chlorides transfer, thus distorting the steel behavior.

Once the corrosion process has started, a low resistivity gives a high **corrosion rate** [191]; when corrosion begins, ρ can represent a "controlling factor" of its speed [194,195]. After depassivation, an inverse relationship between concrete electrical resistivity and corrosion rate has been found [28,175,196,197] but not with a general validity, since it depends on concrete composition [197], age, carbonation, and environmental conditions, especially with low chloride content [74]. Millard and Sadowski [109] adapted the four-point Wenner resistivity method to evaluate the corrosion rate of steel without the requirement of a direct electrical connection to the reinforcement rebars, as necessary for the Linear Polarization Resistance (LPR) method [198]. LPR consists in the introduction of a small perturbative DC signal to a corroding steel rebar and in the measurement of the corresponding electric potential change, to derive the polarization resistance; it requires the knowledge of the area of steel being perturbed. Wenner's method alone cannot directly measure the corrosion rate; however, in this study, the short-circuit effect caused by steel reinforcement rebars was exploited to do measurements both in DC and AC (directly over a steel bar, parallel to the bar itself) to assess the resistive (i.e., the charge transfer resistance) and the capacitive (i.e., the double layer of charged ions on the surface of the steel rebar) components of the corrosion interface. This offers a quick method to quantify the corrosion rate on the surface of the considered rebar, since its instantaneous value is proportional to the charge transfer resistance itself [110].

Sadowski [112] combined the resistivity measurement performed by four-point probe method and the galvanostatic resistivity measurement [110] with the use of a Multi-Layer Perceptron (MLP) **artificial neural network** (ANN) model. A dataset of 70% of the available experimental data was used for training, 15% for testing, and 15% for the MLP verification process. A correlation coefficient $R^2 \approx 0.85$ with a MAPE (mean absolute percentage error) of 0.000266 was obtained for the first set of data (related to a specimen with a high corrosion rate), whereas $R^2 \approx 0.98$ and MAPE error of 0.000027 for the second dataset (related to a specimen with a moderate corrosion rate), indicating very good performance of the neural network in the prediction of the corrosion current rate. In this way, also the influence of environmental conditions on electrical resistivity is considered. It is worth noting how ANNs have spread also in civil engineering applications [199,200], giving important contributions to SHM. Sadowski and Nikoo [111] considered electrical resistivity (both in DC and AC) as input parameters for an artificial neural network base model for the estimation of corrosion current density. They found that imperialist competitive algorithm (ICA) provides higher accuracy and flexibility than genetic algorithms (GA).

Simon and Vass [12] underlined the existence of an inverse correlation between corrosion current density and the electrolyte resistance, as reported in Equation (4):

$$I_{corr} \propto \frac{1}{\rho} \tag{4}$$

where ρ [$\Omega \cdot m$] is the electrical resistivity of concrete. This indicates that the rate of corrosion increases together with decreasing electrical resistivity of concrete. Gulikers [175] stated that the relationship between the corrosion current density and the electrical resistance (and resistivity, if anodic and cathodic sites are considered fixed) can be considered almost linear for a wide range of corrosion current density values. However, many times cathodic activation control (electrical charge transfer) dominates on concrete resistance in driving the corrosion resistance, as well as oxygen diffusion can play a significant role in the cathodic control (particularly in wet environments).

EIS can provide information on the corrosion kinetics, indicating the corrosion mechanism (activation, concentration, or diffusion), monitoring the situation over time [201]. McCarter et al. [202] investigated the electrical resistivity of concrete through the cover region to the reinforcement rebar, at intervals of 5 mm. Considering that the electrical resistivity of pore water, salt solution, and pore fluid mixture can be represented by parallel conduction elements (equivalent electric circuit), the concentration of NaCl in the pore water after salt solution exposure can be estimated. Yu et al. [126] used the Wenner method to measure the electrical impedance of concrete for corrosion detection, without connecting to the rebars, by placing the electrodes just above the reinforcements. They found that the impedance decreases as corrosion proceeds, enabling a clear discrimination between healthy and corroded rebars. Therefore, the impedance difference in a certain frequency range (10-100 Hz) can be considered as a suitable parameter for monitoring reinforced concrete structures. However, they stressed that corrosion detection is more difficult if the rebar is located more deeply and that above 1 kHz the passive film formation on the rebars cause a sharp increase in electrical impedance. Guthrie et al. [139] identified the frequency range between 10 Hz and 1 kHz as the best to investigate the corrosion degree of reinforcing steel. Also, Zhang et al. [116] aligned the electrodes with the reinforcement rebar, stating that polarization resistance and double layer capacitance are clearly visible in the measured electrical impedance. This approach provides results comparable to standard EIS method without the need of contact with the reinforcement rebar and faster thanks to the use of higher frequency values. However, the result heavily depends on the geometry, the electrodes positioning, the concrete cover depth, the diameter of reinforcement rebars and the concrete electrical resistivity. Morris et al. [105] investigated corrosion on concrete specimens exposed to seashore and submerged. In the former scenario, the electrical resistivity values are approximately three times greater than in the latter and, in both cases, electrical resistivity increases over time. It worth noting that chloride concentration does not significantly influence concrete electrical resistivity; reinforcement steel rebars are probably in an active corrosion state when electrical resistivity is <10 k$\Omega \cdot$cm, whereas they remain in their passive state when ρ > 30 k$\Omega \cdot$cm. Electrical resistance is important not only for estimating the corrosion rate but also in the design of cathodic protection systems against corrosion [203].

3.6. Freezing/Thawing

In cold regions, the formation of frost inside cement-based structure is of great concern, since rapid cooling causes significant temperature and moisture gradients, thus resulting in considerable internal pressure caused by water confined in pores, as well as the crystallization pressure, giving possible damages. This is mirrored also by a compressive strength loss and cracks formation. Electrical resistivity, being related to moisture content and temperature, can provide information on the ice content; moreover, ice alters the porosity and connectivity of the material, which induce changes in the electrical resistivity.

Kim et al. [72] studied the influence of freezing/thawing on electrical impedance of concrete. They adopted specific thermal cycling, passing from 120 °C to −70°C (at a rate of 10 °C/h), maintaining the temperature for 3 h to reach thermal equilibrium, then they increased the temperature of 20 °C (at the same rate) till equilibrium; the cycle was repeated twice in a day. Both the real and the imaginary parts of impedance change with temperature; in particular, the impedance increases with the ice formation in the pore network [90,204], making the pore-water volume decrease, besides increasing slightly ionic concentration in the unfrozen pore water [83]. Moreover, the ions mobility decreases when temperature decreases, consequently decreasing electrical impedance as confirmed by Wang et al. [65]. Furthermore, the maximum value of ρ increases with the number of freezing/thawing cycles, indicating progressive damage caused by frost [205,206]). After ice nucleation, a linear relationship between the logarithm of electrical resistivity and temperature can be identified. This curve provides valuable information on the effect of frost on concrete durability [83].

Wang et al. [90] estimated the ice content of mortar through electrical impedance measurements. If the electrical impedance variations are not constant over cycles, it means that some frost damages have occurred. The freezing point is identified at approximately −7 °C; above 0 °C, there is a linear relationship between the logarithm of electrical resistivity and temperature (no ice formation), whereas the increase of electrical impedance is sharp below the freezing point (which increases with cycles). The ice content is higher during thawing than freezing, reasonably because of the natural supercooling effect of the material (during freezing, water cannot form ice because of lack of ice nucleation crystals). Sato and Beaudoin [83] highlighted that, during freezing, ice formation increases salt concentration in the unfrozen pore water, making its electrical resistance decrease. Therefore, another resistance component, linked to the paste-pore water interface, which increases with temperature decrease, appears. Moreover, after melting, there is a residual expansion in cement pastes causing cracking and pore deformation, possibly changing the micro-scale structure. Wang et al. [66] analyzed cement pastes with different moisture contents, highlighting that electrical resistivity is very sensitive to nucleation and growth of ice crystals in pores of cement-based materials. As already said, electrical resistivity increases with temperature decrease (because of a decrease in ion mobility); at first, the decrease proceeds slightly, then rapidly, indicating crystallization of the pore solution. The variation of electrical resistivity is stronger in thawing than in freezing as a consequence of moisture redistribution. When moisture content is higher, both freezing and melting temperatures increase; moreover, electrical resistivity is lower. It is possible to find a log-linear relationship between electrical resistivity and temperature that is due to changes in both pore solution molecular activities and conductive pathways. Finally, Perron and Beaudoin [84] underlined that EIS can provide valuable information on the effects of pore structure in water transfer and ice crystals formation. In fact, temperature decrease causes a reduction in pore solution conductivity and pore water movement, hence resulting in an increased resistance. When freezing begins, the slope of the electrical resistance vs. temperature curve changes; when there is no still pore water freezing, the resistance increases, but with a decreased rate, linked to the only temperature decrease.

3.7. Stress and Strain

Self-sensing construction materials have gained much interest for SHM purposes, enabling the perception of stress/strain by means of electrical resistivity measurements. Indeed, electrical resistivity enables concrete to behave as a smart material for the self-sensing of strain and stresses (piezoresistive effect), especially when conductive additions are used [68].

A good electrical conductivity is a fundamental prerequisite for piezoresistive behavior; it can be stated that cement-based material is the stress sensor, whereas the conductive addition makes the piezoresistive effect significative. Conductive fillers and/or fibers can be used at this aim, such as carbon-based materials (e.g., virgin/recycled carbon fibers, carbon nanotubes [207], carbon nanofibers [208], char, and carbon black [209]), graphene (e.g., graphene nanoplatelets [48]), steel fibers, graphite powder, nickel powder, titanium dioxide, and iron oxide. These functional fillers increase the concrete ability to sense not only stress and strain, but also internal damages (e.g., cracks). When a cement-based material contains a conductive addition, its electrical resistivity depends on its dispersion degree, the electrical conductivity of the addition, and of the interface between the addition and the cement matrix. Also, moisture plays an important role in the piezoresistive ability of concrete [48].

There are manifold applications of piezoresistivity property of concrete, such as SHM [121], traffic monitoring [210] (also including self-sensing pavement applications for vehicle detection [122]), and cement-based sensors [55]. Fractional Change of Resistivity (FCR)—or Fractional Change of Resistance—during loading/unloading phases is commonly considered and the sensitivity is evaluated by means of the Gauge Factor (GF), defined as the ratio of electrical resistivity variation ($\Delta\rho/\rho$) and strain (ε) [33]. Intrinsic self-sensing concrete is advantageous in terms of sensitivity, mechanical performance, durability, ease of installation, and maintenance [211].

Carbon fibers are often used to enhance the piezoresistive behavior; when short carbon fibers are added to the mortar mix, it can be noted that the electrical resistivity varies with the applied stress; in particular, ρ decreases with compression longitudinal loads (fiber push-in), whereas increases upon tension (fiber pull-out) [21]. Wen and Chung [212] observed that the DC electrical resistance increases upon tensile loading in specimens containing carbon fibers, due to the degradation of the fiber-matrix interface. This degradation is partly reversible, meaning that carbon fibers composites behave as strain sensors, in both transverse and longitudinal directions (even if FCR is higher in the case of longitudinal resistivity) [59]. The same authors [57] evaluated the electrical resistivity of cement pastes subjected to uniaxial compression; they considered paste specimens of two types, namely carbon fiber silica fume cement paste and carbon fiber latex cement paste. They evaluated electrical resistivity both in longitudinal and transverse directions, considering cubic (51 mm side) and rectangular specimens (150 × 12 × 11 mm), observing a reversible decrease in both the resistivities, with the exception of some irreversible increase at the end of the first stress cycle due to minor damage.

Carbon fibers were used also in combination with **graphite** [45]: the former form a conductive network, whereas the latter fill the spaces among fibers, increasing the smart agility of smart concrete. Electrical resistivity is negatively correlated with stress: it is possible to establish a quadratic polynomial relationship between the variance ratio of resistance ($\Delta R/R$) and stress (or strain), with $R^2 > 0.95$ for different concrete compositions. The same materials combinations was tested by Liu and Wu [49] in conductive asphalt concrete; they noticed that FCR slows down with more loading cycles, meaning that piezoresistivity tends to decrease because of viscoelasticity.

Also **steel fibers** can be used. According to Teomete and Kocyigit [51], the relationship between the percent change in electrical resistance and tensile strain is almost linear before cracking ($R^2 = 0.92$ and $R^2 > 0.94$ for mixtures without and with fibers, respectively; the correlation coefficient converges at 0.99 approximately at percolation threshold). The gage factor increases with fibers volume, converging approximately in correspondence of the percolation threshold (1 vol.%). Nguyen et al. [53,54] evaluated different types of steel fibers, namely smooth, twisted, and hooked, noticing that the most effective for enhancing mechanical resistance are those twisted (better meso than macro twisted), followed by smooth and hooked ones. Electrical resistivity decreases with tensile strain until post-cracking, due to the formation of multiple cracks due to strain-hardening.

Cementitious strain sensors can be realized to be embedded in structures to be monitored, providing a high monitoring efficiency in continuous, low-cost, high durability, and simple construction technology to be embedded only in critical points of concrete structures, forming a distributed monitoring network. Strain sensing provides valuable information for the control of structural vibrations, traffic monitoring, and weighing [213]. Azhari and Banthia [121] highlighted a non-linear and rate-dependent relationship between change in resistivity and compressive stress, measured in electrically conductive cementitious composites with carbon fibers and CNTs (better when used in conjunction). Also, D'Alessandro et al. [38] used CNTs to realize sensors to be embedded on a concrete beam (250 × 250 × 2200 mm) at a distance of 250 mm from each other, forming a durable distributed SHM solution able to provide results in agreement with traditional strain gauges. Han and Ou [67] employed carbon fibers and carbon black to realize piezoresistive sensors embedded in concrete structures. They obtained a linear relationship between FCR and compressive stress ($y = -1.35x$); another linear relationship was established between FCR and compressive strain (y = −0.0227x). The response of the contact resistance to elastic deformation is reversible [214], whereas becomes irreversible in the case of plastic deformation. Ding et al. [40] developed cementitious sensors, manufactured with the addition of CNT/NCB (nano carbon black) composite filler, to incorporate in concrete columns. These sensors presented high stability and repeatability within the elastic regime, providing a polynomial relationship between stress/strain and FCR (y = $7.743 \cdot 10^{-5} x^3 - 0.005 x^2 + 0.510x$, with R^2 = 0.999 and y = $0.018 x^3 - 0.759 x^2 + 28.00751$ with R^2 = 0.997 for stress and strain, respectively). During loading, FCR decreases; no damages to the sensor are caused by the compressive stress. The change of electrical resistance is reversible upon loading and unloading phases, even if the amplitude of FCR is slightly different in subsequent load cycles; under monotonic loading, when the sensor ultimate stress has been reached, FCR remains quite stable. Similarly, Han et al. [55] developed highly sensitive piezoresistive sensors using nickel powder-filled cement-based composites; they found that electrical resistivity decreases monotonously as compressive load increases, showing a decrease by approximately 63% in the elastic regime (investigated range: 0–12.5 MPa). FCR is highly correlated to compressive force (Boltzmann sigmoidal regression curve, R^2 > 0.99). Nickel powder enables contacting and tunneling conduction effects, causing a decrease in the electrical resistivity, besides enhancing the sensitivity to stress/strain. Han et al. [215] designed a self-sensing pavement (embedded with smart cement-based sensors, realized with the addition of nickel powders enhancing self-sensing capability—[216]) for traffic detection and monitoring. They proposed an FCR equal to approximately 18% with a compressive strength of 0.5 MPa (caused by a small vehicle of approximately 1000 kg on an area of the four tires of approximately 20000 mm^2), thus obtaining a reversible and sensitive response in terms of electrical resistivity. Luo and Chung [52] studied contact electrical resistance under dynamic loading in two overlapped mortar strips; they observed an irreversible increase in electrical resistivity at a stress amplitude of 15 MPa, probably because of the production of severe debris. The inclusion of steel fibers decreases the noise present in FCR data [122]. Contrarily, Lee et al. [122] noticed that FCR cannot predict the compressive behavior of ultra-high-performance concretes containing CNTs, probably because the external load changes the conductive paths only slightly, due to their dense microstructure, whereas it can simulate quite well the tensile behavior (in terms of stress-strain and stress-crack opening displacement). A non-linear relationship between stress and FCR and a linear relationship between strain and FCR (R^2 > 0.9) were found.

3.8. Defects and Damages Detection

Cracks (provoked by internal stresses [217] or external loadings [218], such as mechanical loads [219], drying shrinkage [220], and freezing action [221]) weaken the matrix resistance, besides constituting a preferential ingress path for contaminants [222], fluids and ion flows, which ease corrosion processes; therefore, durability of concrete is significantly affected by cracks and other damages formation. When cracks form, conductive pathways are partially destroyed. As a consequence, electrical resistivity increases. In this case, and consequently, the electrical resistance can be used as a **damage sensor** besides strain sensor. Teomete and Kocyigit [51] observed that electrical resistance has abrupt changes when crack initiates and propagates in steel fiber reinforced cement matrix composites.

Real-time monitoring makes it possible to promptly detect damages when they occur, easing the detection of the causes of these damages and enabling a timely repairing action, hence prolonging structural service life. Electrical resistivity of concrete is sensitive to concrete stress-strain and cracking behavior, with a response mainly depending on the load level and the water saturation degree [76]. In fact, crack formation and propagation cause changes in the path of electric current, hence ρ will change, following their evolution in time. It is worth underling that the effect will be dependent on the cracks conditions, as demonstrated by Lataste et al. [119] with a 4-probe resistivity meter. If cracks are insulating (i.e., dry concrete, where cracks are filled with air or sometimes dust, showing a high electrical resistivity), ρ will increase; if they are water-saturated (i.e., wet concrete), ρ will decrease (since water has a good electrical resistivity) [119]; in fact, possible impurities present in the mixture can alter electrical resistance, increasing or decreasing it depending on their nature (e.g., metallic or dust particles). The same meter was used by Goueygou et al. [118] for the comparison of electrical resistivity measurements and transmission of ultrasonic (US) surface waves in crack detection; both the techniques are suitable to localize the main cracks, but not to measure the crack depths. The information provided by electrical resistivity can be useful to complement US results, distinguishing between wet and dry cracks. However, secondary cracks represent a disturbing factor for electrical resistivity measurements. Wiwattanachang and Giao [98] used the resistivity meter SYSCAL R1 plus, with 24 electrode connected to a multicore cable system, for the evaluation of both artificial cracks (made of plastic sheets) and loading cracks. The wider the crack width, the higher the electrical resistivity. Electric imaging can provide 2D representation useful for crack detection. Peled et al. [75] noticed a dramatic change in impedance values when a sudden growth in crack occurred; the real part of the electrical impedance increases through the fracture process, following the crack opening. Dong et al. [41] monitored the cracks initiation and propagation by means of electrical resistivity measurements; in fact, FCR is dominated by the cracks opening, showing the same trend of opening with load.

Also, cementitious materials with conductive additions were evaluated for damage detection. It is worth noting that conductive fibers/fillers can be used as a reinforcement, providing the so-called "crack-bridging effect". Lim et al. [46] analyzed cement composites containing CNTs, confirming the possibility to detect crack (and to evaluate their width) by means of electrical resistivity measurements, provided that the specimen under test is moist. Otherwise, under dry conditions, the crack breaks the conductive network of CNTs and the electrical conductivity is not influenced by the crack width anymore. Also, Zhang et al. [39] found that reduction in water content significantly decreases the piezoresistive property, both in terms of FCR and sensitivity to stress and strain; this can be attributed to the decrease of contact between eventual fillers (they tested both carbon nanotube and nanocarbon black) and the shrinkage increase [223]. It is worth noting that the water content generally increases with conductive fillers content because of their water absorption capability. The importance of moisture content is highlighted also in other studies. Boulay et al. [50] studied the evolution of electrical resistivity with cracking in concrete specimens saturated with a basic solution, noticing a decrease of electrical resistivity when crack opens, since the ionic solution takes up and fills it. Fan et al. [123] maintained specimens in saturated water condition before tests, to exclude moisture effect on impedance measurement; after damage, the specimens were subjected to wet/dry cycles to enable self-healing. They noticed that both opening and number of cracks influence the measurement results; the relation between electrical impedance and crack opening is frequency-dependent and presents both a resistor and a capacitor effect: the former is linked to the conductive media inside the crack, whereas the latter linearly decreases with increasing crack opening. Zhu et al. [91] considered both plain mortar and ECCs; they highlighted that in fractured mortars the electrical impedance increases several hundred times, resulting in different shape and trend curves representation in Nyquist plot.

4. Let's Wrap Up

4.1. Discussion

This extended literature review presented so far aims to highlight that many different methods are nowadays used to measure the electrical resistivity of concrete. These methods involve different measurement setups (Figure 3) and sensing electrode (Figure 4) configuration, they provide different information, suitable for distinct applications and with different accuracies.

There are many target applications where electrical resistivity measurements can provide valuable information (Figure 5). The approach representing the best compromise between ease of application, time to result, and measurement accuracy should be chosen by considering the target application. For example, the uniaxial method is suitable for laboratory measurements on specimens, but it cannot be exploited for in-field testing, whereas the non-contact method is hardly exploitable for continuous monitoring due to the specific testing configuration (particularly the ring-shaped specimens needed). Surface electrodes are widely used in 4-point probe AC configuration, but this setup is valid only for inspection purposes and not for continuous monitoring. Embedded electrodes are considered more durable and can ensure better bonding with the cement-based material with respect to pasted electrodes [55].

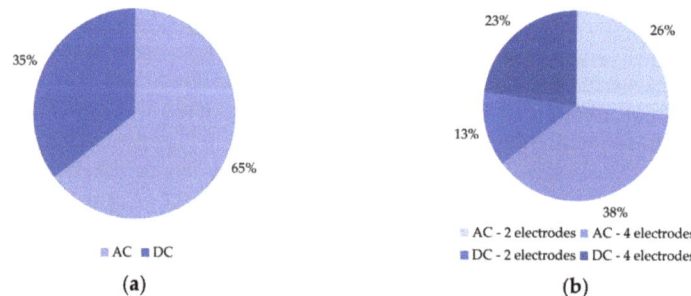

Figure 3. (a) Percentages of papers reporting DC and AC measurements; (b) Percentages of papers reporting 2- and 4-electrode DC/AC measurements.

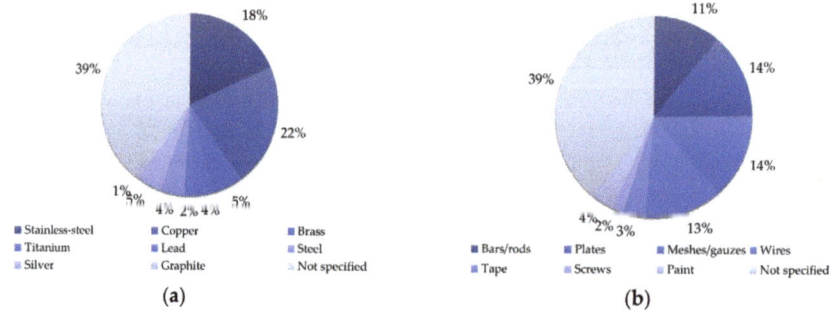

Figure 4. Different electrodes configurations used for electrical resistivity measurement in terms of (a) Materials and (b) Geometries.

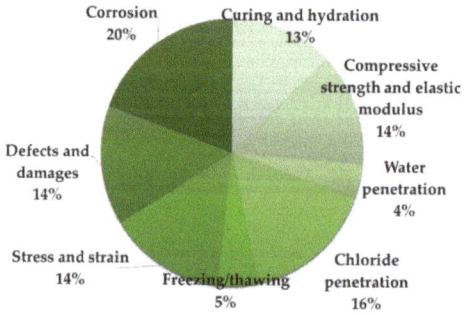

Figure 5. Percentages of papers reporting different application fields.

In electrical impedance testing, it is fundamental to take into account all the **possible side effects** that might increase the uncertainty associated with the measurement. Different issues with **DC measurement** with **two electrodes** are pointed out in the literature, namely possible electrochemical reactions linked to the electrical potential difference applied between the two electrodes, polarization at the electrode/specimen interface, and accuracy affected by the contact resistance between electrodes and the specimen [158]. The excitation signal causes an accumulation of charges at the interface between the electrodes and the material itself (the electrodes behave like capacitors, whose conductive planes are the electrode itself and the underneath material) and such interface polarization gives an important contribution to the measurement result (since it includes the contact impedance). This holding, it becomes clear that measurements performed with two electrodes significantly overestimate the concrete electrical resistivity [80] and it would be necessary to take into account the electrodes polarization contribution, subtracting it from the result [175,224] to get accurate resistivity values. The addition of conductive materials (e.g., carbon fibers) can decrease polarization effect, since conductors cannot support high electric fields [44].

Differentiating between excitation and measurement electrodes significantly avoid polarization phenomena and minimizes the insertion error [44,125]. For these reasons, the **4-electrode** configuration should be preferred since separating the excitation/response electrodes results in increasing the measurement accuracy. Polder et al. [7] gave some recommendations on the approaches for measuring the electrical resistivity of concrete, including the four-point method (i.e., Wenner's array) for the on-site measurement; they also addressed reinforced structures. However, the variability of the measurement results is quite high: 10% variation coefficients are considered good, 20% normal; also values up to 25% in the field are possible [25]. Some commercial devices are available, but they are targeted to inspection purposes and not suitable for continuous monitoring, for which a fix sensor network regularly measuring the electrical resistivity of concrete would be necessary [68].

The use of an **AC** approach in electrical resistivity measurement makes it possible to focus on the material microstructure, thus excluding electrode polarization as a possible interfering input. The measurement of electrical impedance of concrete should be performed according to the EIS method, thus spanning multiple frequencies. This is of utmost importance, since the material electrical properties (both the electrical resistivity and the dielectric permittivity) are frequency-dependent [69] and the analysis in different ranges allows to investigate several different aspects of materials (e.g., the composite resistance of specimens with conductive fibers is visible at a higher frequency, since at low frequency the superficial passive layer makes them non-conductive [22]). The results of an electrical resistivity measurement are strictly linked to the frequency and the shape of the electric signal used for the measurement. Frequencies greater than 1 kHz should be employed to avoid the effect of material polarization (i.e., the orientation of dipoles according to the electromagnetic field originated from the measurement itself) [80,224]. McCarter et al. [77] monitored the electrical impedance of cement pastes in the frequency range of 1–100 kHz, with a sampling time of 10 min in 48 h test period. They highlighted that electrode polarization at the lowest frequency (1 kHz) masks the bulk response from the paste, which instead vanishes with increasing frequency. Moreover, in order to examine composite materials, e.g., concrete with the addition of carbon fibers, higher frequency values, which mirror the interaction between the conductive addition and the cement-based material, should be considered.

To avoid measurement errors, attention should be paid to several aspects, including the specimen geometry, the possible non-homogeneity of the material, the environmental conditions (particularly temperature and relative humidity) and their changes, the interface between electrodes and material (a good electric contact should be maintained), the presence of reinforcing steel rebars or other conductive materials close to the measurement electrodes, the different resistivity of bulk and surface layers, the local variations in concrete, and the electronic noise.

4.2. Conclusions

This review paper aims at providing the interested reader with hints and suggestions on how to approach electrical resistivity/impedance measurements on cement-based structures. Indeed, it was one of the authors' priority to demonstrate that, depending on target applications, several variables should be considered to ensure reliability and accuracy of results, e.g., electrode configuration or excitation type (DC vs. AC), just to cite some. Literature can indeed provide useful recommendations in approaching electrical resistivity/impedance measurements in concrete [106], in particular:

- The use of an AC excitation signal (electric current or potential difference) is important to avoid the Faradic effect of charges (ions) separation in the material;
- Frequencies greater than 1 kHz should be used to fully avoid the alignment of dipoles consequent to the excitation;
- A low-frequency (less than hundreds of MHz or some GHz) AC excitation current should be applied to avoid spurious mode voltage;
- The 4-point probe method should be used to avoid any influence from the contact surface area;
- A good electrical connection between each contact and the concrete surface should be ensured;
- A contact spacing of at least 1.5 times the size of the maximum aggregate size in the concrete should be used;
- A Wenner contact spacing less than or equal to one quarter of the concrete section thickness should be used;
- Concrete resistivity should be measured at a minimum distance of twice the contact spacing from the edge of the concrete section;
- An electromagnetic cover meter or bar locator should be used to locate the reinforcement rebars;
- A contact spacing less than or equal to two thirds of the concrete cover should be used where the proximity of steel reinforcements is unavoidable, in order to minimize the error due to the presence of the rebar;
- A contact spacing of at least 4 cm should be used where surface wetting effect might be expected;
- The measurement should be performed at least 24 h after a rainfall has occurred;
- A contact spacing not less than eight times the thickness of this layer should be used if the presence of a low-resistivity surface layer is unavoidable;
- A temperature compensation of +1 kW·cm per 3 °C fall in ambient temperature might be used to convert resistivity measurements to a standard temperature.

Moreover, it is worth noting that electrical resistivity is a parameter varying within the volume, since there is no homogeneity in concrete elements; in addition, those layers close to the surface are more sensitive to variations of climate conditions (e.g., temperature changes or rain events). For this reason, the measurement should be considered as a local measurement. This suggests that a proper monitoring of a concrete structure should be approached targeting a distributed monitoring system using several sensing nodes. These nodes should be places on those parts of the target structure that are considered as the most critical to concrete degradation. On the other hand, electrical resistivity measurements can be applied also targeting inspection. No matter the final approach, i.e., monitoring or inspection, the addition of self-sensing material would produce positive effects on a resistivity measurement. The calibration of the measurement devices is a key aspect that should always be considered, especially in a measurement for which environmental variables, like temperature and relative humidity, or material characteristics, like moisture content, act as disturbing inputs and consequently contribute also as sources of uncertainty. For in-lab tests, calibration can be performed with liquid of known electrical conductivity and on standard concrete specimens. The procedure is, however, difficult for in-field measurements. Hence, calibration turns out in a check on cabling and connections or a performance verification with reference devices (e.g., electronic circuits, concrete specimens of known resistivity, etc.) Measurement uncertainty would deserve more consideration by the scientific community and should be addressed more often, given the lack of documents available in the literature on this topic. Indeed, only a proper assessment of measurement uncertainty can ensure reliability of results and hence robustness to the inference on the health status of the cement-based structure that is inspected/monitored, also given the heterogeneity of the cement-based materials, surely affecting the measurement reproducibility.

Last but not least, it should be recalled that, since electrical resistivity of concrete is influenced by many factors, e.g., concrete composition, environmental conditions, etc., pre-determined ranges correlating resistivity with corrosion risks of reinforcement, or relationships correlating resistivity with concrete durability, just to cite some, should be considered in a critical manner, as they could not be considered valid in all conditions.

Author Contributions: Conceptualization, G.C. and A.M.; data curation, G.C. and A.M.; writing—original draft preparation, G.C. and A.M.; writing—review and editing, G.C., A.M., P.C., F.T., G.M.R.; visualization, G.C. and A.M.; supervision, P.C. All authors have read and agreed to the published version of the manuscript.

Funding: This research received no external funding.

Conflicts of Interest: The authors declare no conflict of interest.

References

1. Monticelli, C.; Criado, M.; Fajardo, S.; Bastidas, J.M.; Abbottoni, M.; Balbo, A. Corrosion behaviour of a Low Ni austenitic stainless steel in carbonated chloride-polluted alkali-activated fly ash mortar. *Cem. Concr. Res.* **2014**, *55*, 49–58. [CrossRef]
2. European Standard. *EN 1990: Eurocode—Basis of Structural Design*; European Standard: Brussels, Belgium, 2002.
3. Hoła, J.; Bień, J.; Sadowski, Ł.; Schabowicz, K. Non-destructive and semi-destructive diagnostics of concrete structures in assessment of their durability. *Bull. Pol. Acad. Sci. Tech. Sci.* **2015**, *63*. [CrossRef]
4. Sengul, O.; Gjørv, O.E. Electrical resistivity measurements for quality control during concrete construction. *ACI Mater. J.* **2008**, *105*, 541–547. [CrossRef]
5. Sengul, O. Use of electrical resistivity as an indicator for durability. *Constr. Build. Mater.* **2014**, *73*, 434–441. [CrossRef]
6. Gjørv, O.E.; Vennesland, Ø.E.; El-Busaidy, A.H.S. Electrical resistivity of concrete in the oceans. In Proceedings of the Offshore Technology Conference, Houston, TX, USA, 2–5 May 1977.
7. Polder, R.; Andrade, C.; Elsener, B.; Vennesland, Ø.; Gulikers, J.; Weidert, R.; Raupach, M. Test methods for on site measurement of resistivity of concrete. *Mater. Struct.* **2000**, *33*, 603–611. [CrossRef]
8. Bertolini, L.; Elsener, B.; Pedeferri, P.; Polder, R. *Corrosion of Steel in Concrete: Prevention, Diagnosis, Repair*; Wiley-VCH Verlag GmbH & Co. KGaA, Ed.; Wiley Blackwell: Weinheim, Germany, 2005; ISBN 9783527603374.
9. Gorzelańczyk, T. Moisture influence on the failure of self-compacting concrete under compression. *Arch. Civ. Mech. Eng.* **2011**, *11*, 45–60. [CrossRef]
10. Tuutti, K. *Corrosion of Steel in Concrete*; Swedish Cement and Concrete Research Institute: Stockholm, Sweden, 1982.
11. Hou, T.C.; Nguyen, V.K.; Su, Y.M.; Chen, Y.R.; Chen, P.J. Effects of coarse aggregates on the electrical resistivity of Portland cement concrete. *Constr. Build. Mater.* **2017**, *133*, 397–408. [CrossRef]
12. Simon, T.K.; Vass, V. The electrical resistivity of concrete. *Concr. Struct.* **2012**, *13*, 61–64. [CrossRef]
13. Collepardi, M. *The New Concrete*; Tintoretto: Lancenigo, Italy, 2010; ISBN 88-903777-2-0.
14. Dong, B.; Zhang, J.; Wang, Y.; Fang, G.; Liu, Y.; Xing, F. Evolutionary trace for early hydration of cement paste using electrical resistivity method. *Constr. Build. Mater.* **2016**, *119*, 16–20. [CrossRef]
15. López, W.; González, J.A.; Andrade, C. Influence of temperature on the service life of rebars. *Cem. Concr. Res.* **1993**, *23*, 1130–1140. [CrossRef]
16. Ferreira, R.M.; Jalali, S. NDT measurements for the prediction of 28-day compressive strength. *NDT E Int.* **2010**, *43*, 55–61. [CrossRef]
17. Hope, B.B.; Ip, A.C. Corrosion of steel in concrete made with slag cement. *ACI Mater. J.* **1987**, *84*, 525–531. [CrossRef]
18. Bijen, J. Benefits of slag and fly ash. *Constr. Build. Mater.* **1996**, *10*, 309–314. [CrossRef]
19. Adil, G.; Kevern, J.T.; Mann, D. Influence of silica fume on mechanical and durability of pervious concrete. *Constr. Build. Mater.* **2020**, *247*, 118453. [CrossRef]
20. Wee, T.H.; Suryavanshi, A.K.; Tin, S.S. Evaluation of rapid chloride permeability test (RCPT) results for concrete containing mineral admixtures. *ACI Struct. J.* **2000**, *97*, 221–232. [CrossRef]
21. Donnini, J.; Bellezze, T.; Corinaldesi, V. Mechanical, electrical and self-sensing properties of cementitious mortars containing short carbon fibers. *J. Build. Eng.* **2018**, *20*, 8–14. [CrossRef]
22. Berrocal, C.G.; Hornbostel, K.; Geiker, M.R.; Löfgren, I.; Lundgren, K.; Bekas, D.G. Electrical resistivity measurements in steel fibre reinforced cementitious materials. *Cem. Concr. Compos.* **2018**, *89*, 216–229. [CrossRef]

23. Chiarello, M.; Zinno, R. Electrical conductivity of self-monitoring CFRC. *Cem. Concr. Compos.* **2005**, *27*, 463–469. [CrossRef]
24. Whittington, H.W.; McCarter, J.; Forde, M.C. The conduction of electricity through concrete. *Mag. Concr. Res.* **1981**, *33*, 48–60. [CrossRef]
25. Polder, R.B. Test methods for on site measurement of resistivity of concrete—A RILEM TC-154 technical recommendation. *Constr. Build. Mater.* **2001**, *15*, 125–131. [CrossRef]
26. Polder, R. Test methods for on-site measurement of resistivity of concrete—RILEM TC 154-EMC: Electrochemical techniques for measuring metallic corrosion. *Mater. Struct.* **2000**, *33*, 603–611. [CrossRef]
27. Garzon, A.J.; Sanchez, J.; Andrade, C.; Rebolledo, N.; Menéndez, E.; Fullea, J. Modification of four point method to measure the concrete electrical resistivity in presence of reinforcing bars. *Cem. Concr. Compos.* **2014**, *53*, 249–257. [CrossRef]
28. Nguyen, A.Q.; Klysz, G.; Deby, F.; Balayssac, J.P. Evaluation of water content gradient using a new configuration of linear array four-point probe for electrical resistivity measurement. *Cem. Concr. Compos.* **2017**, *83*, 308–322. [CrossRef]
29. Glass, G.K.; Page, C.L.; Short, N.R. Factors affecting the corrosion rate of steel in carbonated mortars. *Corros. Sci.* **1991**, *32*, 1283–1294. [CrossRef]
30. Dérobert, X.; Lataste, J.F.; Balayssac, J.P.; Laurens, S. Evaluation of chloride contamination in concrete using electromagnetic non-destructive testing methods. *NDT E Int.* **2017**, *89*, 19–29. [CrossRef]
31. Lübeck, A.; Gastaldini, A.L.G.; Barin, D.S.; Siqueira, H.C. Compressive strength and electrical properties of concrete with white Portland cement and blast-furnace slag. *Cem. Concr. Compos.* **2012**, *34*, 392–399. [CrossRef]
32. Wen, S.; Chung, D.D.L. Piezoresistivity-based strain sensing in carbon fiber-reinforced cement. *ACI Mater. J.* **2007**, *104*, 171–179. [CrossRef]
33. Lee, S.-J.; You, I.; Zi, G.; Yoo, D.-Y. Experimental investigation of the piezoresistive properties of cement composites with hybrid carbon fibers and nanotubes. *Sensors* **2017**, *17*, 2516. [CrossRef]
34. Hope, B.B.; Ip, A.K.; Manning, D.G. Corrosion and electrical impedance in concrete. *Cem. Concr. Res.* **1985**, *15*, 525–534. [CrossRef]
35. Chen, C.T.; Chang, J.J.; Yeih, W.C. The effects of specimen parameters on the resistivity of concrete. *Constr. Build. Mater.* **2014**, *71*, 35–43. [CrossRef]
36. Schuetze, A.P.; Lewis, W.; Brown, C.; Geerts, W.J. A laboratory on the four-point probe technique. *Am. J. Phys.* **2004**. [CrossRef]
37. ASTM C1760—12 Standard Test Method for Bulk Electrical Conductivity of Hardened Concrete. Available online: https://www.astm.org/Standards/C1760 (accessed on 24 June 2020).
38. D'Alessandro, A.; Meoni, A.; Ubertini, F.; Luigi Materazzi, A. Strain measurement in a reinforced concrete beam using embedded smart concrete sensors. In *Lecture Notes in Civil Engineering*; Springer: Cham, Switzerland, 2020; Volume 42, pp. 289–300.
39. Zhang, L.; Ding, S.; Han, B.; Yu, X.; Ni, Y.Q. Effect of water content on the piezoresistive property of smart cement-based materials with carbon nanotube/nanocarbon black composite filler. *Compos. Part. A Appl. Sci. Manuf.* **2019**, *119*, 8–20. [CrossRef]
40. Ding, S.; Ruan, Y.; Yu, X.; Han, B.; Ni, Y.Q. Self-monitoring of smart concrete column incorporating CNT/NCB composite fillers modified cementitious sensors. *Constr. Build. Mater.* **2019**, *201*, 127–137. [CrossRef]
41. Dong, S.; Dong, X.; Ashour, A.; Han, B.; Ou, J. Fracture and self-sensing characteristics of super-fine stainless wire reinforced reactive powder concrete. *Cem. Concr. Compos.* **2020**, *105*, 103427. [CrossRef]
42. Melugiri-Shankaramurthy, B.; Sargam, Y.; Zhang, X.; Sun, W.; Wang, K.; Qin, H. Evaluation of cement paste containing recycled stainless steel powder for sustainable additive manufacturing. *Constr. Build. Mater.* **2019**, *227*, 116696. [CrossRef]
43. Huang, Y.; Li, H.; Qian, S. Self-sensing properties of engineered cementitious composites. *Constr. Build. Mater.* **2018**, *174*, 253–262. [CrossRef]
44. Wen, S.; Chung, D.D.L. Electric polarization in carbon fiber-reinforced cement. *Cem. Concr. Res.* **2001**, *31*, 141–147. [CrossRef]
45. Chen, M.; Gao, P.; Geng, F.; Zhang, L.; Liu, H. Mechanical and smart properties of carbon fiber and graphite conductive concrete for internal damage monitoring of structure. *Constr. Build. Mater.* **2017**, *142*, 320–327. [CrossRef]

46. Lim, M.-J.; Lee, H.K.; Nam, I.-W.; Kim, H.-K. Carbon nanotube/cement composites for crack monitoring of concrete structures. *Compos. Struct.* **2017**, *180*, 741–750. [CrossRef]
47. Dehghanpour, H.; Yilmaz, K. Investigation of specimen size, geometry and temperature effects on resistivity of electrically conductive concretes. *Constr. Build. Mater.* **2020**, *250*, 118864. [CrossRef]
48. Belli, A.; Mobili, A.; Bellezze, T.; Tittarelli, F.; Cachim, P. Evaluating the self-sensing ability of cement mortars manufactured with graphene nanoplatelets, virgin or recycled carbon fibers through piezoresistivity tests. *Sustainability* **2018**, *10*, 4013. [CrossRef]
49. Liu, X.; Wu, S. Research on the conductive asphalt concrete's piezoresistivity effect and its mechanism. *Constr. Build. Mater.* **2009**, *23*, 2752–2756. [CrossRef]
50. Boulay, C.; Dal Pont, S.; Belin, P. Real-time evolution of electrical resistance in cracking concrete. *Cem. Concr. Res.* **2009**, *39*, 825–831. [CrossRef]
51. Teomete, E.; Kocyigit, O.I. Tensile strain sensitivity of steel fiber reinforced cement matrix composites tested by split tensile test. *Constr. Build. Mater.* **2013**, *47*, 962–968. [CrossRef]
52. Luo, X.; Chung, D.D.L. Concrete-concrete pressure contacts under dynamic loading, studied by contact electrical resistance measurement. *Cem. Concr. Res.* **2000**, *30*, 323–326. [CrossRef]
53. Nguyen, D.L.; Ngoc-Tra Lam, M.; Kim, D.J.; Song, J. Direct tensile self-sensing and fracture energy of steel-fiber-reinforced concretes. *Compos. Part. B Eng.* **2020**, *183*, 107714. [CrossRef]
54. Nguyen, D.L.; Song, J.; Manathamsombat, C.; Kim, D.J. Comparative electromechanical damage-sensing behaviors of six strain-hardening steel fiber-reinforced cementitious composites under direct tension. *Compos. Part. B Eng.* **2015**, *69*, 159–168. [CrossRef]
55. Han, B.G.; Han, B.Z.; Ou, J.P. Experimental study on use of nickel powder-filled Portland cement-based composite for fabrication of piezoresistive sensors with high sensitivity. *Sens. Actuators A Phys.* **2009**, *149*, 51–55. [CrossRef]
56. Wen, S.; Chung, D.D.L. A comparative study of steel- and carbon-fibre cement as piezoresistive strain sensors. *Adv. Cem. Res.* **2003**, *15*, 119–128. [CrossRef]
57. Wen, S.; Chung, D.D.L. Uniaxial compression in carbon fiber-reinforced cement, sensed by electrical resistivity measurement in longitudinal and transverse directions. *Cem. Concr. Res.* **2001**, *31*, 297–301. [CrossRef]
58. Wen, S.; Chung, D.D.L. Damage monitoring of cement paste by electrical resistance measurement. *Cem. Concr. Res.* **2000**, *30*, 1979–1982. [CrossRef]
59. Wen, S.; Chung, D.D.L. Uniaxial tension in carbon fiber reinforced cement, sensed by electrical resistivity measurement in longitudinal and transverse directions. *Cem. Concr. Res.* **2000**, *30*, 1289–1294. [CrossRef]
60. Cao, J.; Chung, D.D.L. Defect dynamics and damage of concrete under repeated compression, studied by electrical resistance measurement. *Cem. Concr. Res.* **2001**, *31*, 1639–1642. [CrossRef]
61. Chen, B.; Liu, J. Damage in carbon fiber-reinforced concrete, monitored by both electrical resistance measurement and acoustic emission analysis. *Constr. Build. Mater.* **2008**, *22*, 2196–2201. [CrossRef]
62. Cao, J.; Chung, D.D.L. Use of fly ash as an admixture for electromagnetic interference shielding. *Cem. Concr. Res.* **2004**, *34*, 1889–1892. [CrossRef]
63. Cao, J.; Chung, D.D.L. Coke powder as an admixture in cement for electromagnetic interference shielding. *Carbon N. Y.* **2003**, *41*, 2433–2436. [CrossRef]
64. Wen, S.; Chung, D.D.L. Partial replacement of carbon fiber by carbon black in multifunctional cement-matrix composites. *Carbon N. Y.* **2007**, *45*, 505–513. [CrossRef]
65. Wang, Z.; Zeng, Q.; Wang, L.; Yao, Y.; Li, K. Electrical resistivity of cement pastes undergoing cyclic freeze-thaw action. *J. Mater. Civ. Eng.* **2015**, *27*, 04014109. [CrossRef]
66. Wang, Z.; Zeng, Q.; Wang, L.; Yao, Y.; Li, K. Effect of moisture content on freeze-thaw behavior of cement paste by electrical resistance measurements. *J. Mater. Sci.* **2014**, *49*, 4305–4314. [CrossRef]
67. Han, B.; Ou, J. Embedded piezoresistive cement-based stress/strain sensor. *Sens. Actuators A Phys.* **2007**, *138*, 294–298. [CrossRef]
68. Cosoli, G.; Mobili, A.; Giulietti, N.; Chiariotti, P.; Pandarese, G.; Tittarelli, F.; Bellezze, T.; Mikanovic, N.; Revel, G.M. Performance of concretes manufactured with newly developed low-clinker cements exposed to water and chlorides: Characterization by means of electrical impedance measurements. *Constr. Build. Mater.* **2020**, 121546. [CrossRef]

69. Taha, H.; McCarte, W.J.; Suryanto, B.; Starrs, G. Frequency- and time-domain dependency of electrical properties of cement-based materials during early hydration. *Adv. Civ. Eng. Mater.* **2017**, *6*, 20140367. [CrossRef]
70. AASHTO TP 119—Standard Method of Test for Electrical Resistivity of a Concrete Cylinder Tested in a Uniaxial Resistance Test. Available online: https://standards.globalspec.com/std/9945672/aashto-tp-119 (accessed on 12 August 2020).
71. Mason, T.O.; Campo, M.A.; Hixson, A.D.; Woo, L.Y. Impedance spectroscopy of fiber-reinforced cement composites. *Cem. Concr. Compos.* **2002**, *24*, 457–465. [CrossRef]
72. Kim, J.; Suryanto, B.; McCarter, W.J. Conduction, relaxation and complex impedance studies on Portland cement mortars during freezing and thawing. *Cold Reg. Sci. Technol.* **2019**, *166*, 102819. [CrossRef]
73. Osterminski, K.; Polder, R.B.; Schießl, P. Long term behaviour of the resistivity of concrete. *Heron* **2012**, *57*, 211–230.
74. Polder, R.B.; Peelen, W.H.A. Characterisation of chloride transport and reinforcement corrosion in concrete under cyclic wetting and drying by electrical resistivity. *Cem. Concr. Compos.* **2002**, *24*, 427–435. [CrossRef]
75. Peled, A.; Torrents, J.M.; Mason, T.O.; Shah, S.P.; Garboczi, E.J. Electrical impedance spectra to monitor damage during tensile loading of cement composites. *ACI Mater. J.* **2001**, *98*, 313–322.
76. Zeng, X.; Liu, H.; Zhu, H.; Ling, C.; Liang, K.; Umar, H.A.; Xie, Y.; Long, G.; Ma, C. Study on damage of concrete under uniaxial compression based on electrical resistivity method. *Constr. Build. Mater.* **2020**, *254*, 119270. [CrossRef]
77. McCarter, W.J.; Chrisp, T.M.; Starrs, G.; Blewett, J. Characterization and monitoring of cement-based systems using intrinsic electrical property measurements. *Cem. Concr. Res.* **2003**, *33*, 197–206. [CrossRef]
78. Rovnaník, P.; Kusák, I.; Bayer, P.; Schmid, P.; Fiala, L. Comparison of electrical and self-sensing properties of Portland cement and alkali-activated slag mortars. *Cem. Concr. Res.* **2019**, *118*, 84–91. [CrossRef]
79. Van Noort, R.; Hunger, M.; Spiesz, P. Long-term chloride migration coefficient in slag cement-based concrete and resistivity as an alternative test method. *Constr. Build. Mater.* **2016**, *115*, 746–759. [CrossRef]
80. McCarter, W.J.; Starrs, G.; Kandasami, S.; Jones, R.; Chrisp, M. Electrode configurations for resistivity measurements on concrete. *ACI Mater. J.* **2009**, *106*, 258–264. [CrossRef]
81. Spragg, R.; Castro, J.; Nantung, T.; Paredes, E.; Weiss, W.J. *Variability Analysis of the Bulk Resistivity Measured Using Concrete Cylinders*; Joint Transportation Research Program, Indiana Department of Transportation and Purdue University: West Lafayette, IN, USA, 2011.
82. Newlands, M.D.; Jones, M.R.; Kandasami, S.; Harrison, T.A. Sensitivity of electrode contact solutions and contact pressure in assessing electrical resistivity of concrete. *Mater. Struct. Constr.* **2008**, *41*, 621–632. [CrossRef]
83. Sato, T.; Beaudoin, J.J. Coupled AC impedance and thermomechanical analysis of freezing phenomena in cement paste. *Mater. Struct.* **2011**, *44*, 405–414. [CrossRef]
84. Perron, S.; Beaudoin, J.J. Freezing of water in portland cement paste—An ac impedance spectroscopy study. *Cem. Concr. Compos.* **2002**, *24*, 467–475. [CrossRef]
85. Zhou, C.; Li, K.; Han, J. Characterizing the effect of compressive damage on transport properties of cracked concretes. *Mater. Struct.* **2012**, *45*, 381–392. [CrossRef]
86. Díaz, B.; Freire, L.; Merino, P.; Nóvoa, X.R.; Pérez, M.C. Impedance spectroscopy study of saturated mortar samples. *Electrochim. Acta* **2008**, *53*, 7549–7555. [CrossRef]
87. Carsana, M.; Canonico, F.; Bertolini, L. Corrosion resistance of steel embedded in sulfoaluminate-based binders. *Cem. Concr. Compos.* **2018**, *88*, 211–219. [CrossRef]
88. Ghosh, P.; Tran, Q. Influence of parameters on surface resistivity of concrete. *Cem. Concr. Compos.* **2015**, *62*, 134–145. [CrossRef]
89. Velay-Lizancos, M.; Azenha, M.; Martínez-Lage, I.; Vázquez-Burgo, P. Addition of biomass ash in concrete: Effects on E-Modulus, electrical conductivity at early ages and their correlation. *Constr. Build. Mater.* **2017**, *157*, 1126–1132. [CrossRef]
90. Wang, Y.; Gong, F.; Zhang, D.; Ueda, T. Estimation of ice content in mortar based on electrical measurements under freeze-thaw cycle. *J. Adv. Concr. Technol.* **2016**, *14*, 35–46. [CrossRef]
91. Zhu, Y.; Zhang, H.; Zhang, Z.; Dong, B.; Liao, J. Monitoring the cracking behavior of engineered cementitious composites (ECC) and plain mortar by electrochemical impedance measurement. *Constr. Build. Mater.* **2019**, *209*, 195–201. [CrossRef]

92. Balestra, C.E.T.; Reichert, T.A.; Pansera, W.A.; Savaris, G. Evaluation of chloride ion penetration through concrete surface electrical resistivity of field naturally degraded structures present in marine environment. *Constr. Build. Mater.* **2020**, *230*, 116979. [CrossRef]
93. ASTM G57—06(2012) Standard Test Method for Field Measurement of Soil Resistivity Using the Wenner Four-Electrode Method. Available online: https://www.astm.org/Standards/G57.htm (accessed on 12 August 2020).
94. Mendes, S.E.S.; Oliveira, R.L.N.; Cremonez, C.; Pereira, E.; Pereira, E.; Medeiros-Junior, R.A. Electrical resistivity as a durability parameter for concrete design: Experimental data versus estimation by mathematical model. *Constr. Build. Mater.* **2018**, *192*, 610–620. [CrossRef]
95. AASHTO T 358. *Standard Method of Test for Surface Resistivity Indication of Concrete's Ability to Resist Chloride Ion Penetration*; AASHTO: Washington, DC, USA.
96. Tibbetts, C.M.; Paris, J.M.; Ferraro, C.C.; Riding, K.A.; Townsend, T.G. Relating water permeability to electrical resistivity and chloride penetrability of concrete containing different supplementary cementitious materials. *Cem. Concr. Compos.* **2020**, *107*, 103491. [CrossRef]
97. Florida Department of Transportation FM 5-578. *Florida Method of Test for Concrete Resistivity as an Electrical Indicator of Its Permeability*; Florida Department of Transportation: Tallahassee, FL, USA, 2004; pp. 4–7.
98. Wiwattanachang, N.; Giao, P.H. Monitoring crack development in fiber concrete beam by using electrical resistivity imaging. *J. Appl. Geophys.* **2011**, *75*, 294–304. [CrossRef]
99. Alhajj, M.A.; Palma-Lopes, S.; Villain, G. Accounting for steel rebar effect on resistivity profiles in view of reinforced concrete structure survey. *Constr. Build. Mater.* **2019**, *223*, 898–909. [CrossRef]
100. Ramezanianpour, A.A.; Pilvar, A.; Mahdikhani, M.; Moodi, F. Practical evaluation of relationship between concrete resistivity, water penetration, rapid chloride penetration and compressive strength. *Constr. Build. Mater.* **2011**. [CrossRef]
101. Tanesi, J.; Ardani, A. Surface resistivity test evaluation as an indicator of the chloride permeability of concrete. *TechBrief* **2012**, 1–6. [CrossRef]
102. Presuel-Moreno, F.; Wu, Y.Y.; Liu, Y. Effect of curing regime on concrete resistivity and aging factor over time. *Constr. Build. Mater.* **2013**, *48*, 874–882. [CrossRef]
103. Medeiros, R.A.; Lima, M.G. Electrical resistivity of unsaturated concrete using different types of cement. *Constr. Build. Mater.* **2016**, *107*, 11–16. [CrossRef]
104. Kessler, R.J.; Powers, R.G.; Paredes, M.A. *Resistivity Measurements of Water Saturated Concrete as an Indicator of Permeability*; Florida Department of Transportation: Gainesville, FL, USA, 2005.
105. Morris, W.; Vico, A.; Vazquez, M.; De Sanchez, S.R. Corrosion of reinforcing steel evaluated by means of concrete resistivity measurements. *Corros. Sci.* **2002**, *44*, 81–99. [CrossRef]
106. Gowers, K.R.; Millard, S.G. Measurement of Concrete Resistivity for Assessment of Corrosion Severity of Steel Using Wenner Technique. *Mater. J.* **1999**, *96*, 536–541. [CrossRef]
107. UNE 83988-2:2014 Concrete Durability. Test Methods. Determination of the Electrical Resistivity. Part 2: Four Points or Wenner Method. Available online: https://www.une.org/encuentra-tu-norma/busca-tu-norma/norma?c=N0052651 (accessed on 12 August 2020).
108. Sengul, O.; Gjørv, O.E. Effect of embedded steel on electrical resistivity measurements on concrete structures. *ACI Mater. J.* **2009**, *106*, 11–18. [CrossRef]
109. Millard, S.; Sadowski, L. Novel method for linear polarisation resistance corrosion measurement. In Proceedings of the NDTCE'09 Non-Destructive Testing in Civil Engineering, Nantes, France, 30 June–3 July 2009.
110. Sadowski, L. New non-destructive method for linear polarisation resistance corrosion rate measurement. *Arch. Civ. Mech. Eng.* **2010**, *10*, 109–116. [CrossRef]
111. Sadowski, L.; Nikoo, M. Corrosion current density prediction in reinforced concrete by imperialist competitive algorithm. *Neural Comput. Appl.* **2014**, *25*, 1627–1638. [CrossRef] [PubMed]
112. Sadowski, L. Non-destructive investigation of corrosion current density in steel reinforced concrete by artificial neural networks. *Arch. Civ. Mech. Eng.* **2013**, *13*, 104–111. [CrossRef]
113. Sadowski, L. Methodology for assessing the probability of corrosion in concrete structures on the basis of half-cell potential and concrete resistivity measurements. *Sci. World J.* **2013**, *2013*. [CrossRef] [PubMed]
114. Zhang, J.; Monteiro, P.J.M.; Morrison, H.F. Noninvasive surface measurement of corrosion impedance of reinforcing bar in concrete—Part 1: Experimental results. *ACI Mater. J.* **2001**, *98*, 116–125. [CrossRef]

115. Zhang, J.; Monteiro, P.J.M.; Morrison, H.F. Noninvasive surface measurement of corrosion impedance of reinforcing bar in concrete—Part 2: Forward modeling. *ACI Mater. J.* **2002**, *99*, 242–249. [CrossRef]
116. Zhang, J.; Monteiro, P.J.M.; Morrison, H.F.; Mancio, M. Noninvasive surface measurement of corrosion impedance of reinforcing bar in concrete—Part 3: Effect of geometry and material properties. *ACI Mater. J.* **2004**, *101*, 273–280. [CrossRef]
117. Nguyen, A.Q.; Klysz, G.; Deby, F.; Balayssac, J.P. Assessment of the electrochemical state of steel reinforcement in water saturated concrete by resistivity measurement. *Constr. Build. Mater.* **2018**, *171*, 455–466. [CrossRef]
118. Goueygou, M.; Abraham, O.; Lataste, J.F. A comparative study of two non-destructive testing methods to assess near-surface mechanical damage in concrete structures. *NDT E Int.* **2008**, *41*, 448–456. [CrossRef]
119. Lataste, J.F.; Sirieix, C.; Breysse, D.; Frappa, M. Electrical resistivity measurement applied to cracking assessment on reinforced concrete structures in civil engineering. *NDT E Int.* **2003**, *36*, 383–394. [CrossRef]
120. Yim, H.J.; Bae, Y.H.; Kim, J.H. Method for evaluating segregation in self-consolidating concrete using electrical resistivity measurements. *Constr. Build. Mater.* **2020**, *232*, 117283. [CrossRef]
121. Azhari, F.; Banthia, N. Cement-based sensors with carbon fibers and carbon nanotubes for piezoresistive sensing. *Cem. Concr. Compos.* **2012**, *34*, 866–873. [CrossRef]
122. Lee, S.H.; Kim, S.; Yoo, D.Y. Hybrid effects of steel fiber and carbon nanotube on self-sensing capability of ultra-high-performance concrete. *Constr. Build. Mater.* **2018**, *185*, 530–544. [CrossRef]
123. Fan, S.; Li, X.; Li, M. The effects of damage and self-healing on impedance spectroscopy of strain-hardening cementitious materials. *Cem. Concr. Res.* **2018**, *106*, 77–90. [CrossRef]
124. Loche, J.M.; Ammar, A.; Dumargue, P. Influence of the migration of chloride ions on the electrochemical impedance spectroscopy of mortar paste. *Cem. Concr. Res.* **2005**, *35*, 1797–1803. [CrossRef]
125. Yim, H.J.; Lee, H.; Kim, J.H. Evaluation of mortar setting time by using electrical resistivity measurements. *Constr. Build. Mater.* **2017**, *146*, 679–686. [CrossRef]
126. Yu, J.; Sasamoto, A.; Iwata, M. Wenner method of impedance measurement for health evaluation of reinforced concrete structures. *Constr. Build. Mater.* **2019**, *197*, 576–586. [CrossRef]
127. FHWA. Long-Term Bridge Performance. Available online: https://highways.dot.gov/research/long-term-infrastructure-performance/ltbp/long-term-bridge-performance (accessed on 12 August 2020).
128. Li, Z.; Wei, X.; Li, W. Preliminary interpretation of portland cement hydration process using resistivity measurements. *ACI Mater. J.* **2003**, *100*, 253–257. [CrossRef]
129. Li, Z.; Tang, S.; Lu, Y. Pore Structure Analyzer Based on Non-Contact Impedance Measurement for Cement-Based Materials. U.S. Patent US9488635B2, 21 December 2011.
130. Tang, S.W.; Li, Z.J.; Chen, E.; Shao, H.Y. Impedance measurement to characterize the pore structure in Portland cement paste. *Constr. Build. Mater.* **2014**, *51*, 106–112. [CrossRef]
131. Tang, S.W.; Li, Z.J.; Zhu, H.G.; Shao, H.Y.; Chen, E. Permeability interpretation for young cement paste based on impedance measurement. *Constr. Build. Mater.* **2014**, *59*, 120–128. [CrossRef]
132. Li, Z.; Li, W. Contactless, Transformer-Based Measurement of the Resistivity of Materials. U.S. Patent US6639401B2, 19 July 2001.
133. Wei, X.; Xiao, L. Influence of the aggregate volume on the electrical resistivity and properties of portland cement concretes. *J. Wuhan Univ. Technol. Mater. Sci. Ed.* **2011**, *26*, 965–971. [CrossRef]
134. Wei, X.; Tian, K.; Xiao, L. Prediction of compressive strength of Portland cement paste based on electrical resistivity measurement. *Adv. Cem. Res.* **2010**, *22*, 165–170. [CrossRef]
135. Wei, X.; Xiao, L.; Li, Z. Prediction of standard compressive strength of cement by the electrical resistivity measurement. *Constr. Build. Mater.* **2012**, *31*, 341–346. [CrossRef]
136. Lianzhen, X.; Xiaosheng, W. Early age compressive strength of pastes by electrical resistivity method and maturity method. *J. Wuhan Univ. Technol. Sci. Ed.* **2011**. [CrossRef]
137. Shao, H.; Zhang, J.; Fan, T.; Li, Z. Electrical method to evaluate elastic modulus of early age concrete. *Constr. Build. Mater.* **2015**, *101*, 661–666. [CrossRef]
138. Xiao, L.; Li, Z. Early-age hydration of fresh concrete monitored by non-contact electrical resistivity measurement. *Cem. Concr. Res.* **2008**, *38*, 312–319. [CrossRef]
139. Guthrie, W.S.; Baxter, J.S.; Mazzeo, B.A. Vertical electrical impedance testing of a concrete bridge deck using a rolling probe. *NDT E Int.* **2018**, *95*, 65–71. [CrossRef]

140. Guthrie, W.S.; Mazzeo, B.A. Vertical impedance testing for assessing protection from chloride-based deicing salts provided by an asphalt overlay system on a concrete bridge deck. In Proceedings of the Cold Regions Engineering, Salt Lake City, UT, USA, 19–22 July 2015; American Society of Civil Engineers: Reston, VA, USA, 2015; Volume 2015, pp. 358–369.
141. Argyle, H.M. Sensitivity of Electrochemical Impedance Spectroscopy Measurements to Concrete Bridge Deck Properties. Master's Thesis, Ira A. Fulton College of Engineering and Technology, Civil and Environmental Engineering, Provo, UT, USA, 20 March 2014.
142. Bartholomew, P.D.; Guthrie, W.S.; Mazzeo, B.A. Vertical impedance measurements on concrete bridge decks for assessing susceptibility of reinforcing steel to corrosion. *Rev. Sci. Instrum.* **2012**, *83*. [CrossRef] [PubMed]
143. Melara, E.K.; Mendes, A.Z.; Andreczevecz, N.C.; Bragança, M.O.G.P.; Carrera, G.T.; Medeiros-Junior, R.A. Monitoring by electrochemical impedance spectroscopy of mortars subjected to ingress and extraction of chloride ions. *Constr. Build. Mater.* **2020**, *242*, 118001. [CrossRef]
144. ASTM G106—89(2015) Standard Practice for Verification of Algorithm and Equipment for Electrochemical Impedance Measurements. Available online: https://www.astm.org/Standards/G106.htm (accessed on 14 May 2020).
145. Page, C.L.; Treadaway, K.W.J.; Bamforth, P.B. *Corrosion of Reinforcement in Concrete*; Elsevier Applied Science Publishers Ltd: Amsterdam, The Netherlands, 1990; ISBN 1851664874.
146. ISO. *GUM ISO/IEC Guide 98-3:2008 Uncertainty of Measurement—Part 3: Guide to the Expression of Uncertainty in Measurement (GUM:1995)*; ISO: Geneva, Switzerland, 2008.
147. Bourreau, L.; Bouteiller, V.; Schoefs, F.; Gaillet, L.; Thauvin, B.; Schneider, J.; Naar, S. Uncertainty assessment of concrete electrical resistivity measurements on a coastal bridge. *Struct. Infrastruct. Eng.* **2019**, *15*, 443–453. [CrossRef]
148. Andrade, C.; Gulikers, J.; Polder, R.; Raupach, M. RILEM TC 154-EMC: Electrochemical techniques for measuring metallic corrosion, half-cell potential measurements—Potential mapping on reinforced concrete structures. *Mater. Struct. Matrriaux Constr.* **2003**, *36*, 461–471.
149. Morris, W.; Moreno, E.I.; Sagüés, A.A. Practical evaluation of resistivity of concrete in test cylinders using a Wenner array probe. *Cem. Concr. Res.* **1996**, *26*, 1779–1787. [CrossRef]
150. Andrade, C.; Polder, R.; Basheer, M. *Non-Destructive Methods to Measure Ion Migration*; Torrent, R., Fernandez Luco, L., Eds.; RILEM Publications SARL: Paris, France, 2007; ISBN 978-2-35158-054-7.
151. Lecieux, Y.; Schoefs, F.; Bonnet, S.; Lecieux, T.; Lopes, S.P. Quantification and uncertainty analysis of a structural monitoring device: Detection of chloride in concrete using DC electrical resistivity measurement. *Nondestruct. Test. Eval.* **2015**, *30*, 216–232. [CrossRef]
152. Su, T.; Wu, J.; Zou, Z.; Wang, Z.; Yuan, J.; Yang, G. Influence of environmental factors on resistivity of concrete with corroded steel bar. *Eur. J. Environ. Civ. Eng.* **2019**. [CrossRef]
153. Su, J.K.; Yang, C.C.; Wu, W.B.; Huang, R. Effect of moisture content on concrete resistivity measurement. *J. Chin. Inst. Eng. Trans. Chinese Inst. Eng. A Chung-kuo K. Ch'eng Hsuch K'an* **2002**, *25*, 117–122. [CrossRef]
154. Zaccardi, Y.A.V.; García, J.F.; Huélamo, P.; Maio, Á.A. Di Influence of temperature and humidity on Portland cement mortar resistivity monitored with inner sensors. *Mater. Corros.* **2009**, *60*, 294–299. [CrossRef]
155. Azarsa, P.; Gupta, R. Electrical resistivity of concrete for durability evaluation: A review. *Adv. Mater. Sci. Eng.* **2017**, *2017*. [CrossRef]
156. Corva, D.M.; Hosseini, S.S.; Collins, F.; Adams, S.D.; Gates, W.P.; Kouzani, A.Z. Miniature resistance measurement device for structural health monitoring of reinforced concrete infrastructure. *Sensors* **2020**, *20*, 4313. [CrossRef] [PubMed]
157. Levita, G.; Marchetti, A.; Gallone, G.; Princigallo, A.; Guerrini, G.L. Electrical properties of fluidified Portland cement mixes in the early stage of hydration. *Cem. Concr. Res.* **2000**, *30*, 923–930. [CrossRef]
158. Tang, S.W.; Cai, X.H.; He, Z.; Zhou, W.; Shao, H.Y.; Li, Z.J.; Wu, T.; Chen, E. The review of pore structure evaluation in cementitious materials by electrical methods. *Constr. Build. Mater.* **2016**, *117*, 273–284. [CrossRef]
159. Wei, X.; Xiao, L. Kinetics parameters of cement hydration by electrical resistivity measurement and calorimetry. *Adv. Cem. Res.* **2014**, *26*, 187–193. [CrossRef]
160. Liu, Z.; Zhang, Y.; Jiang, Q. Continuous tracking of the relationship between resistivity and pore structure of cement pastes. *Constr. Build. Mater.* **2014**, *53*, 26–31. [CrossRef]

161. Zhang, J.; Li, Z. Hydration process of cements with superplasticizer monitored by non-contact resistivity measurement. In Proceedings of the Advanced Testing of Fresh Cementtitious Materials, Stuttgart, Germany, 3–4 August 2006; Reinhardt, H.W., Ed.;
162. Itagaki, M.; Suzuki, S.; Shitanda, I.; Watanabe, K.; Nakazawa, H. Impedance analysis on electric double layer capacitor with transmission line model. *J. Power Sources* **2007**, *164*, 415–424. [CrossRef]
163. Aligizaki, K.K. *Pore Structure of Cement-Based Materials: Testing, Interpretation and Requirements*; CRC Press: Boca Raton, FL, USA, 2019; ISBN 9780367863838.
164. Scott, A.; Alexander, M.G. Effect of supplementary cementitious materials (binder type) on the pore solution chemistry and the corrosion of steel in alkaline environments. *Cem. Concr. Res.* **2016**, *89*, 45–55. [CrossRef]
165. ASTM C1202. *Test Method for Electrical Indication of Concretes Ability to Resist Chloride Ion Penetration*; ASTM: Washington, DC, USA, 2019.
166. Tomlinson, D.; Moradi, F.; Hajiloo, H.; Ghods, P.; Alizadeh, A.; Green, M. Early age electrical resistivity behaviour of various concrete mixtures subject to low temperature cycling. *Cem. Concr. Compos.* **2017**, *83*, 323–334. [CrossRef]
167. Wang, Z.; Zeng, Q.; Wang, L.; Yao, Y.; Li, K. Characterizing blended cement pastes under cyclic freeze-thaw actions by electrical resistivity. *Constr. Build. Mater.* **2013**, *44*, 477–486. [CrossRef]
168. Dinakar, P.; Babu, K.G.; Santhanam, M. Corrosion behaviour of blended cements in low and medium strength concretes. *Cem. Concr. Compos.* **2007**, *29*, 136–145. [CrossRef]
169. Yeh, I.C. Generalization of strength versus water-cementitious ratio relationship to age. *Cem. Concr. Res.* **2006**, *36*, 1865–1873. [CrossRef]
170. EN 12390-8. *Testing Hardened Concrete—Part. 8: Depth of Penetration of Water Under Pressure*; CEN: Brussels, Belgium, 2019.
171. Djerbi Tegguer, A.; Bonnet, S.; Khelidj, A.; Baroghel-Bouny, V. Effect of uniaxial compressive loading on gas permeability and chloride diffusion coefficient of concrete and their relationship. *Cem. Concr. Res.* **2013**, *52*, 131–139. [CrossRef]
172. Soongswang, P.; Tia, M.; Bloomquist, D.G.; Meletiou, C.; Sessions, L.M. Efficient test setup for determining the water-permeability of concrete. *Transp. Res. Rec.* **1988**, *1204*, 77–82.
173. Beushausen, H.; Fernandez Luco, L. *Performance-Based Specifications and Control. of Concrete Durability: State-of-the-Art Report RILEM TC 230-PSC*; Springer: Dordrecht, The Netherlands, 2015; Volume 18, ISBN 9789401773096.
174. Sengül, Ö. Probabilistic design for the durability of reinforced concrete structural elements exposed to chloride containing environments. *Tek. Dergi Tech. J. Turk. Chamb. Civ. Eng.* **2011**, *22*, 5409–5423. [CrossRef]
175. Gulikers, J. Theoretical considerations on the supposed linear relationship between concrete resistivity and corrosion rate of steel reinforcement. *Mater. Corros.* **2005**, *56*, 393–403. [CrossRef]
176. Luping, T. *Guidelines for Practical Use of Methods for Testing the Resistance of Concrete to Chloride Ingress*; Swedish National Testing and Research Institute: Boras, Sweden, 2005.
177. Polder, R.B. Chloride diffusion and resistivity testing of five concrete mixes for marine environment. In *RILEM International Workshop on Chloride Penetration into Concrete*; Nilsson, L.-O., Ollivier, P., Eds.; Springer Nature: St-Remy-les-Chevreuses, France, 1997; pp. 225–233.
178. Layssi, H.; Ghods, P.; Alizadeh, A.R.; Salehi, M. Electrical resistivity of concrete concepts, applications, and measurement techniques. *Concr. Int.* **2015**, *37*, 41–46.
179. ASTM. *New Test Method for Measuring the Surface Resistivity of Hardened Concrete Using the Wenner Four-Electrode Method*; WK37880 ASTM; American Society of Testing and Materials: Washington, DC, USA, 2014.
180. Whiting, D. *Rapid Determination of the Chloride Permeability of Concrete*; The Portland Cement Association: Washington, DC, USA, 1981.
181. Julio-Betancourt, G.A.; Hooton, R.D. Study of the Joule effect on rapid chloride permeability values and evaluation of related electrical properties of concretes. *Cem. Concr. Res.* **2004**, *34*, 1007–1015. [CrossRef]
182. Sharfuddin Ahmed, M.; Kayali, O.; Anderson, W. Chloride penetration in binary and ternary blended cement concretes as measured by two different rapid methods. *Cem. Concr. Compos.* **2008**, *30*, 576–582. [CrossRef]
183. Shi, M.; Chen, Z.; Sun, J. Determination of chloride diffusivity in concrete by AC impedance spectroscopy. *Cem. Concr. Res.* **1999**, *29*, 1111–1115. [CrossRef]

184. NT BUILD 492—Concrete, Mortar and Cement-Based Repair Materials: Chloride Migration Coefficient from Non-Steady-State Migration Experiments. Available online: http://www.nordtest.info/index.php/methods/item/concrete-mortar-and-cement-based-repair-materials-chloride-migration-coefficient-from-non-steady-state-migration-experiments-nt-build-492.html (accessed on 21 May 2020).
185. DURAR. *Manual for Inspecting, Evaluating and Diagnosing Corrosion in Reinforced Concrete Structures*; Rincon, O.T.; Carruyo, A.R., Andrade, C., Helene, P.R.L., Diaz, I., Eds.; CYTED; Ibero-American Program Science and Technology for Development, CYTED Research Network XV.B (DURAR): Maracaibo, Venezuela, 2000; ISBN 980-296-541-3.
186. Reou, J.S.; Ann, K.Y. Electrochemical assessment on the corrosion risk of steel embedment in OPC concrete depending on the corrosion detection techniques. *Mater. Chem. Phys.* **2009**, *113*, 78–84. [CrossRef]
187. Vedalakshmi, R.; Dolli, H.; Palaniswamy, N. Embeddable corrosion rate-measuring sensor for assessing the corrosion risk of steel in concrete structures. *Control. Health Monit.* **2009**, *16*, 441–459. [CrossRef]
188. Yu, B.; Liu, J.; Chen, Z. Probabilistic evaluation method for corrosion risk of steel reinforcement based on concrete resistivity. *Constr. Build. Mater.* **2017**, *138*, 101–113. [CrossRef]
189. ASTM International. *ASTM C876—15 Standard Test Method for Corrosion Potentials of Uncoated Reinforcing Steel in Concrete*; ASTM Int.: Washington, DC, USA, 2015; pp. 1–8. [CrossRef]
190. Bowler, N.; Huang, Y. Electrical conductivity measurement of metal plates using broadband eddy-current and four-point methods. *Meas. Sci. Technol.* **2005**, *16*, 2193. [CrossRef]
191. Polder, R.B. Critical chloride content for reinforced concrete and its relationship to concrete resistivity. *Mater. Corros.* **2009**, *60*, 623–630. [CrossRef]
192. Andrade, C.; Sanjuán, M.A.; Recuero, A.; Río, O. Calculation of chloride diffusivity in concrete from migration experiments, in non steady-state conditions. *Cem. Concr. Res.* **1994**, *24*, 1214–1228. [CrossRef]
193. Poupard, O.; Aït-Mokhtar, A.; Dumargue, P. Corrosion by chlorides in reinforced concrete: Determination of chloride concentration threshold by impedance spectroscopy. *Cem. Concr. Res.* **2004**, *34*, 991–1000. [CrossRef]
194. Alonso, C.; Andrade, C.; González, J.A. Relation between resistivity and corrosion rate of reinforcements in carbonated mortar made with several cement types. *Cem. Concr. Res.* **1988**, *18*, 687–698. [CrossRef]
195. Andrade, C.; Alonso, C.; Gulikers, J.; Polder, R.; Cigna, R.; Vennesland, M.; Salta, M.; Raharinaivo, A.; Elsener, B. Test methods for on-site corrosion rate measurement of steel reinforcement in concrete by means of the polarization resistance method. *Mater. Struct. Constr.* **2004**, *37*, 623–643. [CrossRef]
196. Hope, B.B.; Page, J.A.; Ip, A.K.C. Corrosion rates of steel in concrete. *Cem. Concr. Res.* **1986**, *16*, 771–781. [CrossRef]
197. Bertolini, L.; Polder, R.B. *Concrete Resistivity and Reinforcement Corrosion Rate as a Function of Temperature and Humidity of the Environment*; TNO Report 97-BT-R0574; Netherlands Organisation for Applied Scientific Research: Delft, The Netherlands, 1997; Volume 273, pp. 12466–12475.
198. Millard, S.; Broomfield, J. Measuring the corrosion rate of reinforced concrete using linear polarisation resistance. *Concrete* **2003**, *37*, 36–38.
199. Hoła, J.; Schabowicz, K. New technique of nondestructive assessment of concrete strength using artificial intelligence. *NDT E Int.* **2005**, *38*, 251–259. [CrossRef]
200. He, S.; Zou, Y.; Quan, D.; Wang, H. Application of RBF neural network and ANFIS on the prediction of corrosion rate of pipeline steel in soil. In *Lecture Notes in Electrical Engineering*; Springer: Berlin/Heidelberg, Germany, 2012; Volume 124, pp. 639–644.
201. Ribeiro, D.V.; Abrantes, J.C.C. Application of electrochemical impedance spectroscopy (EIS) to monitor the corrosion of reinforced concrete: A new approach. *Constr. Build. Mater.* **2016**, *111*, 98–104. [CrossRef]
202. McCarter, W.J.; Ezirim, H.; Emerson, M. Properties of concrete in the cover zone: Water penetration, sorptivity and ionic ingress. *Mag. Concr. Res.* **1996**, *48*, 149–156. [CrossRef]
203. Oleiwi, H.M.; Wang, Y.; Curioni, M.; Chen, X.; Yao, G.; Augusthus-Nelson, L.; Ragazzon-Smith, A.H.; Shabalin, I. An experimental study of cathodic protection for chloride contaminated reinforced concrete. *Mater. Struct.* **2018**, *51*, 148. [CrossRef]
204. McCarter, W.J.; Starrs, G.; Chrisp, T.M.; Basheer, P.A.M.; Nanukuttan, S.V.; Srinivasan, S. Conductivity/activation energy relationships for cement-based materials undergoing cyclic thermal excursions. *J. Mater. Sci.* **2015**, *50*, 1129–1140. [CrossRef]
205. Scherer, G.W. Crystallization in pores. *Cem. Concr. Res.* **1999**, *29*, 1347–1358. [CrossRef]

206. Zeng, Q.; Fen-Chong, T.; Dangla, P.; Li, K. A study of freezing behavior of cementitious materials by poromechanical approach. *Int. J. Solids Struct.* **2011**, *48*, 3267–3273. [CrossRef]
207. Ying Li, G.; Ming Wang, P.; Zhao, X. Pressure-sensitive properties and microstructure of carbon nanotube reinforced cement composites. *Cem. Concr. Compos.* **2007**. [CrossRef]
208. Gao, D.; Sturm, M.; Mo, Y.L. Electrical resistance of carbon-nanofiber concrete. *Smart Mater. Struct.* **2009**, *18*, 095039. [CrossRef]
209. Li, H.; Xiao, H.G.; Ou, J. Ping effect of compressive strain on electrical resistivity of carbon black-filled cement-based composites. *Cem. Concr. Compos.* **2006**, *28*, 824–828. [CrossRef]
210. Monteiro, A.O.; Loredo, A.; Costa, P.M.F.J.; Oeser, M.; Cachim, P.B. A pressure-sensitive carbon black cement composite for traffic monitoring. *Constr. Build. Mater.* **2017**, *154*, 1079–1086. [CrossRef]
211. Han, B.; Ding, S.; Yu, X. Intrinsic self-sensing concrete and structures: A review. *Meas. J. Int. Meas. Confed.* **2015**, *59*, 110–128. [CrossRef]
212. Wen, S.; Chung, D.D.L. Piezoresistivity in continuous carbon fiber cement-matrix composite. *Cem. Concr. Res.* **1999**, *29*, 445–449. [CrossRef]
213. Chung, D.D.L. Piezoresistive Cement-Based Materials for Strain Sensing. *J. Intell. Mater. Syst. Struct.* **2002**, *13*, 599–609. [CrossRef]
214. Wen, S.; Chung, D.D.L. Defect dynamics of cement paste under repeated compression studied by electrical resistivity measurement. *Cem. Concr. Res.* **2001**, *31*, 1515–1518. [CrossRef]
215. Han, B.; Zhang, K.; Yu, X.; Kwon, E.; Ou, J. Nickel particle-based self-sensing pavement for vehicle detection. *Meas. J. Int. Meas. Confed.* **2011**, *44*, 1645–1650. [CrossRef]
216. Han, B.; Yu, X.; Kwon, E. A self-sensing carbon nanotube/cement composite for traffic monitoring. *Nanotechnology* **2009**, *20*, 445501. [CrossRef]
217. Acker, P.; Boulay, C.; Rossi, P. On the importance of initial stresses in concrete and of the resulting mechanical effects. *Cem. Concr. Res.* **1987**, *17*, 755–764. [CrossRef]
218. Mehta, P.K.; Monteiro, P.J.M. *Concrete: Microstructure, Properties, and Materials*, 3rd ed.; McGraw-Hill: New York, NY, USA, 2005; ISBN 0071462899.
219. Nemati, K.M.; Monteiro, P.J.M.; Scrivener, K.L. Analysis of compressive stress-induced cracks in concrete. *ACI Mater. J.* **1998**, *95*, 617–630. [CrossRef]
220. Bisschop, J.; Van Mier, J.G.M. How to study drying shrinkage microcracking in cement-based materials using optical and scanning electron microscopy? *Cem. Concr. Res.* **2002**, *32*, 279–287. [CrossRef]
221. Litorowicz, A. Identification and quantification of cracks in concrete by optical fluorescent microscopy. *Cem. Concr. Res.* **2006**, *36*, 1508–1515. [CrossRef]
222. Dal Pont, S.; Durand, S.; Schrefler, B.A. A multiphase thermo-hydro-mechanical model for concrete at high temperatures-Finite element implementation and validation under LOCA load. *Nucl. Eng. Des.* **2007**, *237*, 2137–2150. [CrossRef]
223. Loukili, A.; Khelidj, A.; Richard, P. Hydration kinetics, change of relative humidity, and autogenous shrinkage of ultra-high-strength concrete. *Cem. Concr. Res.* **1999**, *29*, 577–584. [CrossRef]
224. Hou, T.-C. Wireless and Electromechanical Approaches for Strain Sensing and Crack Detection in Fiber Reinforced Cementitious Materials. Ph.D. Thesis, University of Michigan, Ann Arbor, MI, USA, 2008.

Publisher's Note: MDPI stays neutral with regard to jurisdictional claims in published maps and institutional affiliations.

© 2020 by the authors. Licensee MDPI, Basel, Switzerland. This article is an open access article distributed under the terms and conditions of the Creative Commons Attribution (CC BY) license (http://creativecommons.org/licenses/by/4.0/).

Communication

The Sonic Resonance Method and the Impulse Excitation Technique: A Comparison Study

Tomáš Húlan [1], Filip Obert [1], Ján Ondruška [1], Igor Štubňa [1] and Anton Trník [1,2,*]

[1] Department of Physics, Faculty of Natural Sciences, Constantine the Philosopher University in Nitra, Tr. A. Hlinku 1, 94974 Nitra, Slovakia; thulan@ukf.sk (T.H.); filip.obert@ukf.sk (F.O.); jondruska@ukf.sk (J.O.); istubna@ukf.sk (I.Š.)

[2] Department of Materials Engineering and Chemistry, Faculty of Civil Engineering, Czech Technical University in Prague, Thákurova 7, 16629 Prague, Czech Republic

* Correspondence: atrnik@ukf.sk; Tel.: +421-37-6408-616

Abstract: In this study, resonant frequencies of flexurally vibrating samples were measured using the sonic resonant method (SRM) and the impulse excitation technique (IET) to assess the equivalency of these two methods. Samples were made from different materials and with two shapes (prism with rectangular cross-section and cylinder with circular cross-section). The mean values and standard deviations of the resonant frequencies were compared using the t-test and the F-test. The tests showed an equivalency of both methods in measuring resonant frequency. The differences between the values measured using SRM and IET were not significant. Graphically, the relationship between the resonant frequencies is a line with a slope of $0.9993 \approx 1$.

Keywords: sonic resonant method; impulse excitation technique; resonant frequency

1. Introduction

Young's modulus is a mechanical quantity of great importance for solid materials. It depends on different external influences, in addition to the intrinsic properties of the measured material. Therefore, Young's modulus allows an indirect study of, for example, the microstructure (porosity, texture) and the influences of some technological steps (drying or sintering) on ceramic materials. Young's modulus is also a necessary quantity in some engineering calculations, e.g., in the determination of the critical rate of heating a ceramic body.

The most commonly used methods for determining Young's modulus of metals, ceramics, concrete, glass, composites, and biological materials are dynamical methods. Thomaz et al. [1] studied Young's modulus of concrete containing basaltic aggregates using static and dynamic methods, such as the ultrasonic pulse velocity (UPV) and impulse excitation technique (IET). They found out that the dynamic Young's moduli had higher values than the static moduli by approximately 16% for IET and 28% for UPV. Using IET, Quaglio et al. [2] determined Young's modulus of samples from basalt and diabase mines used as aggregates in the construction industry. Their results showed that values of Young's modulus had high repeatability and agreed with those reported in the literature for the same material. Using IET, Guicciardi et al. [3] studied the dynamic Young's modulus of ZrB_2-based composites containing $MoSi_2$ as a secondary phase up to 1430 °C. Duan et al. [4] used IET to compare the microstructures of several glasses by measuring Young's modulus and the internal friction as a function of temperature. Wang et al. [5] investigated the validity of using the frequency and decay rate of free-free beam vibrations, which were measured by IET, to characterize the viscoelastic properties of glass in the temperature range of glass transition. Ligoda-Chmiel et al. [6] used a traditional compression test and the ultrasonic and impulse excitation of vibration methods to compare and analyze Young's modulus, Kirchoff's modulus, and Poisson's ratio using alumina foam/tri-functional epoxy resin composites with an interpenetrating network structure. Radovic et al. [7] compared four different experimental techniques, namely, resonant ultrasound spectroscopy (RUS), impulse excitation, nanoindentation, and the four-point bending test to determine Young's and shear moduli of 99.9% pure Al_2O_3, 7075 aluminum, 4140 steel, and Pyrex glass. They found that dynamic methods (RUS and IET) have superior precision and repeatability, and the differences between the results of RUS and IET were not statistically significant. Haines et al. [8] compared the results from a resonance flexure method and from four-point static flexure tests for wood samples.

Dynamical methods based on measurement of the resonant frequency of a vibrating sample are relatively simple and produce very low mechanical stress that does not initiate inelastic processes in tested material. Under such a low stress, the assumptions of the elastic theory of vibration are well fulfilled. Another advantage of these methods is their applicability for high temperature measurements [9,10]. If resonant frequency is measured in a defined temperature regime, e.g., during heating/cooling with a constant rate, such measurement falls under thermal analysis and is called dynamical thermomechanical analysis (D-TMA).

A rectangular prism or a rod with a circular cross-section, both having free ends, are commonly used for determination of Young's modulus. The longitudinal vibration or flexural vibration of such samples are possible for measurement purposes, but the flexural vibration is preferable because it can be easily excited, gives more intense amplitude, and the resonant frequency is lower compared to the longitudinal vibration. Two kinds of vibrations are used [11,12]:

(a) Driven vibrations with a known frequency. The driven vibrations are the base of the sonic resonance method (SRM).
(b) Free vibrations excited by the mechanical impulse. The free vibrations are the base of the impulse excitation technique (IET).

Historically, the first technique was SRM [10,12]. The equipment used consists of a tunable oscillator with an amplifier connected to an exciter as the source of driven vibrations. The sample is suspended in its nodal points, vibrations are registered using a sensor connected to a preamplifier, and the sensor's output signal is observed. The frequency at which the output signal reaches the maximum value is the resonant frequency. This method can be automated, for example, if the RC oscillator works in a sweeping regime [13,14] or the sample is permanently kept in the resonant vibration with the help of a voltage-controlled oscillator in a feedback loop which contains the sensor [15].

Roebben et al. [16] presented an apparatus to measure elastic properties and the internal friction of materials. Their apparatus excited the sample fixed in the nodal points of the fundamental vibration mode via a light mechanical impact. The response includes many transient frequencies that rapidly die out, and thus there is a natural filtering action leaving the main fundamental resonant vibrations as the only detected signal. Then the apparatus performed a software-based analysis of the resulting vibration, i.e., IET was used. The resonant frequency of the sample was determined and Young's modulus was calculated. Similar techniques were also described in [11,17–20]. The sample vibration is captured by a microphone or a piezo electrical sensor and subsequently analyzed using the fast Fourier transformation (FFT). The result of the FFT is a frequency spectrum of the sample vibrations, where the resonant frequency of the fundamental mode of the vibrations can be found. Another way to extract the resonant frequency from the measured signal is based on measurement of the duration of a selected number of cycles by counting zero-crossings and determining the cycle period. Its duration is directly proportional to the reciprocal value of the resonant frequency.

The free vibrations are naturally suppressed by internal processes in the sample. Therefore, a coefficient of the internal damping (internal friction) can be determined. The mechanical impulse can be realized manually with a hammer if the measurement is conducted at room temperature. The impulse can be generated at an elevated temperature by steel or ceramic balls which fall on the sample, and by an electromagnetic impactor.

In high-temperature measurements, IET can be designed as non-contact, whereas SRM needs two thin wire suspensions located at the antinodal points or at the ends of the sample. These suspensions are a drawback of SRM—they are often the source of spurious resonances, and their strength is limited at high temperatures. Consequently, IET is more reliable at high temperatures.

Both methods, SRM and IET, have the same theoretical basis, which is an equation of the flexural vibration of the rectangular beam or rod with a circular cross-section, and their material is homogenous and isotropic [10,21]. A derivation of the equation of the flexural vibration can be found, for example, in [9,10,21–23]. The relationship for Young's modulus E derived from this simplified equation has a form:

$$E = \left(K\frac{l^2 f_0}{d}\right)^2 \rho T, \tag{1}$$

where f_0 is the resonant frequency of the fundamental mode, ρ is the material bulk density, l is the length of the sample, and d is the diameter of the cylindrical sample or thickness of the prismatic sample in the direction of vibration. If $l/d > 20$, the correction coefficient $T = 1$. If $l/d < 20$, the influence of rotary inertia and shear forces have to be taken into account to obtain correct values of Young's modulus. Two ways are possible: (1) The use of the very complex frequency equation for the so-called Timoshenko beam. When the measured frequency is substituted in this equation, Young's modulus can be calculated using a numerical method. (2) The use of the simplified equation for a slender beam from which the frequency equation and Equation (1) with $T = 1$ can be derived. The correction coefficient $T > 1$ should be used for cases $l/d < 20$. The value of T can be calculated from formulae given in [11,12] or can be found in tables in [10]. This second way is commonly used in experimental practice and is described in standards, e.g., ASTM [11,12].

The values of the constant K are:

$K = 1.12336$ for a cylindrical sample and the fundamental resonant frequency;
$K = 0.97286$ for a prismatic sample and the fundamental resonant frequency.

Theoretically, IET and SRM should give the same resonant frequency for the given sample. The authors have found only one short technical note [24] which confirms this equality on the basis of experimental results obtained on a concrete prism.

The aim of this article is to compare the resonant frequencies of the flexurally vibrating free-free beam measured by IET and SRM on different samples. This frequency can be substituted into a formula for the calculation of Young's modulus (together with dimensions and mass of the sample). When the same sample is used for SRM and IET, the difference of resonant frequencies can be only observed, because the shape, dimensions, and intrinsic properties of the sample, in addition to the boundary conditions (free-free sample), are the same for SRM and IET.

2. Materials and Methods

2.1. Samples

The measured samples were made from metal, ceramics, and glass. Their material, dimensions, and shape (prism or cylinder) are given in Table 1. Every sample was measured 12 times by SRM and IET.

Table 1. Used materials for the measurements by SRM and IET method.

No.	Sample Material	Sample Dimensions [mm]
1	Aluminum	⌀12 × 110 (cylinder)
2	Stainless steel	⌀8 × 110 (cylinder)
3	Carbon steel	8.5 × 8.2 × 163 (prism)
4	Kaolin ceramics [1]	⌀12 × 110 (cylinder)
5	Kaolin ceramics [2]	⌀15 × 110 (cylinder)
6	Alumina porcelain	10 × 11 × 110 (prism)
7	Soda-lime glass	⌀8 × 145 (cylinder)
8	Corundum ceramics	⌀8 × 175 (cylinder)
9	Silicon carbide	⌀14 × 150 (cylinder)

[1] Ceramics based on Sedlec kaolin fired at 1150 °C. [2] Ceramics based on Kemmlitz kaolin fired at 1150 °C.

2.2. Measurement Methods

Two methods were used for this comparison: the sonic resonant method (SRM) and the impulse excitation technique (IET). The sample was placed horizontally on a narrow soft foam pads at a distance 0.224 l from the sample ends, where the nodal points are located.

The SRM apparatus was as follows (Figure 1): the exciter (speaker Tesla ARZ 098, 75 Ω, 0.15 W, frequency range of 300–6000 Hz) was located under the center of the sample. The speaker was fed by a sinusoidal voltage from a PC-controlled oscillator M631 (ETC Žilina, Slovakia) which worked in the sweeping regime. The oscillator changed the frequency by 1 Hz steps with 20 ms dwelling on each step. The sensor was placed on the end of the sample. The sensor was a piezoelectric gramophone cartridge that affected the sample via a very small force with the help of a lever with a counterweight. The output of the sensor was connected to a preamplifier and PC. If the sensor was moved around the nodal point, the output signal, which was visible in the PC monitor, reached the minimum value in the nodal point. This technique helps to confirm a fundamental mode of the flexural vibrations.

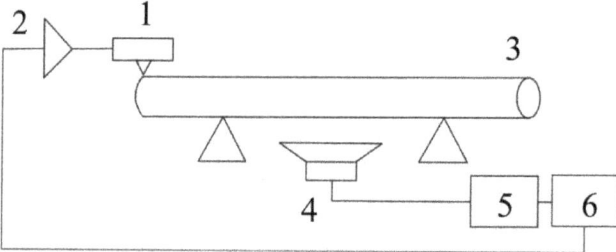

Figure 1. SRM apparatus (1—sensor, 2—preamplifier, 3—sample, 4—exciter, 5—oscillator, 6—personal computer).

The IET apparatus was as follows (Figure 2): The vibrations of the sample were excited by the hit of a small hammer (steel ball glued to a thin wooden stick). The sound was caught by an electric microphone connected to a low-frequency preamplifier and PC, in which the signal was changed into a frequency spectrum using fast Fourier transformation. The sampling frequency was 40 kHz and the period of recording the free vibrations after mechanical impact was 1 s. The resonant frequency was able to be determined with a resolution of 1.221 Hz.

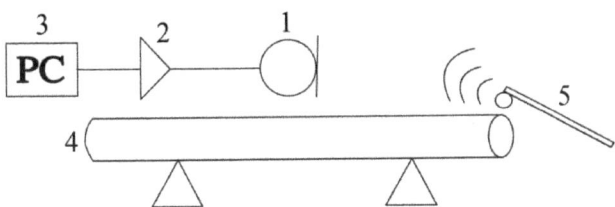

Figure 2. IET apparatus (1—microphone, 2—preamplifier, 3—personal computer, 4—sample, 5—hammer).

The measured sample was placed horizontally on two supports in the nodal points (0.224 l from the both ends) for both the SRM and IET experiments.

3. Results

Because the material, dimensions, mass, and intrinsic properties of the sample were the same for both methods, SRM and IET, only resonant frequencies obtained by SRM and IET could be different. Therefore, only these frequencies were taken into account for the comparison of SRM and IET.

The results obtained by the SRM and IET are shown in Table 2. The resonant frequency of each sample (listed in Table 1) was measured 12 times with both methods and the mean values and standard deviations were calculated. The relative differences between the mean values of the resonant frequencies f_{IET} and f_{SRM}, were calculated according to the equation:

$$\frac{\Delta f}{f_m} = \frac{2(f_{SRM} - f_{IET})}{f_{SRM} + f_{IET}}. \qquad (2)$$

Table 2. Resonant frequency measured using SRM and IET method.

No.	Sample Material	Method	Resonant Frequency [Hz]	Standard Deviation [Hz]	Relative Difference $\Delta f/f_m$	t-Test Score	F-Test Score
1	Aluminum	SRM	4381.75	3.7444	−0.00029	0.88	1.38
		IET	4383.00	3.1856			
2	Stainless steel	SRM	3068.05	2.3448	−0.00003	0.10	1.19
		IET	3068.15	2.5589			
3	Carbon steel	SRM	1553.85	35.0418	0.00158	0.16	1.32
		IET	1551.39	40.2075			
4	Kaolin ceramics [1]	SRM	3374.54	82.9955	0.00562	0.53	1.22
		IET	3355.64	91.7184			
5	Kaolin ceramics [2]	SRM	3421.60	10.6667	−0.00050	0.38	1.16
		IET	3423.31	11.4891			
6	Alumina porcelain	SRM	2952.85	144.0841	−0.00023	0.01	1.03
		IET	2953.52	141.7116			
7	Soda-lime glass	SRM	2372.19	9.3747	−0.00011	0.06	1.22
		IET	2372.45	10.3524			
8	Corundum ceramics	SRM	1813.70	4.2823	0.00006	0.06	1.10
		IET	1813.59	4.4912			
9	Silicon carbide	SRM	2932.75	3.9341	0.00050	1.05	1.91
		IET	2931.28	2.8493			

[1] Ceramics based on Sedlec kaolin fired at 1150 °C. [2] Ceramics based on Kemmlitz kaolin fired at 1150 °C.

Table 2 shows that this difference is low, mostly less than 0.05%, and the highest value of 0.56% was valid for kaolin ceramics. These small differences suggest good agreement between IET and SRM results.

To obtain more reliable information, a t-test (see e.g., [25,26]) was performed to compare mean values. The number of samples was $n_{SRM} = n_{IET} = n = 12$. The degrees of freedom $k = n_{SRM} + n_{IET} - 2 = 22$ and the critical test score $t_{crit} = 2.07$ for the significance level $\alpha = 0.05$ were the same for every comparison. The test scores t for different samples were calculated according to the equation

$$t = |f_{IET} - f_{SRM}| \sqrt{\frac{n}{s_{IET}^2 + s_{SRM}^2}}, \quad (3)$$

where s_{IET} and s_{SRM} are the standard deviations.

Because all t-test scores $< t_{crit} = 2.07$, the mean values obtained by SRM and IET can be considered as equivalent. The differences between the measurement results can be considered in the scope of the measurement errors.

To compare the deviations of measured data, Fisher's test was employed [25,26]. The F-test score was calculated for degrees of freedom $k_{SRM} = n_{SRM} - 1 = 11$ and $k_{IET} = n_{IET} - 1 = 11$ for the significance level $\alpha/2 = 0.025$, for which $F_{crit} = 3.47$. The F-test score was calculated according to the formula:

$$F = \frac{s_1^2}{s_2^2}, \quad (4)$$

where s_1 and s_2 are standard deviations (s_{IET} and s_{SRM}) and $s_1 > s_2$, therefore $F > 1$. The F-test scores (see Table 2) were $< F_{crit} = 3.47$; therefore, it can be considered that standard deviations of SRM and IET are very close to each other. Consequently, both methods are equivalent for measuring the resonant frequency.

The comparative tests confirmed that SRM and IET are equivalent methods that give identical values of the resonant frequency. This is also shown graphically in Figure 3. The relationship between resonant frequencies measured by SRM and IET for different samples must be presented by the line with a slope = 1. As can be seen, the experimental points are lying on the line with a slope $0.9993 \approx 1$, which confirms the equivalence of the SRM and IET methods. The data for samples 6 and 9 are very close to each other and merge to one point in the graph.

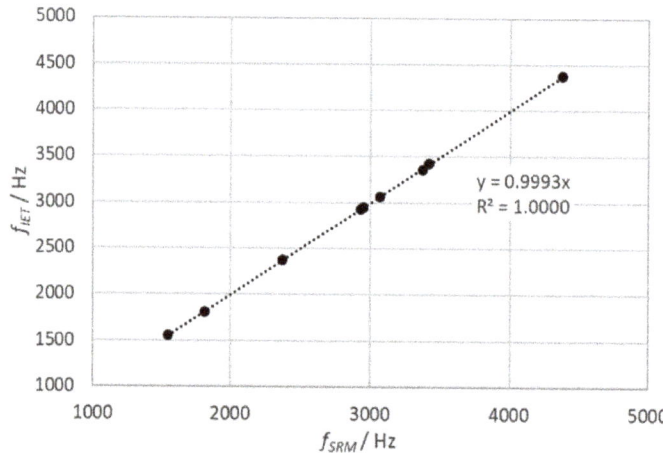

Figure 3. Dependence of resonant frequencies for samples measured by IET and SRM.

Young's moduli calculated from measured resonant frequencies (by SRM and IET) for the tested samples (Table 2) are given in Table 3, where they are compared to Young´s moduli for similar materials from the literature. It can be seen that Young´s moduli obtained by SRM and IET are close to each other, and also are in good agreement with already published results. The differences are mainly caused by different porosity and chemical composition.

Table 3. Young's modulus of used materials determined by SRM and IET methods, and Young's modulus from the literature.

No.	Sample Material	E [GPa] (SRM)	E [GPa] (IET)	E [GPa] (Ref.)
1	Aluminum	73.61	73.65	70 [27]
2	Stainless steel	212.31	212.32	168–206 [28]
3	Carbon steel	226.64	225.92	207 [29]
4	Kaolin ceramics [1]	30.27	29.93	29 [30]
5	Kaolin ceramics [2]	20.57	20.59	29 [30]
6	Alumina porcelain	56.60	56.62	55–85 [31,32]
7	Soda-lime glass	70.06	70.07	73 [33]
8	Corundum ceramics	246.75	246.72	154–377 [34]
9	Silicon carbide	94.00	93.91	100–400 [35]

[1] Ceramics based on Sedlec kaolin fired at 1150 °C. [2] Ceramics based on Kemmlitz kaolin fired at 1150 °C.

4. Conclusions

Samples made from different materials (metal, ceramics, and glass) and with two shapes (a prism with a rectangular cross-section and a cylinder with a circular cross-section) were examined using the sonic resonant method (SRM) and impulse excitation technique (IET) to confirm the equality of the two methods for measuring the resonant frequency f of flexurally vibrating samples. The mean values of the resonant frequencies and standard deviations were compared using the t-test and the F-test. The tests showed that both methods produced the same values of the resonant frequency. Small differences were within the scope of measurement error. This was also confirmed using the graph of the function $f_{IET}(f_{SRM})$, which was a line with the slope of $0.9993 \approx 1$. Young's moduli calculated for the tested samples were compared with those published in the literature and good agreement was found.

Author Contributions: Conceptualization, I.Š.; methodology, T.H., I.Š. and A.T.; investigation, T.H., F.O., J.O. and A.T.; writing—original draft preparation, I.Š., T.H. and A.T.; writing—review and editing, T.H., F.O., J.O., I.Š. and A.T.; visualization, A.T.; supervision, I.Š. and A.T.; project administration, A.T.; funding acquisition, A.T. All authors have read and agreed to the published version of the manuscript.

Funding: This research was funded by Ministry of Education of Slovak Republic, grant number KEGA 027UKF-4/2019 and by RVO: 11000.

Institutional Review Board Statement: Not applicable.

Informed Consent Statement: Not applicable.

Conflicts of Interest: The authors declare no conflict of interest. The funders had no role in the design of the study; in the collection, analyses, or interpretation of data; in the writing of the manuscript, or in the decision to publish the results.

References

1. Thomaz, W.D.; Miyaji, D.Y.; Possan, E. Comparative study of dynamic and static Young's modulus of concrete containing basaltic aggregates. *Case Stud. Constr. Mater.* **2021**, *15*, e00645. [CrossRef]
2. Quaglio, O.A.; da Silva, J.M.; Rodovalho, E.D.; Costa, L.D. Determination of Young's modulus by specific vibration of basalt and diabase. *Adv. Mater. Sci. Eng.* **2020**, *2020*, 4706384. [CrossRef]
3. Guicciardi, S.; Swarnakar, A.K.; Van der Biest, O.; Sciti, D. Temperature dependence of the dynamic Young's modulus of ZrB2-MoSi2 ultra-refractory ceramic composites. *Scr. Mater.* **2010**, *62*, 831–834. [CrossRef]
4. Duan, R.G.; Roebben, G.; Van der Biest, O. Glass microstructure evaluations using high temperature mechanical spectroscopy measurements. *J. Non-Cryst. Solids* **2003**, *316*, 138–145. [CrossRef]
5. Wang, J.B.; Ruan, H.H.; Wang, X.; Wan, J.Q. Investigating relaxation of glassy materials based on natural vibration of beam: A comparative study of borosilicate and chalcogenide glasses. *J. Non-Cryst. Solids* **2018**, *500*, 181–190. [CrossRef]
6. Ligoda-Chmiel, J.; Potoczek, M.; Sliwa, R.E. Mechanical properties of alumina foam/tri-functional epoxy resin composites with an interpenetrating network structure. *Arch. Metall. Mater.* **2015**, *60*, 2757–2762. [CrossRef]
7. Radovic, M.; Lara-Curzio, E.; Riester, L. Comparison of different experimental techniques for determination of elastic properties of solids. *Mater. Sci. Eng. A-Struct. Mater. Prop. Microstruct. Process.* **2004**, *368*, 56–70. [CrossRef]
8. Haines, D.W.; Leban, J.M.; Herbe, C. Determination of Young's modulus for spruce, fir and isotropic materials by the resonance flexure method with comparisons to static flexure and other dynamic methods. *Wood Sci. Technol.* **1996**, *30*, 253–263. [CrossRef]
9. Kashtaljan, J.A. *Elastic Characteristics of Materials at High Temperatures*; Naukova Dumka: Kiev, Ukraine, 1970. (In Russian)
10. Schreiber, E.; Anderson, O.; Soga, N. *Elastic Constants and Their Measurement*; McGraw-Hill Book Co.: New York, NY, USA, 1974.
11. ASTM C 1259-15. *Standard Test Method for Dynamic Young's Modulus, Shear Modulus and Poisson's Ratio for Advanced Ceramics by Impulse Excitation of Vibration*; ASTM International: West Conshohocken, PA, USA, 2015.
12. ASTM E1875-13. *Standard Test Method for Dynamic Young's Modulus, Shear Modulus, and Poisson's Ratio by Sonic Resonance*; ASTM International: West Conshohocken, PA, USA, 2013.
13. Štubňa, I.; Trník, A.; Vozár, L. Determination of Young's modulus of ceramics from flexural vibration at elevated temperatures. *Acta Acust. United Acust.* **2011**, *97*, 1–7. [CrossRef]
14. Lins, W.; Kaindl, G.; Peterlik, H.; Kromp, K. A novel resonant beam technique to determine the elastic moduli in dependence on orientation and temperature up to 2000 °C. *Rev. Sci. Instrum.* **1999**, *70*, 3052–3058. [CrossRef]
15. Landers, H.; Melzer, D.; Klinger, W. A system for measurement of dynamical modulus of elasticity at high temperatures. *Silikattechnik* **1977**, *28*, 275–277. (In German)
16. Roebben, G.; Bollen, B.; Brebels, A.; van Humbeeck, J.; van der Biest, O. Impulse excitation apparatus to measure resonant frequencies, elastic moduli and internal friction at room and high temperature. *Rev. Sci. Instrum.* **1997**, *68*, 4511–4515. [CrossRef]
17. Štubňa, I.; Húlan, T.; Trník, A.; Vozár, L. Uncertainty in the determination of Young's modulus of ceramics using the impulse excitation technique at elevated temperatures. *Acta Acust. United Acust.* **2018**, *104*, 269–276. [CrossRef]
18. Sakata, M.; Kimura, K.; Mizunuma, A. Measurement of elastic moduli from the impact sound of engineering ceramics and composites at elevated temperatures. *J. Am. Ceram. Soc.* **1995**, *78*, 3040–3044. [CrossRef]
19. Miloserdin, J.V.; Baranov, V.M. *High-Temperature Testing of Reactor Materials*; Atomizdat: Moskva, Russia, 1978. (In Russian)
20. Heritage, K.; Frisby, C.; Wolfenden, A. Impulse excitation technique for dynamic flexural measurements at moderate temperature. *Rev. Sci. Instrum.* **1988**, *59*, 973–974. [CrossRef]
21. Štubňa, I.; Trník, A. Equations for the flexural vibration of a sample with a uniform cross-section. *Strojniski Vestn.-J. Mech. Eng.* **2005**, *51*, 90–94.
22. Bosomworth, P.A. Improved frequency equations for calculating the Young's modulus of bars of rectangular or circular cross section from their flexural resonant frequencies. *J. ASTM Int.* **2011**, *7*, 1–14. [CrossRef]
23. Štubňa, I.; Majerník, V. An alternative equation of the flexural vibration. *Acustica* **1998**, *84*, 999–1001.
24. Chu, W.T. A comparison of two test methods for measuring Young's modulus of building materials. *Can. Acoust.* **1996**, *24*, 11.
25. Dowdy, S.; Wearden, S. *Statistics for Research*; John Willey & Sons: New York, NY, USA, 1983.
26. Linczényi, A. *Engineering Statistics*; Alfa: Bratislava, Slovakia, 1973. (In Slovak)
27. Yu, Z.Y.; Tan, Z.Q.; Fan, G.L.; Xiong, D.B.; Guo, Q.; Lin, R.B.; Hu, L.; Li, Z.Q.; Zhang, D. Effect of interfacial reaction on Young's modulus in CNT/Al nanocomposite: A quantitative analysis. *Mater. Charact.* **2018**, *137*, 84–90. [CrossRef]
28. Zhang, R.Z.; Buchanan, C.; Matilainen, V.P.; Daskalaki-Mountanou, D.; Ben Britton, T.; Piili, H.; Salminen, A.; Gardner, L. Mechanical properties and microstructure of additively manufactured stainless steel with laser welded joints. *Mater. Des.* **2021**, *208*, 109921. [CrossRef]
29. Qiu, H.; Ueji, R.; Kimura, Y.; Inoue, T. Grain-to-grain interaction effect in polycrystalline plain low-carbon steel within elastic deformation region. *Materials* **2021**, *14*, 1865. [CrossRef]
30. Antal, D.; Húlan, T.; Štubňa, I.; Záleská, M.; Trník, A. The Influence of texture on elastic and thermophysical properties of kaolin- and illite-based ceramic bodies. *Ceram. Int.* **2017**, *43*, 2730–2736. [CrossRef]
31. Al-Shantir, O.; Trník, A.; Csáki, Š. Influence of firing temperature and compacting pressure on density and Young's modulus of electroporcelain. *AIP Conf. Proc.* **2018**, *1988*, 020001. [CrossRef]
32. Štubňa, I.; Šín, P.; Trník, A.; Podoba, R.; Vozár, L. Development of Young's modulus of the green alumina porcelain raw mixture. *J. Aust. Ceram. Soc.* **2014**, *50*, 36–42.

33. Gong, J.H.; Deng, B.; Jiang, D.Y. A universal function for the description of nanoindentation unloading data: Case study on soda-lime glass. *J. Non-Cryst. Solids* **2020**, *544*, 120067. [CrossRef]
34. Bogdanov, S.P.; Kozlov, V.V.; Shevchik, A.P.; Dolgin, A.S. Young's modulus of corundum ceramics sintered from powders with a core-shell structure synthesized by iodine transport. *Refract. Ind. Ceram.* **2019**, *60*, 405–408. [CrossRef]
35. Pabst, O.; Schiffer, M.; Obermeier, E.; Tekin, T.; Lang, K.D.; Ngo, H.D. Measurement of Young's modulus and residual stress of thin SiC layers for MEMS high temperature applications. *Microsyst. Technol.* **2012**, *18*, 945–953. [CrossRef]

Article

Non-Destructive Multi-Method Assessment of Steel Fiber Orientation in Concrete [†]

Sabine Kruschwitz [1,2,*], Tyler Oesch [1,‡,3], Frank Mielentz [1], Dietmar Meinel [1] and Panagiotis Spyridis [4]

1. Bundesanstalt für Materialforschung und -prüfung (BAM), 12205 Berlin, Germany; frank.mielentz@bam.de (F.M.); dietmar.meinel@bam.de (D.M.)
2. Institute of Civil Engineering, Technische Universität Berlin, 13355 Berlin, Germany
3. Federal Office for the Safety of Nuclear Waste Management (BASE), 11513 Berlin, Germany; tyler.oesch@bfe.bund.de (T.O.)
4. Faculty of Architecture and Civil Engineering South Campus, Technische Universität Dortmund, 44227 Dortmund, Germany; panagiotis.spyridis@tu-dortmund.de (P.S.)
* Correspondence: sabine.kruschwitz@bam.de; Tel.: +49-30-8104-1422 (S.K.)
† This paper is an extended version of paper published in the 17th fib Symposium, Concrete Structures for Resilient Society held in Shanghai, China, 22–24 November 2020 online. ISSN: 2617-4820, ISBN: 978-2-940643-04-2.
‡ Denotes a former employer.

Abstract: Integration of fiber reinforcement in high-performance cementitious materials has become widely applied in many fields of construction. One of the most investigated advantages of steel fiber reinforced concrete (SFRC) is the deceleration of crack growth and hence its improved sustainability. Additional benefits are associated with its structural properties, as fibers can significantly increase the ductility and the tensile strength of concrete. In some applications it is even possible to entirely replace the conventional reinforcement, leading to significant logistical and environmental benefits. Fiber reinforcement can, however, have critical disadvantages and even hinder the performance of concrete, since it can induce an anisotropic material behavior of the mixture if the fibers are not appropriately oriented. For a safe use of SFRC in the future, reliable non-destructive testing (NDT) methods need to be identified to assess the fibers' orientation in hardened concrete. In this study, ultrasonic material testing, electrical impedance testing, and X-ray computed tomography have been investigated for this purpose using specially produced samples with biased or random fiber orientations. We demonstrate the capabilities of each of these NDT techniques for fiber orientation measurements and draw conclusions based on these results about the most promising areas for future research and development.

Keywords: steel fiber reinforced concrete; fiber orientation; non-destructive testing; micro-computed tomography; ultrasound; spectral induced polarization

Citation: Kruschwitz, S.; Oesch, T.; Mielentz, F.; Meinel, D.; Spyridis, P. Non-Destructive Multi-Method Assessment of Steel Fiber Orientation in Concrete. *Appl. Sci.* **2022**, *12*, 697. https://doi.org/10.3390/app12020697

Academic Editors: Dario De Domenico, Jerzy Hoła and Łukasz Sadowski

Received: 29 November 2021
Accepted: 5 January 2022
Published: 11 January 2022

Publisher's Note: MDPI stays neutral with regard to jurisdictional claims in published maps and institutional affiliations.

Copyright: © 2022 by the authors. Licensee MDPI, Basel, Switzerland. This article is an open access article distributed under the terms and conditions of the Creative Commons Attribution (CC BY) license (https://creativecommons.org/licenses/by/4.0/).

1. Introduction

In recent decades, the addition of short fibers in modern structural concrete has been a preferred option for improving the composite material's performance under tensile stress states. In relation to plain concrete, steel fiber-reinforced concrete (SFRC) shows not only higher resistance to crack growth, and hence improved resistance to aggressive substance diffusion and increased durability [1], but also enhanced tensile strength and ductility. Frequent industrial applications of SFRC include high-performance slabs (such as industrial floors, pile-supported foundation rafts and slabs on grade), tunnel linings, refractory structures, silos, containers, impact-proof defence structures and prefabricated—often prestressed—structural components. Because of the increasing trend for the use of SFRC in several types of engineering structures, advanced knowledge and standardisation with respect to this material becomes essential. The increasing interest in design guidance for SFRC is reflected in, besides numerous scientific publications and dedicated conferences, the emerging guidelines and standards, e.g., in the fib Model Code 2010 and 2020 [2,3], the German standard DIN 1045 and the associated DAfStb-Richtlinie Stahlfaserbeton [4], the ACI 544.4 Guide to Design with Fiber-Reinforced Concrete [5] and, most importantly, in the inclusion of provisions for the design with SFRC in the next generation of Eurocode 2 [6]. The design with SFRC strongly depends on the properties of the fiber reinforcement, such as the dosage of the fibers (typically expressed in weight of fibers per total volume of concrete), the material (steel, natural, synthetic), the dimensions (length and cross-section) and the shape of the fibers (straight, crimped, hooked, etc.) Furthermore, modern design concepts strongly rely on consistent and predictable overall distribution and orientation characteristics for fibers in the hardened concrete mix, since the amount and inclination of the fibers relative to the internal stress directions of the structural component have been shown to significantly influence its load-bearing performance [7]. Alignment of fibers in a parallel direction to the prevailing tensile stresses understandably increases the composite matrix's resistance in tensile rupture, and unfavorable fiber distributions and orientations can even hinder the structural performance of concrete by inducing anisotropy and weak regions in the material.

Fiber reinforcement in concrete aims to mainly arrest the concrete crack formation and propagation, which generally results in an increased tensile strength, ductility, and durability of the material. As seen in Figure 1, this effect is strongly dependent on the location of the fibers within the concrete matrix with respect to the cracking, which in turn forms in a nearly perpendicular direction to the tensile stresses in the material. These stresses are in turn bridged through the fibers intersecting the crack. Obviously, this behavior is strongly influenced by the orientation of fibers in relation to the principal tensile stresses (represented by the black arrows) and the resulting cracking planes. The fibers can, for example, have an orientation parallel to the cracking plane (blue), they can have a random orientation distribution in relation to the stress and crack planes (yellow), or be aligned parallel to the tensile forces and perpendicularly to the crack plane (red). In the last case (red), an optimum utilisation of the fibers' cross section is expected, and hence the load-displacement response can exhibit substantial ductility and possibly higher strength than plain concrete, which would otherwise fail in a quasi-brittle manner after the crack initiation. The yellow fibers represent the random distribution, which can still exhibit cracking stress retention and, hence, ductility. In the most suboptimal (blue) case, fibers essentially do not contribute to crack bridging, and, hence, concrete tends to behave similarly to a plain unreinforced material. In fact, fibers in this latter orientation can even act as discontinuities in the matrix and accelerate cracking. The fiber distribution and orientation strongly depends on the casting conditions of the fresh mix as well as geometric constraints, and the effect of casting-induced variations in fiber orientations on SFRC structural performance has been shown in previous research [8–11]. Some methods to control the fiber alignment in fresh concrete have also been proposed, e.g., by [12,13]. The effect of the fiber orientations on the load-bearing performance of concrete is accounted for in current design guidelines [2,14] by introducing a so-called fiber orientation factor, which reduces the nominal tensile strength value of fiber reinforced concrete, accounting for favorable and unfavorable fiber alignment effects as well as isotropy.

Figure 1. Indicative alignment of fibers in relation to tensile stresses and cracking in concrete, from left to right: most favorable, with fibers (red) aligned with the tensile stresses; random (yellow), leading to isotropic performance; and unfavorable (blue) with no fibers (effectively) bridging the crack.

Given the critical effect that fiber orientation and distribution have on the strength and failure pattern of SFRC structural components, NDT methods for evaluating fiber orientation and distribution are needed. In addition to measurement systems that can be applied in the lab on small samples, e.g., in [15], in situ testing techniques for large structures need to be developed.

In this study, we focus on the application of methods, which have already shown some value in this regard but have not yet really been systematically applied on samples with defined amounts of fibers in different orientations.

Previous research has indicated that electrical resistivity could be a suitable tool for the measurement of fiber orientations [16] and fiber-volume fractions in larger SFRC structural components [17,18]. Dry cementitious materials are typically electrical insulators. However, conducting materials added to the concrete, such as steel fibers, significantly decrease its electrical resistivity. Nontheless, further research is still needed in order to precisely define quantitative values of steel fiber distribution and orientation based on electrical resistivity data. The frequency dependency of the electrical resistivity and, hence, the polarizability of SFRC in the low frequency range has not yet been studied. It is conceivable that looking at both the capability of a medium to conduct and to store electrical charges will give deeper insight for material characterization. This can, for example, be done when the spectral induced polarization (SIP) method is used for measuring the electrical properties of a medium.

Recent research has also indicated that ultrasonic techniques may be useful for assessing fiber orientation characteristics [19]. If such an ultrasonic-based method could be validated, it would hold great potential, since ultrasound measurement techniques are already widespread within the civil engineering community and the capability of measuring ultrasonic properties up to several meters in depth has been successfully demonstrated [20]. However, further research is still needed to determine the sensitivity and accuracy of ultrasound for fiber orientation measurement, including ultrasound measurement quality and consistency for varying fiber types, structural shapes, and concrete mixtures.

One of the most popular NDT methods for measuring the spatial distribution and directional orientation of fibers in concrete in the lab is X-ray computed tomography (CT) [21–23]. When used for analysis of SFRC, this method typically relies on the fact that the fibers have significantly higher X-ray attenuations than the surrounding concrete material. The primary drawbacks of this method are, however, that specimen sizes tend to be limited and that X-ray CT cannot easily be used on-site in actual structures. Regardless, the accuracy of fiber measurement using X-ray CT on small specimens makes it an ideal method for calibrating or validating fiber measurements using other NDT techniques.

The study described in this paper was undertaken in order to assess the general sensitivity of the ultrasonic and SIP methods for fiber orientation measurement and to identify promising areas for future research. To accomplish these goals, specimens with varying controlled, casting-induced fiber orientation characteristics were fabricated and subsequently measured using ultrasonic and SIP measurement systems. Following these measurements, the specimens were scanned using X-ray CT and fiber orientation analyses of the resulting CT images were carried out. The goal of these CT-based analyses was to provide precise fiber orientation data for validation of the results of the ultrasonic and SIP measurements.

2. Materials and Methods

2.1. Sample Material

In the framework of the investigations presented herein, five samples with varying fiber alignments were produced and tested. For the sample examinations, it was required to create samples with randomly distributed fibers, and with fibers aligned parallel and transversely to the sample main axis. Depending on the test type and configuration, different sample dimensions were required: the samples used for ultrasound testing were prismatic with dimensions of $80 \times 80 \times 300$ mm, while the SIP- and CT-based evaluations used smaller cylindrical samples with a diameter of 50 mm and a length of 150 mm. The prismatic samples were initially cast as described below. Following ultrasonic measurement, the cylindrical samples were extracted from these prisms through diamond core drilling along their longitudinal axis.

The material used was identical for all samples: a concrete of class C35/45 having a target mean compressive strength of 50 MPa. The mix included CEM I 42.5 R cement from the producer Phoenix Zementwerke Krogbeumker Holding GmbH & Co. KG at 344 kg/m^3, added water at 147 kg/m^3 (w/c = 0.43), aggregates from the Rhine river plant of Hülskens GmbH & Co. KG with grading of 0/2, 2/8 and 8/16 mm at 779, 574, and 571 kg/m^3 (or 40%, 30%, and 30%), respectively, and no further admixtures. Furthermore, hooked-ended 5D Dramix® fibers 5D 65/60 BG (with 62 mm length and 0.9 mm diameter ([24]) and fy = 2300 N/mm^2) were used. Fiber dosages of 80 kg/m^3 (or 1 vol.%) and 40 kg/m^3 (or 0.5 vol.%) were realised in the samples (Figure 2, right). One sample was cast without any fibers for reference. The air void content in all samples was estimated as ca. 1.5% per mix volume.

Figure 2. Photograph of steel fibers Dramix 5D (**left**) and photograph during the sequential concrete and fiber laying for orthotropic samples at the laboratory of the TU Dortmund (**right**).

In order to produce the samples, the concrete material was mechanically mixed in a small laboratory concreting barrel without fibers. A portion was separated and kept without fibers, then fibers were added in the rest of the mix and it was further mixed in order to obtain a homogeneous spatial and orientational distribution of the fibers. This latter material was then directly cast in prismatic moulds. In order to produce the samples with orthotropically aligned fibers, the withheld plain concrete portion was cast sequentially in layers, followed by a manual placement of the fibers in the desired alignment (Figure 2, right). At intervals, the samples were compacted on a vibrating plate.

A photo of all prepared samples is shown in Figure 3. However, in this paper, we will investigate and explain in detail the potential of the NDT methods using five selected specimens as examples. These are three specimens with 40 kg/m^3 fiber content, where the fibers are distributed longitudinally, transversely as well as randomly. Furthermore, the results of a sample with 80 kg/m^3 fibers arranged longitudinally to the sample axis and a sample without fibers are discussed. The samples with random fiber distribution, and with fibers aligned parallel and transversely to the longitudinal sample axis are denoted throughout the paper using the specifications *, | | |, and —, respectively. The fiber dosage is stated in the sample name as either 40 or 80 according to their amount of fibers in kg/m^3. The specimen preparation was performed in the concreting facilities of the Building Research Lab at the TU Dortmund. All samples discussed in the following are colour coded in Figure 3. This colour coding will be further used throughout the paper consistently in the presentation of results.

Ultrasonic (US) measurements generally have to be performed on relatively large specimens in order to allow undisturbed wave propagation. Hence, the US tests were performed first on the largest available test specimens, which in our case where the prisms with 300 mm length and 40 mm edge length. Subsequently, cylindrical sub-samples were extracted from these prisms, which were then used for the investigations with X-ray CT and

SIP. Table 1 summarizes the properties of all samples discussed in this paper. Again, it has to be noted that the cylinders are all sub-samples of the corresponding prisms (Figure 3).

Table 1. Summary of the SFRC samples used in this study, their geometric properties, fiber dosages and fiber inclinations relatie to the longitudinal sample axis. Cylinders are sub-samples of prisms, as shown in Figure 3.

Sample	Fiber Dosage [kg/m³]	Length [mm]	Width/Diameter [mm]	Fiber Inclination [deg]
Q40—Prism	40	300	80	90
Q40 ⏐⏐⏐ Prism	40	300	80	0
Q40 B-40* Cylinder	40	150	50	random
Q40—Cylinder	40	150	50	90
Q40 ⏐⏐⏐ Cylinder	40	150	50	0
Q0F Cylinder	0	150	50	-
Q80 ⏐⏐⏐ Cylinder	80	150	50	0

Figure 3. Photograph of the cast prismatic specimens (following being cut halfway along their length) and the extracted cylindrical sub-samples (**left**). The samples discussed in further detail are color-coded. Close-up on a selection of cylindrical sub-samples (**right**).

2.2. Methods

2.2.1. X-ray Computed Tomography

The CT examinations were performed using a micro-CT system designed by BAM. The system has a 225 kV microfocus X-ray tube and a flat detector with 2048 × 2048 pixels (Figure 4). From the 2000 projection images, taken during a 360-degree rotation of the sample, a volume data set with 1001 × 1001 × 1981 voxels (i.e., 3D-pixels) is generated in the subsequent image reconstruction. During each CT scan, an entire specimen was imaged, which, given the specimen size, resulted in a resolution (voxel size) of 80 µm. The X-ray CT measurements were collected on all cylindrical specimens presented in Table 1.

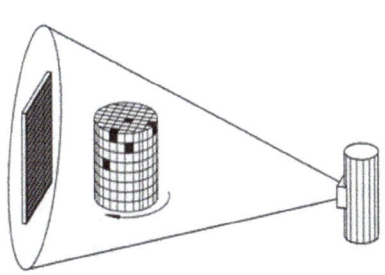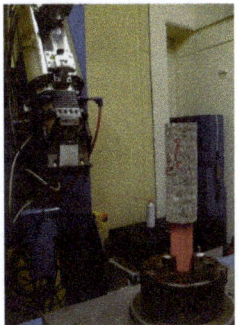

Figure 4. Schematic of the X-ray CT measurement principle (**left**) and photograph of the used CT system with SFRC sample placed on the rotating table (**right**).

2.2.2. Spectral Induced Polarization

The resistivity measurements or, as typically referred to in geophysics, spectral induced polarization (SIP) measurements were performed using a SIP-ZEL (Zentralinstitut für Elektronik) device [25]. Electrical four-point measurements in Wenner configuration were performed over a frequency range from 1 mHz to 45 kHz. A sample holder similar to the one described in [26] was used, the only difference being that it was larger and could accommodate cylindrical samples of 50 mm diameter and 150 mm length (Figure 5, right). An alternating electrical current was introduced across the outer steel caps and the potential decay across the inner ring wires was recorded. The measured electrical resistivities are complex values and can either be expressed as spectra of amplitude and phase or real and imaginary parts. In the following sections, both amplitude and phase of the resistivity as well as real and imaginary parts of the conductivity of our samples will be presented (Figure 5, left). As the drilling of the cylindrical sub-samples from the originally longer, prismatic samples required water cooling, all samples were dried at 40 °C for 24 h prior to the measurement. For the galvanic coupling between the electrodes and the sample in the SIP sample holder, we used 1.5% agar–agar gel. The SIP data were measured on all cylindrical samples given in Table 1.

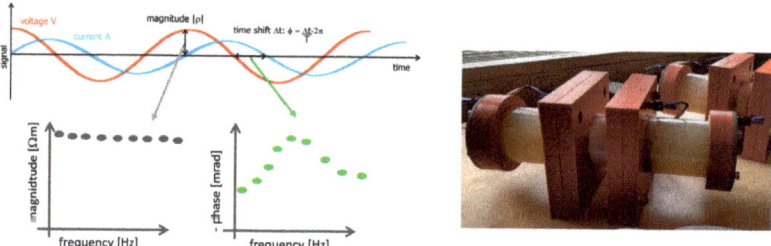

Figure 5. SIP signal for a distinct frequency (**top left**). The applied voltage V causes a subsequent current A signal, which is shifted in time. The amplitude ratio of the two signals is the resistivity magnitude $|\rho|$ (in Ωm due to geometrical considerations of sample diameter and length) and their time shift is related to as phase angle Φ (in mrad or °) (**bottom left**). SIP 4-point measurement cell, where the sample is placed in the middle and the electrodes are galvanically coupled with gel (**right**).

2.2.3. Ultrasound

An ultrasonic measuring system, consisting of a PC, a rectangular transmitter, a DAQ-Pad (manufactured by NI) for data acquisition and two dry contact probes (manufactured by ACS Group) for generation and reception of ultrasonic transverse waves was used. The ultrasonic transmission pulse is emitted by the transmitter (T), passes through the test body and is received at the receiver (R)s (see Figure 6). The signal is then digitized using an analogue-to-digital converter (A/D) and transferred to the PC via the USB interface. The probes require no coupling agent and are pressed against the concrete surface of the test specimens by spring force alone. The polarization of the ultrasonic wave in the concrete test specimen can be altered by rotating the probes, as shown schematically for two polarization states in Figure 6 (right). The probes have a centre frequency of $f_M = 50$ kHz and are excited with a bipolar square wave signal of $u_{pp} = 300$ V. The measurements were carried out on all the prismatic specimens shown in Table 1. Each specimen is measured with different angles of polarization (0°, 45°, 90°, 180°, 225° and 270°). In particular, variations in the signal spectra due to changes in the polarization of the ultrasonic waves were studied.

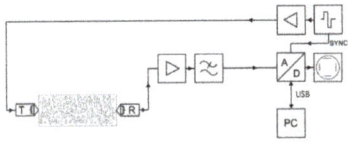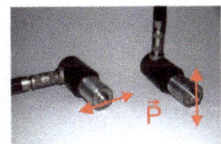

Figure 6. Ultrasonic measuring system: block diagram (**left**), photo of the measurement set-up (**centre**) and close-up, of the ultrasonic probes (**right**). In the close-up the polarization $\vec{P}$ of the transmitters is shown in red.

3. Results

3.1. X-ray Computed Tomography (CT)

Fiber identification and orientation analysis was completed using the program VGStudioMAX 3.3 (Volume Graphics GmbH). The results of the analysis are presented in terms of spherical coordinates (Figure 7, top left). As the fibers are oriented in all three spatial directions, this representation is the most suitable. In our opinion, a cylindrical coordinate system (using polar coordinates supplemented by a Cartesian coordinate for the height) would have been better to represent the data layer by layer focussing then only on the fiber orientation within one respective cylindrical section in 2D. In this study, though, we aim to analyse all fiber orientations in one sample in 3D.

In the spherical coordinate system we use, orientations are characterized by angles Θ and Φ. In the Cartesian coordinate system used for these measurements, the cylindrical axis of the sample is denoted as the z-axis. The angle Θ represents the azimuthal angle in the x-y plane from the x-axis, with $0 < \Theta < 360°$. The angle Φ represents the polar angle from the positive z-axis, with $0 < \Phi < 180°$.

Figure 7 provides fiber renderings and orientation histograms for all five cylindrical samples shown in Table 1. The orientation histograms are shown in the form of orthographic projections, where the centre of the projection is the z-axis (i.e., the cylindrical axis). Thus, these projections represent, effectively, the portion of the fiber orientation histogram plotted on the upper half of the sphere shown in the upper left of Figure 7. Since the fibers are considered to be symmetric about their lengths, it is assumed that differences between the lower and upper halves of the histogram sphere are negligible, which has also been verified visually during data analysis.

During fiber orientation analysis, the fibers were not individually identified, separated and analysed. Rather, composite orientation information was calculated for all steel fiber material in the specimens. This means that the resulting orientation measurements contain not only orientation data related to the primary axis of the fibers, but also orientation data related to the hooked ends of the fibers. Although this effect introduces a partial distortion of the data, the hook-based effects are considered to be acceptable given the relatively small size of the fiber hooks in comparison to the overall fiber dimensions.

The "Fiber Composite Analysis" module automatically scales the colour coding in the polar colour plot to the deviations found from the orientation with the highest frequency. Since there is no fibre material in sample Q0F, the module scales the colour coding to the orientation values determined on dense aggregates. These aggregate related orientation values are no longer visible in the samples containing fibres with the same parameter setting. Therefore, the colour coding for Q0F was manually adjusted so that the colour scale is comparable with the other polar plot representations. Clear differences in the histogram characteristics can, for instance, be observed between the three samples Q40—, Q40||| and Q40* in Figure 7.

3.2. Spectral Induced Polarization

The SIP data collected on all five cylindrical samples listed in Table 1 and the measured real and imaginary electrical conductivities along with the amplitude and phase spectra are plotted in Figure 8. Though all samples were measured in a dry state (meaning that there is no significant pore fluid and electrical conductivity was not possible due to ion movement), we measured relatively low electrical resistivities between 10 and 800 Ohm*m (Figure 8, left). Most of the samples, besides Q0F which does not contain steel fibers, moreover showed distinct polarization behaviour with absolute phase values well above 20 mrad. As could be expected, the sample Q80 ⊥⊥⊥ showed, due to its high fiber content, the lowest resistivity magnitude and highest polarizability. These characteristics were less pronounced for the three samples with 40 kg/m³ fiber content. When comparing the results of these latter three samples, Q40 ⊥⊥⊥, where the steel fibers are oriented along the sample axis (and along the electrically induced current flow within the SIP sample holder), features the lowest electrical resistivity amplitude. For the sample Q40*, where the steel fibers are randomly distributed, the electrical resistivity increases and for sample Q40—, where steel fibers are oriented perpendicularly to the induced current flow, we measure the highest resistivity values for a sample containing fibers. Furthermore, from their electrical polarization behavior, these three samples can clearly be differentiated. The sample Q40 ⊥⊥⊥ Cylinder is most polarizable, with up to −400 mrad. When the steel fibers are randomly distributed (i.e., Q40* Cylinder), the electrical polarizability decreases to about −350 mrad and when the fibers are perpendicular to the sample axis (and induced electrical field) (i.e., Q40—Cylinder), the polarizability becomes less than −50 mrad.

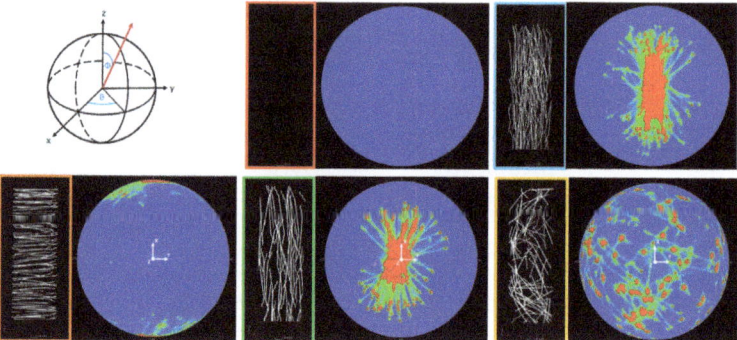

Figure 7. Spherical coordinate system , reproduced from Oesch et al. 2018 under the terms of the Creative Commons Attribution 4.0 International License (https://creativecommons.org/licenses/by/4.0/, accessed on 28 November 2021), (**top left**)). Fiber renderings and orientation histograms for sample Q0F Cylinder (**top centre**), Q80 ⊥⊥⊥ Cylinder (**top right**), Q40—Cylinder (**bottom left**), Q40 ⊥⊥⊥ Cylinder (**bottom centre**), Q40* Cylinder (**bottom right**). Histogram radius: 0 < Φ < 90°; histogram circumference: 0 < Θ < 360°.

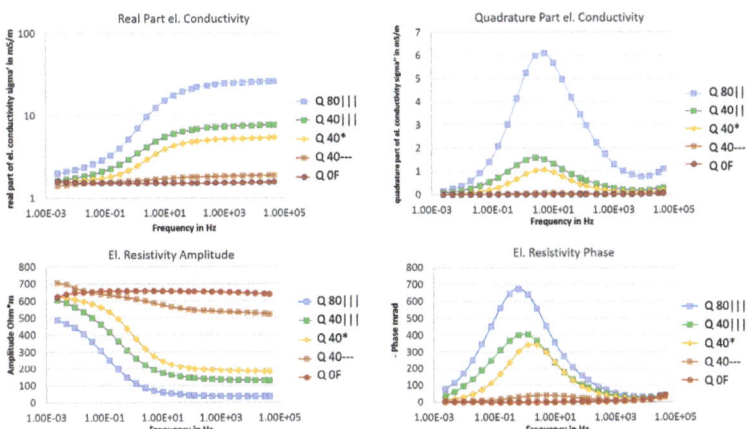

Figure 8. Spectral Induced Polarization spectra of steel fiber reinforced concrete samples Q80 ||| Cylinder, Q40 ||| Cylinder, Q40* Cylinder, Q40—Cylinder and Q0F Cylinder. Real and quadrature parts of electrical conductivity (**top**) and resistivity amplitudes and phases (**bottom**).

Given that the concrete mix and the dosage of steel fibers within these three samples was constant, we can assume that the observed differences in the electrical resistivity behavior are due to the fiber orientation. Thus, we can observe, for instance, that the position of the phase peak slightly shifts towards higher frequencies with increasing amount of steel fibers tilted away from the direction of the induced electrical current flow (which coincides with the long axis of the cylindrical samples).

3.3. Ultrasound

The greatest influence on the ultrasonic signals was expected from the specimen where the steel fibres were oriented perpendicular to the sound propagation (specimen Q40—, Figure 9 top). During the first measurement, the polarization of the ultrasonic waves is 0° and the fibre orientation is 90° (Figure 9, top left). The spectrum shows the centre frequency of the probe at about 50 kHz and strong indications at about 42 kHz and 55 kHz. In this figure, |S(f)| denotes the magnitude of the spectral density in arbitrary units (a.u.). Then, the polarization of the ultrasonic waves was rotated by 90° from 0° (i.e., probe polarization 90°, fibre orientation 90°, Figure 9, top right). Thus, the polarization of the ultrasonic waves and the steel fibres are oriented parallel to each other. The magnitude of the spectrum at about 50 kHz is lower and the higher frequency components are more attenuated. In the Q40 ||| test specimen, the steel fibers are oriented in the direction of sound propagation (Figure 9, bottom). By using a rotation of the polarization, changes in the spectrum are also visible, but the influence on the test frequency and on the higher frequency components is lower. While these measurements do suggest that fibre orientation may have an influence on polarization-induced ultrasonic signal variations, the observed correlations have a limited transferability. A more general validation would require a more thorough measurement program.

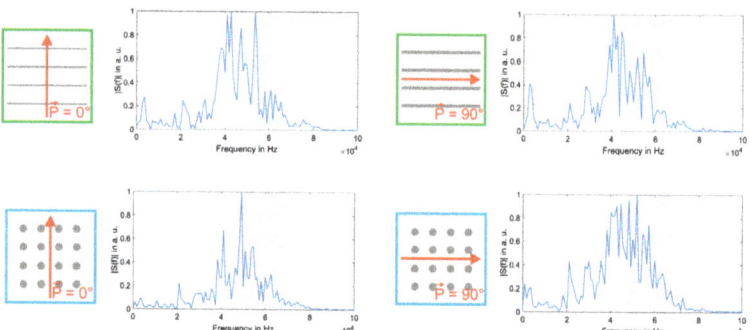

Figure 9. Probe polarization and fibre alignment (schematic), spectrum of the corresponding received signals for the samples Q40—Prism (**top**) and Q40 ∣ ∣ ∣ Prism (**bottom**).

4. Discussion

The X-ray CT results provide both a qualitative and a quantitative demonstration of the variations in fiber orientation characteristics amongst the specimens. In Figure 7, one can visually observe that nearly all fibers within specimen Q40—Cylinder are well aligned with the x-axis of the sample (i.e., perpendicular to the cylindrical axis). This result is confirmed by the concentrations of fiber orientations measured at $(\Phi, \Theta) = (0°, 90°)$ and $(\Phi, \Theta) = (180°, 90°)$, which are the spherical coordinate points corresponding to alignment along the x-axis. Similarly, Figure 7 indicates that the fibers in specimen Q40 ∣ ∣ ∣ Cylinder are well aligned with the z-axis of the specimen (i.e., the cylindrical axis). For specimen Q40* Cylinder, a relatively random scattering of different fiber orientations can be observed (Figure 7). These analysis results, thus, confirm the successfulness of the casting techniques in producing the desired controlled fiber orientation characteristics and can be used as a baseline of fiber orientation information, against which the SIP and ultrasonic measurement results can be evaluated.

In principle, it is conceivable that the measurement results of the non-destructive testing methods are influenced by systematically different (air) pore contents in the specimens due to the manufacturing process and that the observed properties could be distorted as a result. Therefore, the pore radius distributions and the air void contents of the five specimens were determined and compared. Air voids were considered to be those with pore diameters greater than 160 μm, twice the voxel size of the X-ray CT studies performed here. Mercury intrusion porosimetry (MIP) was used as a reference method (compare also [27,28]). For this purpose, 20–30 g of sample material was manually chipped off each of the remnants of the prisms (after the cylinder samples had been taken), dried at 105 °C, and examined by MIP. MIP is used to detect pore sizes down to approximately 3.6 nm. Due to its underlying physics, however, only pore throat diameters can be determined and the volume of wide pore areas (which can only be reached via narrow pore throats) is attributed to the pore throats.

The pore throat diameter distributions are shown in Figure 10, the corresponding total porosities in Table 2. The total porosities of all samples range between 12.14% and 15.19%, but no systematic differences are observed: neither directly between the three samples with 40 kg/m^3 fibre content, where the fibres were manually placed in different orientations, nor with respect to the samples with 0 or 80 kg/m^3 fibre content. Thus, it can be concluded that the casting process had neither a systematic influence on the overall porosity nor on the pore size distribution of the samples.

The MIP method is not suited, though, for the detection of larger pores, as they are usually accessed via narrow pore throats. However, these might be assessed using X-ray CT. To determine the pore content within the volume examined with CT, a low image grey value threshold for pores is used. The boundary between pore (air) and material (cement) is defined by the determined material surface. The ISO threshold value air/cement, automatically determined by the software from the histogram analysis, serves as the starting value for the automatic surface detection with local threshold values. In the CT analysis, however, only objects consisting of two or more adjacent voxels are recognized as pores in order to exclude noise ("false pores"). Consequently, only pores with a diameter larger than 160 µm can be detected. The results are shown in Table 2. It is striking that the two samples with longitudinally aligned fibers (Q80 ||| and Q40 |||) show approximately the same and, compared to the other fiber reinforced samples, somewhat lower amount of large pores. This could, however, be due to the better radiolucency of the fibers and the associated reduction of artefacts that are incorrectly recognized as pores by the automatic analysis software. In the sample without fibres, the amount of large pores is only slightly lower. The results may, therefore, be not very reliable due to the interference caused by the steel fibres. Strictly speaking, the pore content can only be reliably compared if the fibre orientation is the same. Thus, it is not possible to deduce the effect of the casting process on the volume of large pores using the measurement data listed here. However, given that the total porosities measured using MIP did not vary strongly in relation to fiber distribution or orientation characteristics, it appears unlikely that porosity variations had a major influence on the results of the SIP and US investigations.

Table 2. Porosities determined by X-ray CT and MIP in vol.% and BET surface area in m^2/g.

Sample	X-ray CT-Por. ≥ 160 µm	MIP-Por. ≥ 160 µm	MIP-Por. ≥ 3.6 nm	BET Surface Area
Q 80 |||	1.9	0.6	12.22	3.89
Q 40 |||	1.8	0.5	15.19	5.05
Q 40*	3.6	0.4	13.25	4.43
Q 40—	3.4	0.8	12.14	3.44
Q 0F	1.5	0.6	12.98	4.20

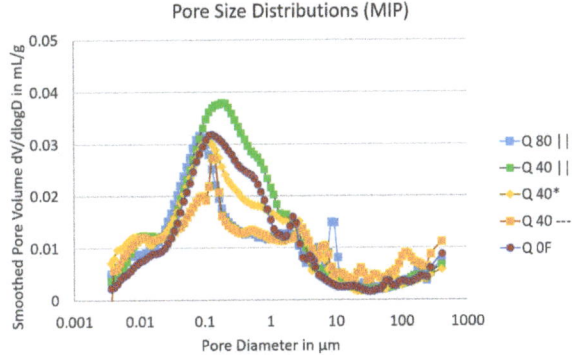

Figure 10. Pore throat size distributions of all five samples determined with MIP.

The SIP measurements showed that our results support the findings of [16], which showed that the main orientation of steel fibers in test specimens coincided with the direction of the lowest resistivity. In the present study, we measured, however, not only the resistivity, which has previously been evaluated by [16], but also the frequency dependency of the complex resistivity by evaluating the polarization behavior of our samples. Our data indicate that the phase spectra might be an even more sensitive quantity for measuring fiber orientation than the amplitude spectra as the differences between samples with parallel and perpendicular oriented steel fibers were even more pronounced. The works of [17,18] both focus on the influence of fiber dosage (not their orientation) on the electrical resistivity for samples prepared with random fiber distributions. The fibre dosage of our samples (for instance, 0.5 vol% for Q40—, Q40||| and Q40*) is at the lower end of what the other working groups used and it is conceivable that by evaluating both resistivity amplitude and phase, we could extract information about fibre orientation and content at the same time.

The observed changes in the ultrasound signal are not very strong and experience has shown that this can also be caused by changes in the coupling. The test specimens are relatively small compared to the ultrasonic wavelength. Due to reflections of the ultrasonic waves at the sides of the test specimen and the superposition of different wave modes, the results are not clear. Thus, further ultrasonic measurements on larger specimens are necessary to confirm these initial promising results. The measurement of steel fiber content and/or orientation with ultrasonic dry contact probes would have the advantage that the concrete surface does not have to be prepared for a measurement. Other influencing variables like the grading curve, the pore structure, cracks or variations in the moisture content of the samples were not regarded in this study. All these parameters can significantly influence the results and sensitivity of SIP or US measurements. Therefore, further research is needed.

5. Conclusions

SFRC is still a relatively new building material, which holds great potential for innovative use and design approaches for future concrete structures. The assessment of SFRC properties, such as fiber orientation and dosage, is still difficult, often requiring coring and elaborate lab analysis. Our study focusses on the effectiveness of alternative methods to extract this information non-destructively.

Evaluating the electrical behavior of SFRC samples, we found that both the resistivity magnitude and phase (i.e., the material's capability to polarize) are sensitive to the main fiber orientation. It is conceivable that when both are measured, fiber orientation and content can be assessed. Likewise, for the measured samples we saw that the transmission amplitude of ultrasonic shear waves is slightly diminished and that higher frequency components are attenuated when the polarization of the wave is parallel to the fibers' direction. However, the dimensions of the test specimens have the same order of magnitude as the ultrasound wavelength itself. This results in reflections of the ultrasonic waves from the test body sides and a superposition of different wave modes in the received signal, making the interpretation of the results difficult. In addition, variations in probe coupling have an influence on the received signals. For this reason, further ultrasonic measurements are necessary, using a larger number of test specimens with bigger dimensions, to further investigate these first indications of an influence of the fibres on the ultrasonic signals.

The X-ray CT analysis provided quantitative measurements of fiber orientation as well as visualizations of actual fiber distributions within the specimens. These measurements not only verified the successfulness of the specimen casting technique in producing specimens with controlled fiber orientations. They also can serve as a quantitative basis for assessing correlations between ultrasonic or SIP measurements and actual fiber conditions. An analysis of the statistical correlation between the quantitative fiber orientation data measured using X-ray CT and the phase variations measured using SIP could yield useful mathematical relationships that could be utilized as a basis for expanding fiber orientation assessment using SIP to further applications (such as measurement of larger components or of different fiber types). Such a statistical analysis is the planned topic of the authors' future work in this area.

Author Contributions: Conceptualization, S.K., T.O. and P.S.; methodology, all; validation, T.O., D.M., F.M. and S.K.; investigation, T.O., D.M., F.M. and S.K.; resources, P.S.; data curation, T.O., D.M., F.M. and S.K.; writing—original draft preparation, all; writing—review and editing, all; visualization, all; supervision, S.K., T.O. and P.S.; project administration, all. All authors have read and agreed to the published version of the manuscript.

Funding: This research received no external funding.

Acknowledgments: The authors express their gratitude to company SV Bekaert SA for the provision of steel fibers as material used in this investigation, and the valuable scientific feedback. Moreover, we thank Heiko Stolpe and Marco Lange from BAM for technical support and sample preparation. The opinions expressed in this publication are those of the authors and they do not purport to reflect opinions or views of the supporting organisations.

Conflicts of Interest: The authors declare no conflict of interest.

References

1. Boshoff, W.P.; Altmann, F.; Adendorff, C.J.; Mechtcherine, V. *A New Approach for Modelling the Ingress of Deleterious Materials in Cracked Strain Hardening Cement-Based Composites*; Springer Science and Business Media LLC: Berlin, Germany, 2015; Volume 49, Chapter A, pp. 2285–2295.
2. di Prisco, M.; Colombo, M.; Dozio, D. Fiber-reinforced concrete infibModel Code 2010: principles, models and test validation. *Struct. Concr.* **2013**, *14*, 342–361. [CrossRef]
3. Matthews, S.; van Vliet, A.B.; Walraven, J.; Mancini, G.; Dieteren, G. fib Model Code 2020: Towards a general code for both new and existing concrete structures. *Struct. Concr.* **2018**, *19*, 969–979. [CrossRef]
4. Mark, P.; Oettel, V.; Look, K.; Empelmann, M. Neuauflage DAfStb-Richtlinie Stahlfaserbeton. *Beton- Stahlbetonbau* **2020**, *116*, 19–25. [CrossRef]
5. ACI Committee 544 4R. *Guide to Design with Fiber-Reinforced Concrete*; Technical Report; American Concrete Institute: Farmington Hills, MI, USA, 2018.
6. Goodchild, C. *Eurocodes Recision—An Update*; Technical Report; The Concrete Centre: London, UK, 2016.
7. Holschemacher, K.; Dehn, F.; Müller, T.; Lobisch, F. *Grundlagen des Faserbetons*; Wiley: Hoboken, NJ, USA, 2016.
8. Ferrara, L.; Park, Y.D.; Shah, S.P. Correlation among Fresh State Behavior, Fiber Dispersion, and Toughness Properties of SFRCs. *J. Mater. Civ. Eng.* **2008**, *20*, 493–501. [CrossRef]
9. Gröger, J.; Nehls, N.; Silbereisen, R.; Tue, N.V. Einfluss der Einbau- und der Betontechnologie auf die Faserverteilung und -orientierung in Wänden aus Stahlfaserbeton. *Beton- Stahlbetonbau* **2011**, *106*, 45–49. [CrossRef]
10. Pujadas, P.; Blanco, A.; Cavalaro, S.H.P.; de la Fuente, A.; Aguado, A. Multidirectional double punch test to assess the post-cracking behavior and fiber orientation of FRC. *Constr. Build. Mater.* **2014**, *58*, 214–224. [CrossRef]
11. Lusis, V.; Krasinovs, A.; Konovova, O.; Lapsa, V.A.; Stonys, R.; Macanovskis, A.; Lukasenoks, A. Effect of Short Fibers Orientation on Mechanical Properties of Composite Material–Fiber Reinforced Concrete. *J. Civ. Eng. Manag.* **2017**, *23*, 1091–1099. [CrossRef]
12. Villar, V.P.; Medina, N.F. Alignment of hooked-end fibers in matrices with similar rheological behavior to cementitious composites through homogeneous magnetic fields. *Constr. Build. Mater.* **2018**, *163*, 256–266. [CrossRef]
13. Lederose, L.; Lehmberg, S.; Budelmann, H.; Kloft, H. Robotergestützte, magnetische Ausrichtung von Mikro-Stahldrahtfasern in dünnwandigen UHPFRC-Bauteilen. *Beton- Stahlbetonbau* **2019**, *114*, 33–42. [CrossRef]
14. Heek, P.; Look, K.; Oettel, V.; Mark, P. Bemessung von Stahlfaserbeton und stahlfaserbewehrtem Stahlbeton. *Beton- Stahlbetonbau* **2021**, *116*, 2–12. [CrossRef]
15. Wichmann, H.J.; Holst, A.; Budelmann, H. Ein praxisgerechtes Messverfahren zur Bestimmung der Fasermenge und - orientierung im Stahlfaserbeton. *Beton- Stahlbetonbau* **2013**, *108*, 822–834. [CrossRef]
16. Lataste, J.F.; Behloul, M.; Breysse, D. Characterisation of fibers distribution in a steel fiber reinforced concrete with electrical resistivity measurements. *NDT E Int.* **2008**, *41*, 638–647. [CrossRef]
17. Solgaard, A.O.S.; Geiker, M.; Edvardsen, C.; Küter, A. Observations on the electrical resistivity of steel fiber reinforced concrete. *Mater. Struct.* **2013**, *47*, 335–350. [CrossRef]
18. Fiala, L.; Toman, J.; Vodička, J.; Ráček, V. Experimental Study on Electrical Properties of Steel-fiber Reinforced Concrete. *Procedia Eng.* **2016**, *151*, 241–248. [CrossRef]
19. Gebretsadik, B. Ultrasonic Pulse Velocity Investigation of Steel Fiber Reinforced Self-Compacted Concrete. Master's Thesis, University of Nevada, Reno, NV, USA, 2013.
20. Wiggenhauser, H.; Niederleithinger, E.; Milmann, B. Zerstörungsfreie Ultraschallprüfung dicker und hochbewehrter Betonbauteile. *Bautechnik* **2017**, *94*, 682–688. [CrossRef]
21. Liu, J.; Li, C.; Liu, J.; Cui, G.; Yang, Z. Study on 3D spatial distribution of steel fibers in fiber reinforced cementitious composites through micro-CT technique. *Constr. Build. Mater.* **2013**, *48*, 656–661. [CrossRef]
22. Herrmann, H.; Pastorelli, E.; Kallonen, A.; Suuronen, J.P. Methods for fiber orientation analysis of X-ray tomography images of steel fiber reinforced concrete (SFRC). *J. Mater. Sci.* **2016**, *51*, 3772–3783. [CrossRef]
23. Oesch, T.; Landis, E.; Kuchma, D. A methodology for quantifying the impact of casting procedure on anisotropy in fiber-reinforced concrete using X-ray CT. *Mater. Struct.* **2018**, *51*, 1–13. [CrossRef]
24. N.V. Bekaert S.A. EC Declaration of Performance Dramix® 5D 65/60BG (Creation Date 2019/07/10). Technical Report. 2019. Available online: https://www.bekaert.com/en/product-catalog/content/dop/dramix-5d-technical-documents (accessed on 28 November 2021).
25. Zimmermann, E.; Kemna, A.; Berwix, J.; Glaas, W.; Münch, H.M.; Huisman, J.A. A high-accuracy impedance spectrometer for measuring sediments with low polarizability. *Meas. Sci. Technol.* **2008**, *19*, 105603. [CrossRef]

26. Kruschwitz, S. Assessment of the Complex Resistivity Behavior of Salt Affected Building Materials. Ph.D. Thesis, Bundesanstalt für Materialforschung und -prüfung (BAM), Berlin, Germany, 2008.
27. Kruschwitz, S.; Halisch, M.; Dlugosch, R.; Prinz, C. Toward a better understanding of low-frequency electrical relaxation—An enhanced pore space characterization. *Geophysics* **2020**, *85*, MR257–MR270. [CrossRef]
28. Washburn, E.W. The Dynamics of Capillary Flow. *Phys. Rev.* **1921**, *17*, 273–283. [CrossRef]

Article

Analysis of the Applicability of Non-Destructive Techniques to Determine In Situ Thermal Transmittance in Passive House Façades

Blanca Tejedor *, Kàtia Gaspar, Miquel Casals and Marta Gangolells

Group of Construction Research and Innovation (GRIC), Universitat Politècnica de Catalunya (UPC), C/Colom, 11, Ed. TR5, 08222 Terrassa, Spain; katia.gaspar@upc.edu (K.G.); miquel.casals@upc.edu (M.C.); marta.gangolells@upc.edu (M.G.)
* Correspondence: blanca.tejedor@upc.edu; Tel.: +34-937-398-919

Received: 26 October 2020; Accepted: 20 November 2020; Published: 24 November 2020

Abstract: Within the European framework, the passive house has become an essential constructive solution in terms of building efficiency and CO_2 reduction. However, the main approaches have been focused on post-occupancy surveys, measurements of actual energy consumption, life-cycle analyses in dynamic conditions, using simulation, and the estimation of the thermal comfort. Few studies have assessed the in situ performance of the building fabric of passive houses. Hence, this paper explores the applicability of non-destructive techniques—heat flux meter (*HFM*) and quantitative infrared thermography (*QIRT*)—for assessing the gap between the predicted and actual thermal transmittance of passive house façades under steady-state conditions in the Mediterranean climate. Firstly, the suitability of in situ non-destructive techniques was checked in an experimental mock-up, and, subsequently, a detached house was tested in the real built environment. The findings revealed that both Non-Destructive Testing (NDT) techniques allow for the quantification of the gap between the design and the actual façades U-value of a new passive house before its operational stage. *QIRT* was faster than the *HFM* technique, although the latter was more accurate. The results will help practitioners to choose the most appropriate method based on environmental conditions, execution of the method, and data analysis.

Keywords: Nearly Zero Energy Buildings (NZEB); passive house (PH); heat flux meter (*HFM*); quantitative infrared thermography (*QIRT*); building thermal performance; U-value; Mediterranean climate

1. Introduction

Within the context of European regulation on energy efficiency and energy conservation, the passive house (PH) concept has emerged as a global quality assurance standard [1–3]. Nearly Zero Energy Buildings (NZEBs) have become an essential element in developed countries to achieve a reduction in energy consumption and CO_2 in the construction sector [4–6], using efficient systems of HVAC (Heating, Ventilating, and Air-Conditioning) and increasing the thermal insulation of buildings [7]. Indeed, PH requires 80–90% less heating energy than conventional buildings to provide optimal thermal comfort conditions, while the initial investment only represents an increase of 5–10% [8]. However, the main barriers for this type of construction are the high performance building materials and the cost of adoption (training and certification) [9,10].

Some authors stated that a PH should be defined by six principles: (i) a high level of thermal insulation and thermal capacity of opaque walls [3,5–7,11], (ii) minimization of thermal bridges [3,5,11,12], (iii) high efficient windows [5,12], (iv) high levels of airtightness [3,5–7], (v) passive solar gains [7,11,12], and (vi) efficient mechanical ventilation with heat recovery system [3,5–7].

For this reason, researchers put their efforts on the assessment of the energy performance of PH dwellings. The main approaches consisted of post-occupancy surveys, measurements of the actual energy consumption, life-cycle analyses (LCA) in dynamic conditions, using simulation (i.e., EcoHestia, Computation Fluid Dynamics (CFD) model, or EnergyPlus) [6,11], and the determination of the thermal comfort by the PMV (Predicted Mean Vote) model according to ASHRAE 55 and EN 15,251 [4,5,7–9,11,13]. As regards the design parameters, recent studies were focused on building performance optimization (BPO) through the implementation of meta-models (i.e., multiple linear regression, support vector machines (SVM), and artificial neural networks (ANN)) [8]. Nevertheless, a lack of systematic optimization methods for façades is still detected in the research field [8]. Besides this, few studies assessed the in situ performance of the building fabric of PH dwellings in comparison with the theoretical design [3,14–18]. Along this line, Johnston et al. [3] highlighted that most of in situ tests were conducted on prefabricated timber frames or externally insulated thin-joint blockworks, using heat flux meter (*HFM*) and qualitative infrared thermography (*IRT*) techniques. No references were found for heavy multi-leaf walls and quantitative internal IRT.

The heat flux meter (*HFM*) method is widely used as a non-destructive method for measuring the actual thermal transmittance of façades [19–25]. However, the literature review revealed that the *HFM* average method is not frequently applied to verify compliance with technical specifications of projects in façades with low U-value in PH on site.

It is challenging to obtain accurate results during in situ measuring of façades with low thermal transmittance. Very few initiatives have conducted in situ measurements on façades with low U-values, using the *HFM* average method. Furthermore, in most of these studies, the relative deviation of measured U-values from theoretical values was significant. Asdrubali et al. [21] measured the actual thermal transmittance of façades, with theoretical U-values ranging from 0.23 up to 0.33 W/m^2·K. Their results showed relative deviations of between 4 to 75%. Mandilaras et al. [26] obtained a relative deviation between theoretical and measured U-values of 28% when they monitored, in situ, a building envelope with theoretical thermal transmittance of 0.20 W/m^2·K. Albatici et al. [27] and Nardi et al. [28] used the *HFM* average method to validate results obtained by using the quantitative infrared thermography technique. Researchers obtained relative deviations between U-values from 1 up to 6% and of 83%, respectively. Bros-Williamson et al. [29] calculated the actual U-value of two façades with theoretical U-values of 0.10 and 0.23 W/m^2·K. Results had relative deviations between theoretical and measured U-values from 10 up to 65%. Finally, Samardzioska and Apostolska [30] studied façades with a theoretical thermal transmittance of 0.22 W/m^2·K, with relative deviations between U-values from 3 up to 59%.

Infrared thermography (*IRT*) is a widely accepted NDT technique that allows users to inspect entire wall areas [31]. The employment of thermographic inspection can be divided into qualitative and quantitative studies [32–34]. In fact, Tejedor et al. [35] highlighted that most studies were qualitative, to discover heterogeneities due to anomalies (moisture, thermal bridges, cracks, air leakages, etc.) below the plaster [36–40] and to define the geometry of a masonry, among other purposes [41]. Besides this, international standards (i.e., ISO 6781:1983 [42]), UNE EN 13187:1998 [43], ASTM E1311 [44], ASTM E1862 [45], and guidelines (i.e., RESNET [32]) only recommend boundary conditions for the use of qualitative *IRT* tests to carry out energy audits in buildings [33,46,47]. In the last decade, few systematic attempts have emerged for developing accurate approaches related to the estimation of in situ U-values of façades by quantitative *IRT*. Nevertheless, the studies tended to analyze the convective and radiative heat-transfer processes of the wall from outside the building [27,28,48–56] instead of inside the building [57–65]. Some researchers pointed out two main constraints in the use of external thermography: (i) the tabulated value of the external convective heat transfer coefficient is considered a precautionary value, since it is computed to estimate the heat loss during the design stage of the building [51]; (ii) the calculated value of the convective heat-transfer coefficient based on Jürge's equation may present a greater variability due to wind characteristics (angle, intensity, and direction) [66–70]; (iii) uncontrolled reflections indexes of surroundings may be given on the target [33,61]. As regards internal thermography, Madding et al. [71] proposed a numerical model where the equation of the specific heat flux by radiation was linearized. Moreover, the convective heat-transfer coefficient was focused on a dimensional approach that was reported in Earle's study [72] and Holman's study [73]. According to Sham et al. [69], the calculation procedure with a dimensionless approach is more accurate. Along this line, Bienvenido–Huertas et al. [74] carried out a comparative analysis among different numerical models based on quantitative internal *IRT* in terms of radiative heat flux: Madding [71], Fokaides et al. [57], and Tejedor et al. [61]. The results did not show significant differences in the use of the three equations. For determining the influence of internal convective heat transfer coefficients (hc) on the thermal characterization of building envelopes using *QIRT*, a cluster analysis was drawn up. The study took into account 25 correlations of temperature differences and 20 correlations of dimensionless numbers. The outcomes demonstrated that the use of dimensionless numbers for computing hc was the most efficient approach for the *QIRT* from inside the building. Concerning the precision of the method, researchers obtained relative deviations between theoretical and measured U-values from 10 to 60% for external thermography and from 2 to 12% for internal thermography. Dall'O et al. [51] stated that the deviation between *HFM* and *QIRT* measured data was greater for well-insulated walls, reaching >50%. As seen, the main benefit of internal *QIRT* is that the practitioner can work under controllable test conditions, obtain a better precision, and avoid unknown reflections of surroundings on the target. Hence, it might be interesting to observe whether passive house façades can be evaluated by internal thermography and whether the results are similar to *HFM* measurements.

Based on the outlined background, the aim of the paper was to assess the use of non-destructive techniques (*HFM* and *QIRT*) for determining the gap between the predicted and actual thermal transmittance of passive house façades in steady-state conditions. With this purpose, two measurement campaigns were carried out. The first one was conducted on an experimental mock-up with controlled climatic conditions (February 2017), to define good practices in the monitoring process for applying the *HFM* average method and the quantitative internal *IRT* technique. The conclusions obtained from this preliminary study were taken into account for the second measurement campaign in a detached house under real environmental conditions (February 2019). This research will help practitioners to select the most appropriate method for determining in situ the actual thermal transmittance of façades with a low U-value.

This paper is organized as follows. Section 2 specifies the research methodology implemented in this paper and describes the opaque walls to be assessed. Section 3 discusses the use of *HFM* and *QIRT* to determine building thermal performance of façades with a high level of insulation. Finally, Section 4 highlights the major contributions of this research.

2. Materials and Methods

The research methodology is represented in Figure 1. Firstly, an experimental mock-up was designed and constructed with prefabricated panels in February 2017, incorporating the passive-house concept in the Mediterranean climate. The façade's U-value was determined by using the heat flux meter (*HFM*) and quantitative internal infrared thermography (*QIRT*) techniques under controlled climatic conditions. Once the applicability of in situ NDT for walls with low U-value had been checked, a passive house was built in 2018 and monitored in February 2019, to carry out a post-construction evaluation of the façade before the operational stage. In both case studies, the in situ measured thermal transmittance was computed following the methods reported in Gaspar et al. [75] and Tejedor et al. [61]. Furthermore, the two measurement campaigns took place during winter, to guarantee a thermal gradient from 10 to 15 °C across the building envelopes. Subsequently, a comparative analysis between theoretical and measured U-values was carried out, considering the coefficient of variation as the common statistical parameter, to evaluate the dispersion of the measurements.

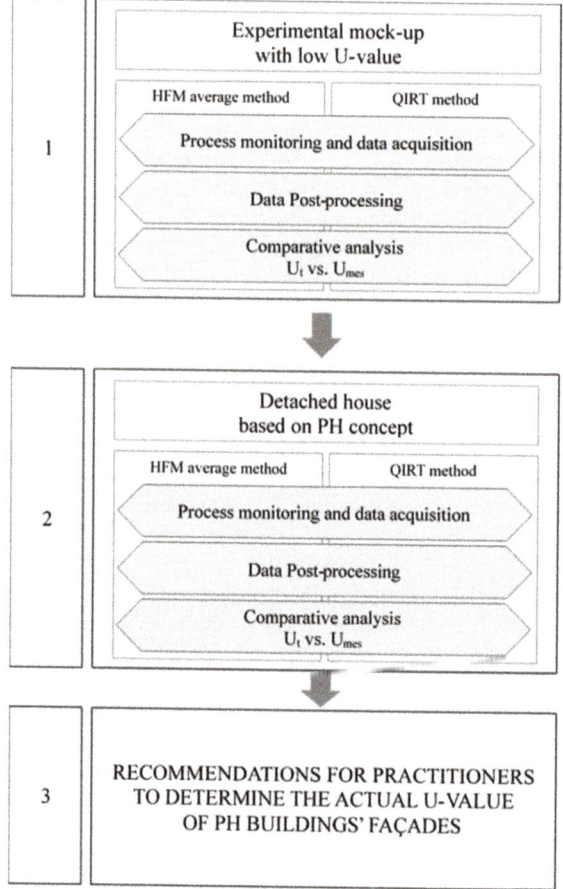

Figure 1. Flowchart of the research methodology.

This section fully describes the measurement setup (Section 2.1), the measuring equipment (Section 2.2), and both case studies (Section 2.3).

2.1. Measurement Setup

The measurement campaigns, which took place in the Mediterranean climate during February 2017 and February 2019, were conducted in an experimental mock-up and a detached house based on the passive-house concept. To ensure the same operating conditions, HFM tests and quantitative IRT tests were performed simultaneously, although with different test durations, depending on the method.

2.1.1. HFM Average Method

The *HFM* average method is a non-destructive method for measuring the thermal transmittance of opaque, plane building elements perpendicular to the heat flow with no significant lateral heat flow. It consists of monitoring the heat flux rate passing through the element (q) and the indoor (T_{IN}) and outdoor (T_{OUT}) environmental temperatures [76]. The *HFM* standardized average method is widely used. It is considered to estimate steady-state conditions well, by monitoring the heat-flow rate and temperatures over an adequately extended duration. According to this method, defined in

ISO 9869-1:2014 [76] and described in Gaspar et al. [75], the thermal transmittance and its combined standard uncertainty ($u_c(U)$) can be obtained by Equations (1) and (2):

$$U_{mes\,HFM}\left[\frac{W}{m^2 \cdot K}\right] = \frac{\sum_{j=1}^{n} q_j}{\sum_{j=1}^{n}(T_{INj} - T_{OUTj})} \qquad (1)$$

$$u_{c\,HFM}^2(U) = \left(\frac{\delta U}{\delta q}\right)^2 \cdot u_c^2(q) + \left(\frac{\delta U}{\delta T_{IN}}\right)^2 \cdot u_c^2(T_{IN}) + \left(\frac{\delta U}{\delta T_{OUT}}\right)^2 \cdot u_c^2(T_{OUT}) =$$
$$= \left(\frac{1}{T_{IN}-T_{OUT}}\right)^2 \cdot u_c^2(q) + \left(\frac{-q}{(T_{IN}-T_{OUT})^2}\right)^2 \cdot u_c^2(T_{IN}) + \left(\frac{q}{(T_{IN}-T_{OUT})^2}\right)^2 \cdot u_c^2(T_{OUT}) \qquad (2)$$

where q is the density of the heat-flow rate per unit area in [W/m²], T_{IN} is the environmental indoor temperature in [K], T_{OUT} is the environmental outdoor temperature in [K], the index j enumerates the individual measurements, $u(q)$ is the uncertainty associated with the heat-flow-rate-measuring equipment, $u(T_{IN})$ is the uncertainty associated with the interior environmental temperature measuring equipment, and $u(T_{OUT})$ is the uncertainty associated with the exterior environmental temperature measuring equipment.

The HFM average method is a standardized method described in ISO 9869-1:2014 [76]. This standard establishes that the duration of the test depends on the values obtained during the course of the test. In this sense, the standard defines three conditions that must be met simultaneously to end the monitoring process: (a) the test must have a minimum duration of 72 h or more, (b) the thermal transmittance obtained when the test finalized must not differ more than 5% from the value obtained 24 h before, and (c) the thermal transmittance obtained by analyzing data from the first time period of INT(2·DT/3) days must not differ more than 5% from the value obtained from the data for the last period of the same duration, where DT is the lasting of the test in days.

Moreover, the variability of results (random error $e(\%)$) was estimated with a 95.4% confidence level, according to Atsonios et al. [77]. The coefficient of variation of the resulting U-values obtained following the HFM average method was calculated with Equations (3) and (4) [77,78]:

$$CV_{HFM}[\%] = \sqrt{\frac{\sum_{i}^{n}(U\_HFM\_mes_i - \overline{U\_HFM\_mes})^2}{n-1}} \times \frac{1}{\overline{U\_HFM\_mes}} \times 100 \qquad (3)$$

$$e(\%) = 2 \times CV_{HFM}(\%) \qquad (4)$$

where $U\_HFM\_mes_i$ is the value of thermal transmittance of the façade in the cycle i, $\overline{U\_HFM\_mes}$ is the average of Um-values of the façade during n cycles, and n is the number of cycles (n = 3). According to the ASTM C1155-95 Standard [78], the coefficient of variation is expected to be less than 10% for the HFM average method.

2.1.2. Quantitative Internal IRT Method

In contrast to the HFM average method, quantitative internal IRT assumes that the façade is crossed by one-dimensional horizontal specific heat flux (q) resulting from radiation (q_r) and convection (q_c) processes under a stationary regime. According to the method extensively reported in Tejedor et al. [61], the instantaneous and average measured U-values [W/m²·K] can be defined by Equations (5) and (6):

$$U_{mes_i\,IRT}\left[\frac{W}{m^2 \cdot K}\right] = \frac{\left\{0.825 + \frac{0.387 \cdot R_a^{\frac{1}{6}}}{\left[1+\left(\frac{0.492}{Pr}\right)^{\frac{9}{16}}\right]^{\frac{8}{27}}}\right\}^2 \cdot \lambda_{air}}{L} \cdot [T_{IN} - T_{WALL}] + \varepsilon_{WALL} \cdot \sigma \cdot [T_{REF}^4 - T_{WALL}^4]}{(T_{IN} - T_{OUT})} \qquad (5)$$

$$U_{mes_{avg}\ IRT}\left[\frac{W}{m^2 \cdot K}\right] = \frac{\sum_{i=1}^{n}(q_{ri} + q_{ci})}{\sum_{i=1}^{n}(T_{INi} - T_{OUTi})} = \frac{\sum_{i=1}^{n} U_{mesi}}{n}. \tag{6}$$

where T_{IN} and T_{OUT} denote the inner and outer ambient air temperatures in [K], respectively, T_{REF} is the reflected ambient temperature in [K], T_{WALL} represents the wall-surface temperature in [K], ε_{WALL} refers to the wall surface emissivity, σ is the Stefan–Boltzmann's constant with a value of 5.67×10^{-8} [W/m^2·K^4], λ_{air} is defined as the air thermal conductivity measured in [W/m·K], L is the wall height seen from the internal side of the prefabricated wall in meters, and n is the total number of thermograms. As regards Rayleigh (Ra) and Prandtl (Pr) numbers, they are dimensionless parameters for a laminar flow assuming that the wall is a vertical plate. It should be pointed out that Pr is set at 0.73 for dry air under atmospheric pressure and T_{IN} = 20–25 °C, while the Rayleigh number is computed by Equation (7).

$$Ra = G_r \cdot P_r = \frac{g \cdot \beta \cdot (T_{IN} - T_{WALL}) \cdot L^3}{v^2} \cdot P_r \tag{7}$$

The parameters that define Ra are the following: the Grashof number (Gr), the gravitation ($g = 9.8$ m/s^2), the volumetric temperature expansion coefficient ($\beta = 1/T_m$ where T_m is the average value of T_{IN} and T_{WALL}), and the air viscosity (v $1.4 \cdot 10^{-5}$ m^2/s for T_{IN} = 0–15 °C or v=1.5·10^{-5} m^2/s for T_{IN} = 15–25 °C).

In terms of operating conditions, the recommendations defined by Tejedor et al. [61] should be taken into account, to guarantee the implementation of the quantitative internal IRT method in the mock-up and the detached house. Based on this, the measurements were performed in the early morning (from 6 a.m. to 9 a.m.), to ensure the temperature difference through the wall and to avoid the incident solar radiation as an external stimulus. For a thermal gradient between 7 and 16 °C, Tejedor et al. [63] demonstrated that the variance of the measured thermal transmittance could be only predicted by changes of T_{OUT}. The parameters T_{IN}, T_{REF}, and T_{WALL} often remain constant during the monitoring process.

Finally, the variability of the results was assessed by calculating the standard deviation (SD) and the coefficient of variation (CV) among the instantaneous measurements of each test for the wall of the experimental mock-up W1 and the wall of the detached house W2 (Equations (8) and (9)). Gaspar et al. [75] stated that the impact of CV was greater on walls with low U-values. Subsequently, the 95% confidence intervals were also computed following Equation (10), to analyze whether there was a relevant difference among $U_{mes\ avg}$ resulting from quantitative internal IRT.

$$SD\left[\frac{W}{m^2 \cdot K}\right] = \sqrt{\frac{\sum_{i=1}^{n}(U_{mesi\ IRT} - U_{mes_{avg}\ IRT})^2}{n}} \tag{8}$$

$$CV_{IRT}(\%) = (SD/U_{mes\ avg\ IRT}) \cdot 100 \tag{9}$$

$$CI(95\%) = \overline{U} \pm 1.96 \cdot \frac{\sigma}{\sqrt{n}} \tag{10}$$

2.1.3. Validation of Methods

To validate the implementation of both methods, all measurements were compared with the theoretical thermal transmittance (also termed as nominal design data—Equation (11) in accordance with previous studies [21,26–30,61,63,75,79]. In fact, Ficco et al. [23] defined four approaches to estimate U_t and to check the QIRT results: (i) use data from historical analysis, (ii) calculate nominal design data, (iii) determine the actual U-value through endoscopy, and (iv) collect in situ data by using a standardized method (i.e., HFM). For this purpose, the Spanish Technical Building Code [80] and European Standards such as UNE EN ISO 10456:2012 [81] and UNE EN ISO 6946:2012 [82] were considered. Therefore, the adjustment between theoretical and measured U-values can be expressed as the absolute value of the relative difference between theoretical and measured U-values (Equation (12)):

$$U_t = \frac{1}{R_t} = \frac{1}{R_{Si} + \sum_{i=1}^{n} \frac{\Delta x_i}{\lambda_i} + R_{Se}} \tag{11}$$

$$\Delta U/U_t \, (\%) = |(U_{mes} - U_t)/U_t| \cdot 100 \tag{12}$$

where R_T is the theoretical total thermal resistance [m²·K/W], R_{si} and R_{se} refer to the interior and exterior superficial resistance for horizontal heat flux (0.13 and 0.04 m²·K/W, respectively), Δxi is the thickness of the layer in meters, λi is the thermal conductivity of the layer [W/(m·K)], U_t is the theoretical thermal transmittance of the façade [W/m²·K], and U_{mes} is the measured thermal transmittance of the façade [W/m²·K].

2.2. Measuring Equipment

The technical specifications of the measuring equipment used for NDT tests are presented in Table 1. For the *HFM* tests, the inner equipment layout included three heat flux meter plates (HFP01, Hukseflux) with a thickness of 5.0 mm, a diameter of 80.0 mm, and a guard made of a ceramic–plastic composite, with an inside air temperature sensor (107, Campbell Scientific Inc., Barcelona, Spain) consisting of a thermistor encapsulated in epoxy-filled aluminum housing, both connected to an acquisition system (CR850, Campbell Scientific Inc., Barcelona, Spain) consisting of measurement electronics encased in a plastic shell with an integrated wiring panel with external power supply. The outer equipment consisted of an air temperature sensor and its acquisition system (175T1, Instrumentos Testo, SA, Barcelona, Spain). The optimal location of sensors was investigated with qualitative *IRT*, according to the procedures reported in ISO 6781:2015 [42] and UNE EN 13187:1998 [43]. In this way, it was possible to avoid unknown heterogeneities or disturbances (i.e., corners, the vicinity of junctions, direct solar radiation, and the direct influence of a heating unit) [22,34,61,76]. The transducers were installed on the internal side of the prefabricated panels, to achieve the most stable operating conditions before and during the tests. Data loggers were configured to store the 30-minute averaged data in their memories, considering a sampling frequency of 1 s and a total test duration of 7 days.

Table 1. Main technical specifications of the equipment.

Equipment	Output	Measuring Range	Resolution	Accuracy
Heat flux meter plate	q [W/m²]	±2000 W/m² Sensitivity of HFM1 61.68 μV/(W/m²) Sensitivity of HFM2 61.29 μV/(W/m²) Sensitivity of HFM3 63.07 μV/(W/m²)	–	±5%
Inner air temperature sensor	T_{IN} (K)	−35 to +50 °C	0.1 °C	±0.5 °C
Inner acquisition system		Input ±5 Vdc at 0 to 40 °C	–	±0.06% of reading
Infrared camera	T_{WALL} (K) T_{REF} (K)	Temperature: −20 to +120 °C CFOV: 25° × 19°; IFOV: 1.36 mrad Spectral Range: 7.5–13 μm Thermal sensitivity: <0.045 °C, at 30 °C Sensor: FPA, uncooled microbolometer	320 × 240 pixels	±2 °C or ±2% reading
Outer air temperature sensor	T_{OUT} (K)	−35 to +55 °C	0.1 °C	±0.5 °C

HFM, heat flux meter; q, the element; T_{IN}, the indoor environmental temperature; T_{WALL}, the wall-surface temperature; T_{REF}, the reflected ambient temperature; T_{OUT}, the outdoor environmental temperature.

Quantitative infrared data (instantaneous wall surface temperature–T_{WALL}- and reflected ambient temperature–T_{REF}-) were recorded by using an IR camera (FLIR E60bx) and FLIR TOOLS+ Software [83].

Hence, each IRT test involved the post-processing of a sequential video with 120 or 180 thermograms. The wall surface emissivity (ε_{WALL}) was estimated at 0.88, for all walls, by means of aluminum crinkled foil (0.20 m × 0.15 m). The IRT measurements were performed over a period of 2–3 h, with a data acquisition interval of 1 min. For both NDTs, the inner and outer ambient air temperatures were monitored by using the same integrated sensors (107, Campbell Scientific, Inc.; 175T1, Instrumentos Testo SA). Notably, all the equipment was placed at 1.5 m above the ground floor, and the distance between the IR camera and target was set at 1 m. According to Tejedor et al. [61], the height of the walls is around of 2.5–3 m. Hence, a height of 1.5 m could be acceptable to obtain a mean value of the ambient air temperature inside the building and to avoid unknown reflection indexes attributed to the furniture or the ground. The instrumentation used is shown in Figure 2.

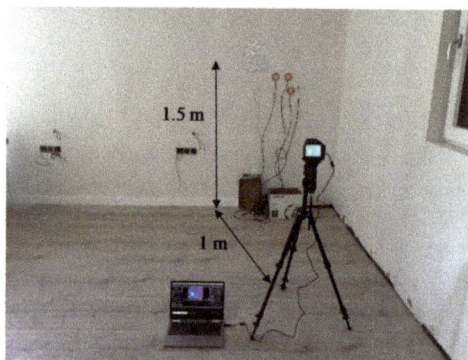

Figure 2. View of the measuring equipment used during the acquisition-of-data process.

2.3. Case Studies

The following procedure was devised to illustrate the applicability of in situ measurement techniques to determine the thermal transmittance of highly insulated façades. Firstly, the experimental mock-up ($A = 12$ m^2) with controlled climatic conditions was erected in January 2017, in order to determine good practices in the monitoring process of constructive solutions in the Mediterranean climate (Figure 3). In this way, it was possible to test a specimen that could ensure four of the six principles that define a passive house (PH): a high level of insulation, a minimum number of thermal bridges, highly efficient windows, and a high level of airtightness. The façade construction system consisted of a galvanized steel structure with prefabricated panels whose internal configuration was (from outside to inside) as follows: mortar, lightweight concrete, polyisocyanurate insulation (PIR), non-ventilated air cavity, lightweight concrete, and gypsum plaster. The main thermo-physical parameters are reported in detail in Table 2. Once the mock-up was erected, installations and finishes were carefully incorporated.

Table 2. Configuration and technical features of the experimental mock-up (from outside to inside).

	N#	Material Layer	Δx_i (m)	λ_i [W/(m·K)]	$R_{t\,i}$ [(m^2·K)/W]	L (m)	U_t [W/(m^2·K)]
	1	Mortar	0.015	0.550	0.027		
W1	2	Lightweight concrete	0.060	0.160	0.375		
Experimental	3	PIR insulation	0.080	0.028	2.857	2.64	0.245
mock-up	4	Non-ventilated air cavity	0.060	—	0.180		
	5	Lightweight concrete	0.070	0.160	0.438		
	6	Gypsum plaster	0.015	0.430	0.035		

Δx_i, thickness of the layer; λ_i, thermal conductivity of the layer; $R_{t\,i}$, theoretical thermal resistance of the layer; L, height of the wall; U_t, theoretical thermal transmittance of the building façade.

Figure 3. Experimental mock-up based on the passive house concept.

Having implemented the *HFM* and quantitative *IRT* as a tool of decision-making in terms of built quality, and taking into account the conclusions of the preliminary study, the wall was reproduced in the real built environment in February 2019 (Figure 4). The differences between specimens *W1* (experimental mock-up) and *W2* (detached house) were given by the thickness and the thermal resistance of the material layers. The technical features are shown below (Table 3). The detached house ($A = 322$ m^2) was also characterized by high efficient doors ($U_D = 0.93$ W/m^2·K) and low-emissivity triple-glazing windows ($U_W = 0.99$ W/m^2·K) that allowed us to achieve an air renovation <0.56 h^{-1}. As regards the building facilities, the mechanical ventilation was focused on a heat-recovery system (model ComfoAir Q450, Zehnder) with an efficiency of 95%, and the indoor environmental conditions were guaranteed by an aerothermal equipment of 6 kW (model HPSU, ROTEX).

Figure 4. Detached house based on the passive house concept.

Table 3. Configuration and technical features of the detached house (from outside to inside).

	No.	Material Layer	Δx_i (m)	λ_i [W/(m·K)]	$R_{t\,i}$ [(m^2·K)/W]	L (m)	U_t [W/(m^2·K)]
W2 Detached house	1	Mortar	0.015	0.550	0.027	2.50	0.233
	2	Lightweight concrete (EVOin)	0.045	0.160	0.281		
	3	PIR insulation	0.080	0.028	2.857		
	4	Cavity (EPS + EVOin)	0.060	0.140	0.429		
	5	Lightweight concrete (EVOin)	0.080	0.160	0.500		
	6	Gypsum plaster	0.015	0.430	0.035		

Δx_i, thickness of the layer; λ_i, thermal conductivity of the layer; $R_{t\,i}$, theoretical thermal resistance of the layer; L, height of the wall; U_t, theoretical thermal transmittance of the building façade.

3. Results

Two heavyweight walls with low U-value were monitored and evaluated in a stationary regime. The comparative analysis of techniques took into account the following: (i) the theoretical U-value, (ii) the U-value measured by the heat flux meter, (iii) the U-value measured by the quantitative internal infrared thermography, (iv) the test duration, and (v) the temperature gradient between inside and outside the building. The following statistical parameters were also estimated: the average value of thermal transmittance [W/(m²·K)], the standard deviation (SD) [W/(m²·K)], the coefficient of variation (CV) (%) of the measurements, and the absolute value of the relative difference between theoretical and measured U-values (%).

3.1. HFM Average Method

Both the experimental mock-up and the detached house were monitored for 168 h, obtaining 337 datasets of readings for each case. The experimental campaign of the building mock-up was conducted with a heater, with an average air temperature difference much higher than 15 °C [79]. However, the experimental campaign of the detached house was performed under actual environmental conditions, which ensured that the average temperature difference was greater than 10 °C [20,21] (Figure 5). Hence, the temperature during process monitoring was not a significant factor that influenced the results of the second case study. Measurement periods, as well as the average temperatures differences, are detailed in Table 4.

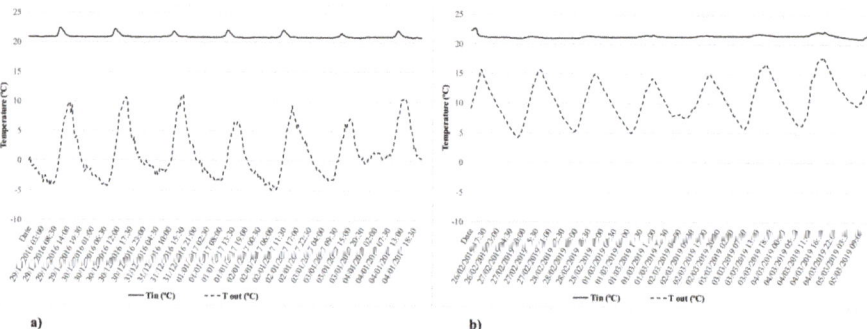

Figure 5. Temperature data obtained from the process of monitoring for both (**a**) the experimental mock-up and (**b**) the detached house.

Table 4. Measured U-values and their associated combined uncertainties, deviation between theoretical and measured U-value, and coefficient of variation of the measured thermal transmittance in both the experimental mock-up and the detached house, using the *HFM* average method.

		Experimental Mock-Up			Detached House		
		$U_{mHFM} \pm u_c(U)$ [W/m²·K]	$\left\|\frac{(U_t - U_m)}{U_t}\right\|$ (%)	CV (%)	$U_{mHFM} \pm u_c(U)$ [W/m²·K]	$\left\|\frac{(U_t - U_m)}{U_t}\right\|$ (%)	CV (%)
Number of cycles (duration)	1 (24 h)	0.246 ± 0.018	0.33	–	0.212 ± 0.018	8.84	–
	2 (48 h)	0.244 ± 0.013	0.35	–	0.225 ± 0.014	3.47	–
	3 (72 h)	0.248 ± 0.011	1.05	1.39	0.225 ± 0.012	3.06	6.80
	4 (96 h)	0.246 ± 0.009	0.44	1.40	0.227 ± 0.010	2.37	1.15
	5 (120 h)	0.249 ± 0.008	1.59	1.13	0.227 ± 0.009	2.33	0.84
	6 (144 h)	0.250 ± 0.008	1.98	1.57	0.230 ± 0.008	1.13	1.43
	7 (168 h)	0.251 ± 0.007	2.41	0.81	0.225 ± 0.008	3.09	2.02
Measurement period		From 29 December 2016 to 4 January 2017			From 26 February 2019 to 4 March 2019		
ΔT average (°C)		19.7			10.8		

The measured thermal transmittances and their related uncertainty were calculated in 7 consecutive cycles of 24 h (Equations (1) and (2)), using accumulative data. Thus, the first cycle was calculated with data obtained during the first 24 h tested, the second cycle with data from the first 48 h tested, and successively for each cycle, up to the seventh day. The results obtained are depicted in Figure 6 and detailed in Table 4, where Um-HFM ± U(C) is the measured thermal transmittance, using the HFM average method and its associated uncertainty. It is noteworthy that uncertainties of measurement decreased as the tests were extended and when the average air temperature difference increased, which is in line with the results obtained by Asdrubali et al. [21] and Nardi et al. [28].

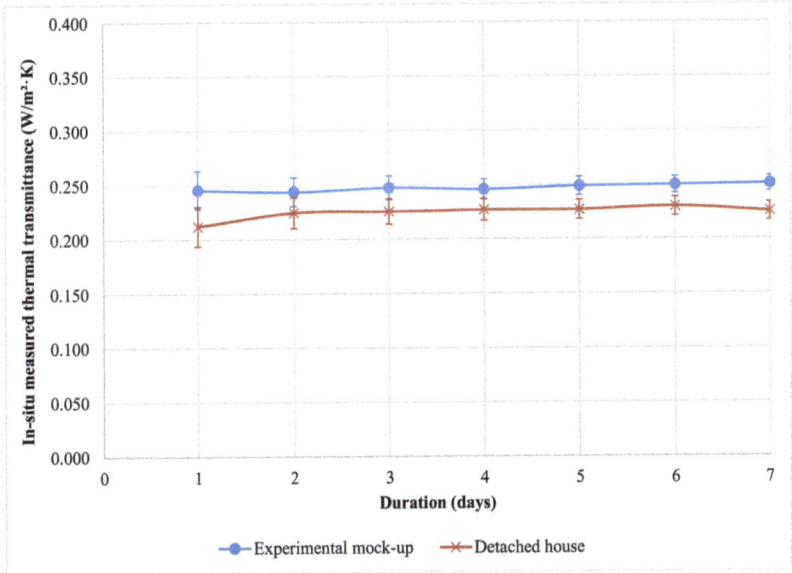

Figure 6. Measured U-values and their associated combined uncertainties in the experimental mock-up and the detached house for the seven cycles of test duration, using the HFM average method.

The minimum test duration for the experimental mock-up and the detached house was 72 h, in view of the fact that the second condition consisting of limiting the deviation between the U-value obtained at the end of the test from the value obtained 24 h earlier to 5% was fulfilled (1.4% and 0.4% respectively, as shown in Table 5), and that the third condition consisting of limiting the deviation between the U-value obtained by analyzing data from an initial period from the value obtained from data for the last period of the same duration to 5% was also fulfilled in both cases (2.3% and 1.5%, respectively, as shown in Table 6). Then, the variability of the results was verified, in both cases, through the coefficient of variation of the resulting U-values (Equations (3) and (4)). It was confirmed that the test could be ended at 72 h in both cases, since the coefficients of variation were lower than expected from 72 h of testing (Figure 7).

Table 5. Deviation of the U-value from the value obtained 24 h earlier, for both the experimental mock-up and the detached house, using the *HFM* average method.

Number of Cycle Evaluated	Related Cycles in the Evaluation	Deviation in $U_{m\text{-}HFM}$ (%)	
		Experimental Mock-Up	Detached House
3 (72 h)	3 vs. 2	1.4	0.4
4 (96 h)	4 vs. 3	−0.6	0.7
5 (120 h)	5 vs. 4	1.1	0.0
6 (144 h)	6 vs. 5	0.4	1.2
7 (168 h)	7 vs. 6	0.4	−2.0

Table 6. Deviation of U-values between the initial and final analysis period for each test duration for both the experimental mock-up and the detached house, using the *HFM* average method.

Number of Cycle Evaluated	Duration of the Analysis Period (Days)	Analysis Period	Experimental Mock-UP		Detached House	
			$U_{m\text{-}HFM}$ [W/m²·K]	Deviation of $U_{m\text{-}HFM}$ between Periods of Analysis (%)	$U_{m\text{-}HFM}$ [W/m²·K]	Deviation of $U_{m\text{-}HFM}$ between Periods of Analysis (%)
3 (72 h)	2	Initial test period	2.34	2.3	2.34	1.5
		Final test period	2.40		2.37	
4 (96 h)	2	Initial test period	2.34	0.8	2.34	0.6
		Final test period	2.33		2.35	
5 (120 h)	3	Initial test period	2.36	0.3	2.35	1.1
		Final test period	2.37		2.37	
6 (144 h)	4	Initial test period	2.33	2.4	2.34	1.7
		Final test period	2.39		2.38	
7 (168 h)	4	Initial test period	2.33	2.2	2.34	2.0
		Final test period	2.39		2.39	

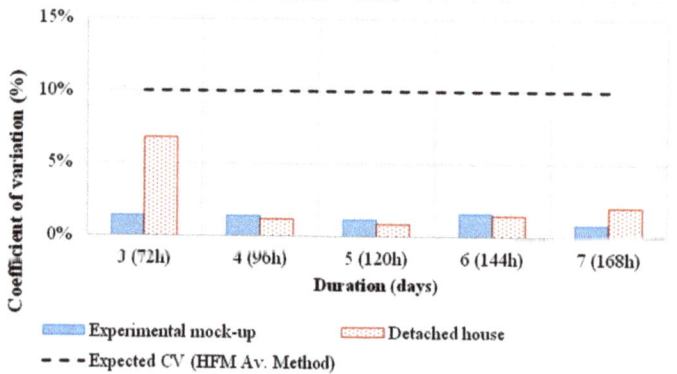

Figure 7. Coefficient of variation of both the experimental mock-up and the detached house, using the HFM average method.

The measured thermal transmittance in both the experimental mock-up and the detached house was closely adjusted to the theoretical transmittance value, with absolute values of relative differences of less than 5%. However, it can be observed that the measured U-value of the experimental mock-up fit better with the theoretical U-value (1.05%) than the one of the detached house (3.06%) (Table 4). This difference in fitting could be due to two situations. Firstly, the theoretical thermal transmittance of the experimental mock-up was greater than that of the detached house. Secondly, conditions for conducting in situ tests were more optimal during the monitoring process of the experimental mock-up than those of the detached house. Consequently, the average temperature difference between indoor and outdoor environments was almost 9 °C greater during the testing process of the experimental mock-up than that of the detached house.

3.2. Quantitative Internal IRT Method

Previous studies have demonstrated that highly insulated walls (U-value < 0.400 W/m²·K) and low heat capacity per unit of area (~200 kJ/m²·K) might be more difficult to evaluate [63]. However, prefabricated panels based on the passive house concept had not been assessed by means of quantitative internal *IRT*. As seen in Section 2.3, the external opaque walls of the detached house were designed practically equal to the building envelopes of the experimental mock-up. In fact, the theoretical thermal transmittances were 0.245 and 0.233 W/m²·K, respectively. Concerning post-construction evaluation by thermography, the results revealed that the gap between the design nominal data and the real thermal performance was found to be 5% in both case studies (Table 7 and Figure 8). Hence, the proposed façade-building system of the mock-up fulfilled the expected thermal behavior in the real built environment. Despite this, a greater fluctuation of the measured thermal transmittance was observed for the detached house (Figure 8).

Table 7. Measured U-values, deviation between theoretical and measured U-value, and coefficient of variation of the measurements in the experimental mock-up and the detached house, using internal quantitative infrared thermography (QIRT).

Parameters	Experimental Mock-Up	Detached House
U_t [W/m^2·K]	0.245	0.233
U_{mes_IRT} [W/m^2·K]	0.257	0.245
Test duration (hours)	2	2
ΔT Range (°C)	8.2–10.4	13.17–15.76
SD [W/m^2·K]	0.021	0.024
CV (%)	8.06	9.69
$\Delta U_{IRT}/U_t$ (%)	4.90	5.15

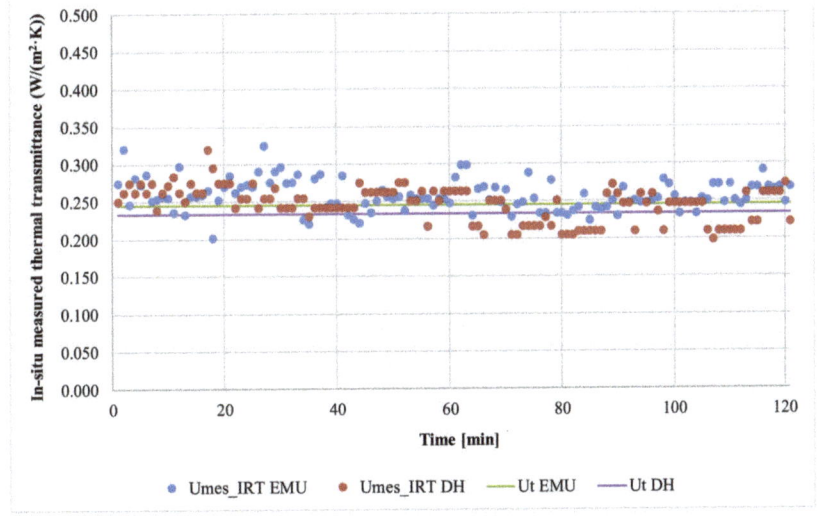

Figure 8. In situ measured thermal transmittance over time for W1 and W2 and comparison with theoretical values.

From the analysis of variability, it can be extrapolated that standard deviations were slightly similar (0.021 W/m^2·K for W1 and 0.024 W/m^2·K for W2) and the coefficients of variation (*CV*) were less than 10%. However, the plot of confidence intervals at 95% for *IRT* measurements (Figure 9) shows some differences between *W1* and *W2*. It should be noted that a box plot indicates if a statistical datum is normally distributed (the median of the data is located in the middle of the box, and the quartiles are symmetric) or skewed. (the median of the data is closer to the bottom or to the top of the box, and the distribution of instantaneous measurements is asymmetric). In the case of the experimental mock-up, the minimum and maximum U-values were 0.202 and 0.324 W/m^2·K (including outliers). A positive skew was also observed, since most of the instantaneous measurements fell in the upper quartile. As regards the three outliers (18, 2, and 27), they corresponded to three instantaneous measurements that were numerically distant from the other observations. Despite being outside of the whiskers of the box plot, these outliers would not affect the results because 121 data points are obtained for each test.

In the case of the detached house, the minimum and maximum U-values were similar to *W1* (0.198 and 0.320 W/m^2·K, respectively). Nevertheless, the box of *W2* had a wider interquartile range (*IQR*), and the distribution of the data was negatively skewed because the measurements were concentrated in the lower quartile. Taking into account that the internal assembly of the specimens was slightly similar, the spread of the data could be attributed to changes in the temperature gradient between inside and outside the building.

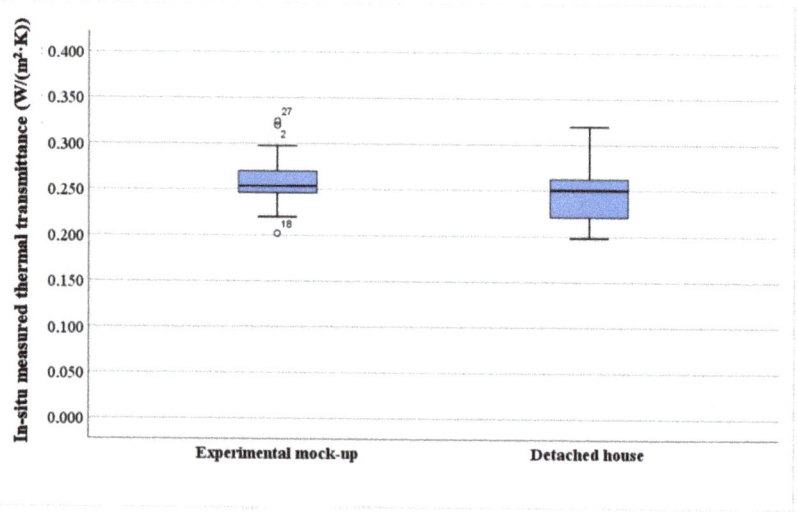

Figure 9. Confidence intervals at 95% for quantitative IRT measurements.

4. Discussion

The results of this research revealed that in situ measurement techniques can be useful to assess the build quality of a new passive house before its operational stage. Indeed, they allow us to quantify the gap between the design or modeling and the actual work. However, it is a proof of concept for both non-invasive techniques, and, consequently, further research is required. For the first time, thermography was implemented in a quantitative way, to determine the in situ thermal transmittance of passive house façades. Previous studies demonstrated that the optimal ΔT of quantitative *IRT* ranged between 7 and 10 °C for conventional heavy multi-leaf walls (U > 0.362 W/m^2·K) [63]. In the current research, the above statement could also be assumed. Besides this, this technique is characterized to be faster than *HFM*, since the *IRT* test only takes 2 h. In contrast, *HFM* offers better results for PH façades in terms of accuracy and dispersion of the instantaneous measurements. The deviation between theoretical and measured U-values was found to be 1.05% (*W1*) and 3.06% (*W2*) for the *HFM*, compared to ~5% for *QIRT*. The coefficient of variation ranged from 1.39 (*W1*) to 6.80 (*W2*) for *HFM*, while the values for the thermographic analysis were found to be 8.06% (*W1*) and 9.69% (*W2*). Notably, the good *HFM* results of the experimental mock-up could be attributed to the high temperature gradient (~19 °C) that was achieved during the measurement campaign.

Determining the actual thermal transmittance of low U-value façades when weather conditions are favorable and temperature differences between the inner and outer environments are greater than 10 °C requires a minimum test duration of 72 h, using the *HFM* average method, and 2 h of using the *QIRT* method. The first condition implies that practitioners should check the short-term weather forecast before starting the test. The second condition represents a limitation in the period of time in which the test can be performed, especially when the *HFM* method is used. In the Mediterranean climate, due to the need to obtain large temperature differences between environments, the test should be scheduled for the coldest months (normally from October to April).

In terms of execution of the method, the practitioner should have experience in conducting measurement campaigns in the real built environment. In both methods, the tests should be performed on northern façades and preferably in the early morning, with a low wind speed (<1 m/s), since wind and incident solar radiation are adopted as external thermal stimuli that could alter convective factors or lead to a time lag of a few hours [61,75]. Despite this, some differences between *HFM* and *QIRT* were highlighted. The installation of the equipment for the *HFM* monitoring process may be easier for professionals to carry out by following the ISO recommendations. However, a prior inspection, with the help of qualitative *IRT*, is needed to detect the ideal location of heat-flow meters. In addition, *HFM* could be considered more invasive. To guarantee proper thermal contact between the transducer and the wall, a layer of thermal interface material (with a thickness of less than 0.01 m) should be applied. As regards the installation of equipment for the *QIRT*, the monitoring process requires practitioners with specific training to calibrate the IR camera in accordance with the target and to achieve a reliable sequential video recorded during the test. Parameters such as the emissivity of the wall or the reflected ambient temperature can be adjusted during post-processing with the specific software, but the wall area or the distance between the IR camera and the specimen cannot be changed later. Furthermore, the practitioner should follow some recommendations: (i) avoid measuring near corners since heated adjacent walls could impact on the results; (ii) pay attention to internal climatic conditions, because some air current peaks can lead to a non-homogeneity of heat flux and temperature on the wall; and (iii) leave the wall free of furniture, to remove uncontrolled reflection indexes.

In terms of data analysis, the *HFM* average method can be performed by using tools readily available to technicians, such as a simple spreadsheet. In contrast, advanced tools are required for data processing when the *QIRT* method is used. The technician needs to assess the 120 thermal images with the specific software (i.e., FLIR TOOLS+) and transfer the readings of wall parameters (T_{WALL}, T_{REF}, and ε_{WALL}) to a spreadsheet, where the numerical model is computed, taking into account measured environmental parameters, among other aspects. Nevertheless, some soft skills are common to both NDTs: (i) thermophysical knowledge about building materials (with or without anomalies), especially to implement the method or to interpret thermograms; and (ii) knowledge of data processing and statistical treatment for the visualization of results.

5. Conclusions

Passive construction is expected to increase due to European environmental regulations and the sensibility of police makers related to building energy consumption. For this reason, the use of NDT as a diagnosis tool could be essential to verify the compliance with design technical specifications. Along this line, the main contribution of this research is the analysis of the applicability of non-destructive techniques for the post-construction evaluation of thermal transmittance of passive house façades in steady-state conditions.

According to the literature review, the heat flux meter method is a non-destructive technique standardized by ISO 9869-1:2014 [76]. Currently its use in PH is not very widespread, even in the Mediterranean climate. Indeed, the literature review showed that some constraints of applying the *HFM* average method can arise when the practitioner wants to verify the compliance with technical specifications of projects in façades with low U-value. One of the main constraints is the high deviations obtained between the measured and its corresponding theoretical U-value, as can be seen in previous studies. According to Ficco et al. [23] and Nardi et al. [28], these deviations are more pronounced when climatic conditions are not stable during the monitoring process. Another constraint is the fact that some testing parameters for the monitoring process are not fully specified for measuring façades with low U-value by the Standard ISO 9869-1:2014 [76], such as temperature difference between indoor and outdoor environment and the test duration. Concerning the quantitative *IRT*, a significant reduction of the prices of IR cameras [33] led the researchers to put their efforts in the development of accurate approaches related to the thermal characterization of building envelopes. Nevertheless, a gap still exists in the implementation of *QIRT* in passive houses.

In the current research, the assessment of an experimental mock-up with controlled climatic conditions allowed to check the possible suitability of the NDT in highly insulated walls and to define good practices for a detached house based on PH concept under real environmental conditions. Based on the analysis of results and the respective discussion, this study demonstrated that the *HFM* average method and quantitative internal *IRT* can be executed successfully for determining the in situ U-value of passive house façades. Despite this, neither one of the NDTs was implemented in warmer climates or conducted in summer. Hence, the applicability of the *HFM* average method and the quantitative internal thermography has not been demonstrated under these mentioned boundary conditions, and, consequently, further research is needed. Furthermore, the practitioner should choose the method, depending on test duration, thermal gradient, experience in measurement campaigns, and knowledge in data analysis.

Author Contributions: Conceptualization, K.G. and B.T.; methodology, K.G. and B.T.; software, B.T.; validation, K.G. and B.T.; formal analysis, K.G. and B.T.; investigation, K.G. and B.T.; resources, M.C.; data curation, M.C.; writing—original draft preparation, K.G. and B.T.; writing—review and editing, M.G.; visualization, M.G.; supervision, M.C.; project administration, M.C; funding acquisition, M.C. All authors have read and agreed to the published version of the manuscript.

Funding: This research was partially funded by the Government of Catalonia, Research Grant 2017–SGR–227.

Acknowledgments: The authors acknowledge Evowall Technology for providing the experimental mock-up and the detached house and facilitating the monitoring process.

Conflicts of Interest: The authors declare no conflict of interest. The funders had no role in the design of the study; in the collection, analyses, or interpretation of data; in the writing of the manuscript; or in the decision to publish the results.

References

1. Kylili, A.; Fokaides, P.A.; Jimenez, P.A.L. Key Performance Indicators (KPIs) approach in buildings renovation for the sustainability of the built environment: A review. *Renew. Sustain. Energy Rev.* **2016**, *56*, 906–915. [CrossRef]
2. Wang, Y.; Du, J.; Kuckelkorn, J.M.; Kirschbaum, A.; Gu, X.; Li, D. Identifying the feasibility of establishing a passive house school in Central Europe: An energy performance and carbon emissions monitoring study in Germany. *Renew. Sustain. Energy Rev.* **2019**, *113*, 109256. [CrossRef]
3. Johnston, D.; Siddall, M.; Ottinger, O.; Peper, S.; Feist, W. Are the energy savings of the passive house standard reliable? A review of the as-built thermal and space heating performance of passive house dwellings from 1990 to 2018. *Energy Effic.* **2020**, *13*, 1605–1631. [CrossRef]
4. Rohdin, P.; Molin, A.; Moshfegh, B. Experiences from nine passive houses in Sweden—Indoor thermal environment and energy use. *Build. Environ.* **2014**, *71*, 176–185. [CrossRef]
5. Dalbem, R.; Grala da Cunha, E.; Vicente, R.; Figueiredo, A.; Oliveira, R.; Baptista da Silva, A.C.S. Optimisation of a social housing for south of Brazil: From basic performance standard to passive house concept. *Energy* **2019**, *167*, 1278–1296. [CrossRef]
6. Piccardo, C.; Dodoo, A.; Gustavsson, L. Retrofitting a building to passive house level: A life cycle carbon balance. *Energy Build.* **2020**, *223*, 110135. [CrossRef]
7. Mihai, M.; Tanasiev, V.; Dinca, C.; Badea, A.; Vidu, R. Passive house analysis in terms of energy performance. *Energy Build.* **2017**, *144*, 74–86. [CrossRef]
8. Wang, R.; Lu, S.; Feng, W. A three-stage optimization methodology for envelope design of passive house considering energy demand, thermal comfort and cost. *Energy* **2020**, *192*, 116723. [CrossRef]
9. Tokarik, M.S.; Richman, R.C. Life cycle cost optimization of passive energy efficiency improvements in a Toronto house. *Energy Build.* **2016**, *118*, 160–169. [CrossRef]
10. Ekström, T.; Bernardo, R.; Blomsterberg, A. Cost-effective passive house renovation packages for Swedish single-family houses from the 1960s and 1970s. *Energy Build.* **2018**, *161*, 89–102. [CrossRef]
11. Kylili, A.; Ilic, M.; Fokaides, P.A. Whole-building Life Cycle Assessment (LCA) of a passive house of the sub-tropical climatic zone. *Resour. Conserv. Recycl.* **2017**, *116*, 169–177. [CrossRef]

12. Wang, Y.; Kuckelborn, J.; Zhao, F.-Y.; Spliethoff, H.; Lang, W. A state of art of review on interactions between energy performance and indoor environment quality in Passive House buildings. *Renew. Sustain. Energy Rev.* **2017**, *72*, 1303–1319. [CrossRef]
13. Liang, X.; Wang, Y.; Royapoor, M.; Wu, Q.; Roskilly, T. Comparison of building performance between Conventional House and Passive House in the UK. *Energy Procedia* **2017**, *142*, 1823–1828. [CrossRef]
14. Guerra-Santin, O.; Tweed, C.; Jenkins, H.; Jiang, S. Monitoring the performance of low energy dwellings: Two UK case studies. *Energy Build.* **2013**, *64*, 32–40. [CrossRef]
15. Johnston, D.; Farmer, D.; Brooke-Peat, M.; Miles-Shenton, D. Bridging the domestic building fabric performance gap. *Build. Res. Inf.* **2014**, *44*, 147–159. [CrossRef]
16. Stamp, S. Assessing Uncertainty in Co-Heating Test: Calibrating a Whole Building Steady State Heat Loss Measurement Method. Ph.D. Thesis, University College London (UCL), London, UK, 2015.
17. Johnston, D.; Siddall, M. The building fabric thermal performance of Passivhaus dwellings—Does it do what it says on the tin? *Sustainability* **2016**, *8*, 97. [CrossRef]
18. Gupta, R.; Kotopouleas, A. Magnitude and extent of building fabric thermal performance gap in UK low energy housing. *Appl. Energy* **2018**, *222*, 673–686. [CrossRef]
19. Baker, P. U-values and traditional buildings. In situ measurements and their comparisons to calculated values. *Hist. Scotl. Alba Aosmhor Tech. Pap.* **2011**, *10*, 70.
20. Desogus, G.; Mura, S.; Ricciu, R. Comparing different approaches to in situ measurement of building components thermal resistance. *Energy Build.* **2011**, *43*, 2613–2620. [CrossRef]
21. Asdrubali, F.; D'Alessandro, F.; Baldinelli, G.; Bianchi, F. Evaluating in situ thermal transmittance of green buildings masonries: A case study. *Case Stud. Constr. Mater.* **2014**, *1*, 53–59. [CrossRef]
22. Evangelisti, L.; Guattari, C.; Gori, P.; De Lieto Vollaro, R. In situ thermal transmittance measurements for investigating differences between wall models and actual building performance. *Sustainability* **2015**, *7*, 10388–10398. [CrossRef]
23. Ficco, G.; Iannetta, F.; Ianniello, E.; d'Ambrosio Alfano, F.R.; Dell'Isola, M. U-value in situ measurement for energy diagnosis of existing buildings. *Energy Build.* **2015**, *104*, 108–121. [CrossRef]
24. Bienvenido-Huertas, D.; Rodríguez-Álvaro, R.; Moyano, J.J.; Rico, F.; Marín, D. Determining the U-value of façades using the thermometric method: Potentials and limitations. *Energies* **2018**, *11*, 360. [CrossRef]
25. Evangelisti, L.; Guattari, C.; De Lieto Vollaro, R.; Asdrubali, F. A methodological approach for heat-flow meter data post-processing under different climatic conditions and wall orientations. *Energy Build.* **2020**, *223*, 110216. [CrossRef]
26. Mandilaras, I.; Atsonios, I.; Zannis, G.; Founti, M. Thermal performance of a building envelope incorporating ETICS with vacuum insulation panels and EPS. *Energy Build.* **2014**, *85*, 654–665. [CrossRef]
27. Albatici, R.; Tonelli, A.M.; Chiogna, M. A comprehensive experimental approach for the validation of quantitative infrared thermography in the evaluation of building thermal transmittance. *Appl. Energy* **2015**, *141*, 218–228. [CrossRef]
28. Nardi, I.; Ambrosini, D.; Rubeis, T.; de Sfarra, S.; Perilli, S.; Pasqualoni, G. A comparison between thermographic and flow-meter methods for the evaluation of thermal transmittance of different wall constructions. *J. Phys. Conf. Ser.* **2015**, *655*, 012007. [CrossRef]
29. Bros-Williamson, J.; Garnier, C.; Currie, J.I. A longitudinal building fabric and energy performance analysis of two homes built to different energy principles. *Energy Build.* **2016**, *130*, 578–591. [CrossRef]
30. Samardzioska, T.; Apostolska, R. Measurement of heat-flux of new type façade walls. *Sustainability* **2016**, *8*, 1031. [CrossRef]
31. Qu, Z.; Jiang, P.; Zhang, W. Development and application of infrared thermography non-destructive testing techniques. *Sensors* **2020**, *20*, 3851. [CrossRef]
32. RESNET—Residential Energy Services Network. RESNET Interim Guideline for Thermographic Inspections of Buildings. 2012. Available online: http://www.resnet.us/standards/RESNET_IR_interim_guidelines.pdf (accessed on 4 October 2020).
33. Nardi, I.; Lucchi, E.; De Rubeis, T.; Ambrosini, D. Quantification of heat energy losses through the building envelope: A state-of-the-art with critical and comprehensive review on infrared thermography. *Build. Environ.* **2018**, *146*, 190–205. [CrossRef]
34. Bienvenido-Huertas, D.; Moyano, J.; Marín, D.; Fresco-Contreras, R. Review of in situ methods for assessing the thermal transmittance of walls. *Renew. Sustain. Energy Rev.* **2019**, *102*, 356–371. [CrossRef]

35. Tejedor, B.; Barreira, E.; Almeida, R.M.S.F.; Casals, M. Thermographic 2D U-value map for quantifying thermal bridges in building façades. *Energy Build.* **2020**, *224*, 110176. [CrossRef]
36. Rocha, J.H.A.; Santos, C.F.; Póvoas, Y.V. Evaluation of the infrared thermography technique for capillarity moisture detection in buildings. *Procedia Struct. Integr.* **2018**, *11*, 107–113. [CrossRef]
37. Nardi, I.; Perilli, S.; De Rubeis, T.; Sfarra, S.; Ambrosini, D. Influence of insulation defects on the thermal performance of walls. An experimental and numerical investigation. *J. Build. Eng.* **2019**, *21*, 355–365. [CrossRef]
38. Garrido, I.; Solla, M.; Lagüela, S.; Fernández, N. IRT and GPR techniques for moisture detection and characterization in buildings. *Sensors* **2020**, *20*, 6421. [CrossRef]
39. Huang, Y.; Shih, P.; Hsu, K.-T.; Chiang, C.-H. To identify the defects illustrated on building façades by employing infrared thermography under shadow. *NDT & E Int.* **2020**, *111*, 102240. [CrossRef]
40. Martens, U.; Schröder, K.-U. Evaluation of infrared thermography methods for analyzing the damage behavior of adhesively bonded repair solutions. *Compos. Struct.* **2020**, *240*, 111991. [CrossRef]
41. Lucchi, E. Thermal transmittance of historical stone masonries: A comparison among standard, calculated and measured data. *Energy Build.* **2017**, *151*, 393–405. [CrossRef]
42. International Organization for Standardization. *Performance of Buildings. Detection of Heat, Air and Moisture Irregularities in Buildings by Infrared Methods—Part 3: Qualifications of Equipment Operators, Data Analysts and Report Writers*; ISO 6781-3:2015; ISO: Geneva, Switzerland, 2015.
43. International Organization for Standardization. *Thermal Performance of Buildings. Qualitative Detection of Thermal Irregularities in Building Envelopes. Infrared Method*; UNE EN 13187:1998; ISO: Geneva, Switzerland, 1998.
44. American Society for Testing Materials. *Standard Test Method for Minimum Detectable Temperature Difference for Thermal Imaging Systems*; ASTM E1311; American Society for Testing Materials: West Conshohocken, PA, USA, 2004.
45. American Society for Testing Materials. *Standard Test Method for Measuring and Compensating for Reflected Ambient Temperature Using Infrared Imaging Radiometers*; ASTM E1862; American Society for Testing Materials: West Conshohocken, PA, USA, 1997.
46. Lucchi, E. Applications of the infrared thermography in the energy audit of buildings: A review. *Renew. Sustain. Energy Rev.* **2018**, *82*, 3077–3090. [CrossRef]
47. Tejedor, B.; Casals, M.; Macarulla, M.; Giretti, A. U-value time series analyses: Evaluating the feasibility of in-situ short-lasting IRT tests for heavy multi-leaf walls. *Build. Environ.* **2019**, *159*, 1–19. [CrossRef]
48. Albatici, R.; Tonelli, A. Verifica sperimentale in situ, con analisi termografiche e algoritmi di calcolo della transmittanza termica di un elemento costruttivo. In *Annali Museo Civico di Rovereto. Sezione: Archeologia, Storia, Scienze Naturali*, 23rd ed.; Museo Civico di Rovereto: Rovereto, Italy, 2008.
49. Albatici, R.; Tonelli, A.M. Infrared thermovision technique for the assessment of thermal transmittance value of opaque building elements on site. *Energy Build.* **2010**, *42*, 2177–2183. [CrossRef]
50. Vavilov, V.P. A pessimistic view of the energy auditing of building structures with the use of infrared thermography. *Russ. J. Nondestruct. Test.* **2010**, *46*, 906–910. [CrossRef]
51. Dall'O, G.; Sarto, L.; Panza, A. Infrared screening of residential buildings for energy audit purposes: Results of a field test. *Energies* **2013**, *6*, 3859–3878. [CrossRef]
52. Ham, Y.; Golparvar-Fard, M. Automated cost analysis of energy loss in existing buildings through thermographic inspections and CFD analysis. In Proceedings of the ISARC-30th International Symposium on Automation and Robotics in Construction and Mining, Montreal, QC, Canada, 11–15 August 2013. [CrossRef]
53. Nardi, I.; Sfarra, S.; Ambrosini, D. Quantitative thermography for the estimation of the U-Value: State of art and a case study. *J. Phys. Conf. Ser.* **2014**, *547*, 012016. [CrossRef]
54. Kim, J.; Lee, J.; Jang, C.; Jeong, H.; Song, D. Appropriate conditions for determining the temperature difference ratio via infrared camera. *Build. Serv. Eng.* **2015**, *37*, 1–16. [CrossRef]
55. Ibos, L.; Monchau, J.-P.; Feuillet, V.; Candau, Y. A comparative study of in-situ measurement methods of a building wall thermal resistance using infrared thermography. *Int. Conf. Qual. Control Artif. Vis.* **2015**. [CrossRef]
56. Marino, B.M.; Muñoz, N.; Thomas, L.P. Estimation of the surface thermal resistance and heat loss by conduction using thermography. *Appl. Therm. Eng.* **2017**, *114*, 1213–1221. [CrossRef]

57. Fokaides, P.A.; Kalogirou, S.A. Application of infrared thermography for the determination of the overall heat transfer coefficient (U-Value) in building envelopes. *Appl. Energy* **2011**, *88*, 4358–4365. [CrossRef]
58. Ham, Y.; Golpavard-Fard, M. 3D Visualization of thermal resistance and condensation problems using infrared thermography for building energy diagnostics. *Vis. Eng.* **2014**, *2*, 12. [CrossRef]
59. Tzifa, V.; Papadakos, G.; Papadopoulou, A.G.; Marinakis, V.; Psarras, J. Uncertainty and method limitations in a short-time measurement of the effective thermal transmittance on a building envelope using an infrared camera. *Int. J. Sustain. Energy* **2017**, *36*, 28–46. [CrossRef]
60. Danielski, I.; Fröling, M. Diagnosis of buildings' thermal performance—A quantitative method using thermography under non-steady state heat flow. *Energy Procedia* **2015**, *83*, 320–329. [CrossRef]
61. Tejedor, B.; Casals, M.; Gangolells, M.; Roca, X. Quantitative internal infrared thermography for determining in-situ thermal behaviour of façades. *Energy Build.* **2017**, *151*, 187–197. [CrossRef]
62. Marshall, A.; Francou, J.; Fitton, R.; Swan, W.; Owen, J.; Benjaver, M. Variation of the U-value measurement of a whole dwelling using infrared thermography under controlled conditions. *Buildings* **2018**, *8*, 46. [CrossRef]
63. Tejedor, B.; Casals, M.; Gangolells, M. Assessing the influence of operating conditions and thermophysical properties on the accuracy of in-situ measured U-values using quantitative internal infrared thermography. *Energy Build.* **2018**, *171*, 64–75. [CrossRef]
64. Bienvenido-Huertas, D.; Oliveira, M.; Rubio-Bullido, C.; Marín, D. A Comparative analysis of the international regulation of thermal properties in building envelope. *Sustainability* **2019**, *11*, 5574. [CrossRef]
65. Lu, X.; Memari, A. Application of infrared thermography for in-situ determination of building envelope termal properties. *J. Build. Eng.* **2019**, *26*, 100885. [CrossRef]
66. Hoyano, A.; Asano, K.; Kanamaru, T. Analysis of the heat flux from the exterior surface of buildings using time sequential thermography. *Atmos. Environ.* **1999**, *33*, 3941–3951. [CrossRef]
67. Emmel, M.G.; Abadie, M.O.; Mendes, N. New external convective heat transfer coefficient correlations for isolated low-rise buildings. *Energy Build.* **2007**, *39*, 335–342. [CrossRef]
68. Rabadiya, A.V.; Kirar, R. Comparative analysis of wind loss coefficient (wind heat transfer coefficient) for solar plate collector. *Int. J. Emerg. Technol. Adv. Eng. (IJETAE)* **2012**, *2*, 2250–2459.
69. Sham, J.F.C.; Lo, T.Y.; Memon, S.A. Verification and application of continuous surface temperature monitoring technique for investigation of nocturnal sensible heat release characteristics by building fabrics. *Energy Build.* **2012**, *53*, 108–116. [CrossRef]
70. Liu, J.; Heidarinejad, M.; Gracik, S.; Srebric, J. The impact of exterior surface convective heat transfer coefficients on the building energy consumption in urban neighbourhoods with different plan area densities. *Energy Build.* **2015**, *86*, 449–463. [CrossRef]
71. Madding, R. Finding R-Values of stud frame constructed houses with IR Thermography. In Proceedings of the InfraMation Conference, Reno, NV, USA, 3–7 November 2008. Available online: https://www.researchgate.net/publication/285737245_Finding_R-values_of_Stud-Frame_Constructed_Houses_with_IR_Thermography (accessed on 10 November 2020).
72. Earle, R.L.; Earle, M.D. *Unit Operations in Food Processing*; Elsevier: Amsterdam, The Netherlands, 1983.
73. Holman, J.P. *Heat Transfer*, 8th ed.; McGraw Hill: New York, NY, USA, 1997; ISBN 978-0078447860.
74. Bienvenido-Huertas, D.; Bermúdez, J.; Moyano, J.J.; Marín, D. Influence of ICHTC correlations on the thermal characterization of façades using the quantitative internal infrared thermography method. *Build. Environ.* **2019**, *149*, 512–525. [CrossRef]
75. Gaspar, K.; Casals, M.; Gangolells, M. A comparison of standardized calculation methods for in situ measurements of façades U-value. *Energy Build.* **2016**, *130*, 592–599. [CrossRef]
76. International Organization for Standardization. Thermal Insulation. Building Elements. In *Situ Measurement of Thermal Resistance and Thermal Transmittance. Part 1: Heat Flow Meter Method*; ISO 9869:2014; ISO: Geneva, Switzerland, 2014.
77. Atsonios, I.A.; Mandilaras, I.D.; Kontogeorgos, D.A.; Founti, M.A. A comparative assessment of the standarized methods for the in-situ measurement of the thermal resistance of buildings walls. *Energy Build.* **2017**, *154*, 198–206. [CrossRef]
78. American Society for Testing Materials. *Standard Practice for Determining Thermal Resistance of Building Envelope Components from the In-Situ Data*; ASTM C1155-95; American Society for Testing Materials: West Conshohocken, PA, USA, 2007.

79. Gaspar, K.; Casals, M.; Gangolells, M. In-situ measurement of façades with a low U-value: Avoiding deviations. *Energy Build.* **2018**, *170*, 61–73. [CrossRef]
80. Spain, Royal Decree 314/2006 Approving the Spanish Technical Building Code CTE-DB-HE1. 2006. Available online: http://www.boe.es/boe/dias/2006/03/28/pdfs/A11816-11831.pdf (accessed on 4 October 2020).
81. International Organization for Standardization. *Building Materials and Products—Hygrothermal Properties—Tabulated Design Values and Procedures for Determining Declared and Design Thermal Values*; UNE EN ISO 10456:2012; ISO: Geneva, Switzerland, 2012.
82. International Organization for Standardization. *Building Components and Building Elements. Thermal Resistance and Thermal Transmittance. Calculation Method*; UNE EN ISO 6946:2012 (ISO 6946:2007); ISO: Geneva, Switzerland, 2012.
83. FLIR Systems. *FLIR TOOLS+ Software*; FLIR Systems: Wilsonwille, OR, USA, 2015.

Publisher's Note: MDPI stays neutral with regard to jurisdictional claims in published maps and institutional affiliations.

© 2020 by the authors. Licensee MDPI, Basel, Switzerland. This article is an open access article distributed under the terms and conditions of the Creative Commons Attribution (CC BY) license (http://creativecommons.org/licenses/by/4.0/).

Article

Combining Signal Features of Ground-Penetrating Radar to Classify Moisture Damage in Layered Building Floors

Tim Klewe [1,*], Christoph Strangfeld [1], Tobias Ritzer [2] and Sabine Kruschwitz [1,3]

1. Bundesanstalt für Materialforschung und -Prüfung (BAM), 12205 Berlin, Germany; christoph.strangfeld@bam.de (C.S.); sabine.kruschwitz@bam.de (S.K.)
2. Ingenieurbüro Tobias Ritzer GmbH, Lindenbachstrasse 29, 91126 Schwabach, Germany; tobias.ritzer@ritzergmbh.com
3. Technische Universität Berlin, 10623 Berlin, Germany
* Correspondence: tim.klewe@bam.de

Citation: Klewe, T.; Stragfeld, C.; Ritzer, T.; Kruschwitz, S. Combining Signal Features of Ground Penetrating Radar to Classify Moisture Damage in Layered Building Floors. *Appl. Sci.* **2021**, *11*, 8820. https://doi.org/10.3390/app11198820

Academic Editors: Jerzy Hoła and Łukasz Sadowski

Received: 29 July 2021
Accepted: 15 September 2021
Published: 23 September 2021

Publisher's Note: MDPI stays neutral with regard to jurisdictional claims in published maps and institutional affiliations.

Copyright: © 2021 by the authors. Licensee MDPI, Basel, Switzerland. This article is an open access article distributed under the terms and conditions of the Creative Commons Attribution (CC BY) license (https://creativecommons.org/licenses/by/4.0/).

Abstract: To date, the destructive extraction and analysis of drilling cores is the main possibility to obtain depth information about damaging water ingress in building floors. The time- and cost-intensive procedure constitutes an additional burden for building insurances that already list piped water damage as their largest item. With its high sensitivity for water, a ground-penetrating radar (GPR) could provide important support to approach this problem in a non-destructive way. In this research, we study the influence of moisture damage on GPR signals at different floor constructions. For this purpose, a modular specimen with interchangeable layers is developed to vary the screed and insulation material, as well as the respective layer thickness. The obtained data set is then used to investigate suitable signal features to classify three scenarios: dry, damaged insulation, and damaged screed. It was found that analyzing statistical distributions of A-scan features inside one B-scan allows for accurate classification on unknown floor constructions. Combining the features with multivariate data analysis and machine learning was the key to achieve satisfying results. The developed method provides a basis for upcoming validations on real damage cases.

Keywords: non-destructive testing; ground-penetrating radar; signal features; material moisture; classification; machine learning; moisture measurements; building floors; civil engineering

1. Introduction

More than half of the building insurance claims in Germany (53%) are caused by piped water damage, which entailed costs of over 3 billion Euro in 2019 alone [1]. One reason for this, apart from generally ageing pipe systems, is that water leakage often remains unrecognized until signs of degradation become noticeable. At that point the extent of damage is already critical, which underlines the demand of an accurate determination and localization of water ingress.

Neutron probes [2] are already successfully applied on building floors to localize the source of damage and to identify affected areas. The radiated fast neutrons lose most of their kinetic energy when colliding with low-mass atoms. This is especially true for hydrogen. As a result, the fast neutrons are transformed into slow (thermal) neutrons, which are then detected by a counter tube inside the probe. Given that, the method is highly sensitive to moisture, however it cannot distinguish between chemically bound or fluid water. Therefore, a calibration must be done by the destructive extraction of drilling cores. These cores are also the only possibility to obtain additional information about the depth of moisture penetration. This is a time- and cost-intensive procedure, especially for building floors, where knowledge about the affected layer is essential to plan and perform efficient renovations. Here, ground-penetrating radar (GPR) can serve as a suitable addition to the neutron probe in order to classify common moisture damages in layered building floors in a non-destructive way.

The sensitivity of GPR for water has already been proven in many publications, especially in geophysics [3,4]. However, in civil engineering (CE), GPR is also increasingly being applied for non-destructive moisture measurements on building materials like asphalt, concrete, and screed [5–10]. Here, various methods have already been established. However, their adequate use and suitability highly depends on the particular case. Due to numerous possible uncertainties, like the given structure, installed materials, and layer thicknesses, interpreting GPR results is not straightforward and requires the expertise of trained personnel. These uncertainties often influence the same signal features that are used for moisture measurements (see Section 1.1). Here, relying on only one feature, as it is done in most of the related publications [11], can lead to high uncertainty.

In contrast, this work pursues the strategy of combining different features, which allows the use of multivariate data analysis. It aims to achieve an automated classification of three scenarios: (1) the dry state, (2) damaged insulation, and (3) a damaged screed, all of them on unknown floor constructions. This is accomplished by a machine learning approach trained with novel radargram features that consider the spatial continuity of the present damage. The features are extracted from an experimentally measured data set, including varying materials and layer thicknesses. Before discussing the methodology in Section 2, a short introduction to moisture measurements with GPR is given.

1.1. Moisture Measurement with GPR

Besides the mostly negligible conductivity and magnetic permeability, the electric permittivity ε is the governing material parameter for moisture measurements with GPR [12,13]. This gets particularly clear by comparing ε for dry concrete and water. Whereas the former lies between 2 to 9 [14], the latter shows values around 81. This difference causes a significant rise for wet concrete (between 10 to 20), which influences various propagation characteristics of the electromagnetic (EM) waves. By analyzing specific time-, amplitude-, or frequency-based features of the received signals, these water-related influences become measurable. A detailed review of those features typically used for moisture measurement with GPR in CE is presented in [11]. However, a short overview is given in the following. First, the velocity v of an EM wave is directly related to ε. For non-magnetic conditions, as it is usually the case in building materials, it can simply be calculated as follows [14,15]:

$$v = \frac{c}{\sqrt{\varepsilon}} = \frac{2D}{T}, \qquad (1)$$

where c is the velocity of EM waves in free space, and T the two-way travel time in a material with the thickness D. Comparing the dry state of a material, sent and reflected pulses are received later for rising moisture content. Furthermore, the intensities and thus the measured amplitudes are reduced due to higher attenuations, caused by generally increased conductivity and more frequently occurring scattering events on water-filled pores. Filled pores also lead to Rayleigh scattering [16], which is one way to explain the observable shift of the received signals to lower frequencies for higher moisture content. Another explanation is given with the presence of dielectric dispersion, presented in the popular models of Debye [17] and Cole–Cole [18]. It describes the rising imaginary part of ε and the resulting absorption of higher-frequency components close to the relaxation frequency, which is 10 GHz to 20 GHz for free water [15,19], but can be smaller for porous materials. Another important characteristic of EM waves is the occurrence of reflection and transmission on material boundaries with different permittivities. With ε_1 and ε_2 of two mediums, an EM wave travelling from medium 1 to medium 2 is reflected by the amount of the reflection coefficient $r \in [-1, 1]$, which is calculated as follows [20]:

$$r = \frac{\sqrt{\varepsilon_1} - \sqrt{\varepsilon_2}}{\sqrt{\varepsilon_1} + \sqrt{\varepsilon_2}} \qquad (2)$$

Therefore, the amplitude of a reflection wave (RW) is highly influenced by the boundary's permittivity contrast, from which it originates. Figure 1 shows this simplified ray-based principle with an exemplary screed plate above air, forming such a permittivity contrast. It also presents the usually performed collection of multiple reflection signals (A-scans) along a survey line, whereas the offset between the transmitter (T) and receiver (R) stays constant (common-offset configuration). The recorded A-scans can then be combined in a radargram (B-scan) that offers the opportunity to visualize spatial deviations caused by inhomogeneities, like the presence of reinforcements or water-damaged areas.

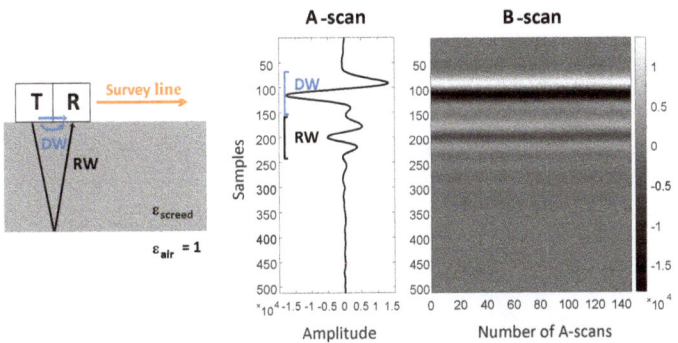

Figure 1. Principle of GPR. Multiple A-scans collected along a survey line form a B-scan.

The most dominant wave-type in an A-scan is the direct wave (DW), which travels the shortest path between T and R and is therefore recorded first. As shown, it is a superposition of an air and a ground wave and is generally used as a time reference for the following RW, since the moment of emitting the pulse (time zero) is unknown [21].

Typical signals and their respective features measured on layered floor constructions are discussed in Section 2.5.

2. Materials and Methods

Figure 2 shows the general procedure of the work presented in this section. After introducing the designed modular test specimen in Section 2.1, the conducted experiments to obtain a dataset of three damage scenarios are discussed in the Sections 2.1–2.4. In Section 2.5, various features are extracted to train and test different classifiers, which are shown in Section 2.6.

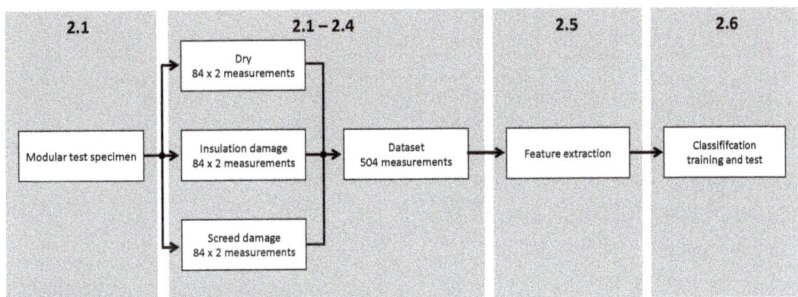

Figure 2. Schematic of the work steps presented in Section 2 divided by their respective subsections.

2.1. Modular Test Specimen

To study multiple different floor constructions, we designed a modular specimen (Figure 3), in which the screed and insulation layer can be exchanged in various ways according to the requirements of the experiment. The inner dimensions of 84 cm length,

84 cm width, and 30 cm height ensure sufficient space for the individual square-shaped parts with an edge length of 80 cm. Table 1 shows the variations of the chosen materials and their thicknesses that are believed to cover most floor setups in practice. Polyethylene (PE) foil is used to create a moisture barrier above and below the insulation. The influence of the laminate flooring and the concrete base layer on the presented classification method is considered to be negligible compared to the screed and insulation layer. Therefore, and with regard to the experimental effort, the flooring and base layer remained unchanged for the entire test series.

The cement and anhydrite screed were both chosen with the popular compressive strength C25 and the consistency class F5. The production process was carried out as instructed by the manufacturer. To guarantee efficient handling of the 60 kg to 100 kg heavy specimens, threaded sleeves were embedded in each corner. This allowed the temporary use of ring bolts to lift the plates.

The amount of different materials and thicknesses (Table 1) allows for the simulation of 84 different floor constructions for each of the three scenarios (252 setups in total). The experimental implementation of water damage in the insulation and screed layer is described in the following sections.

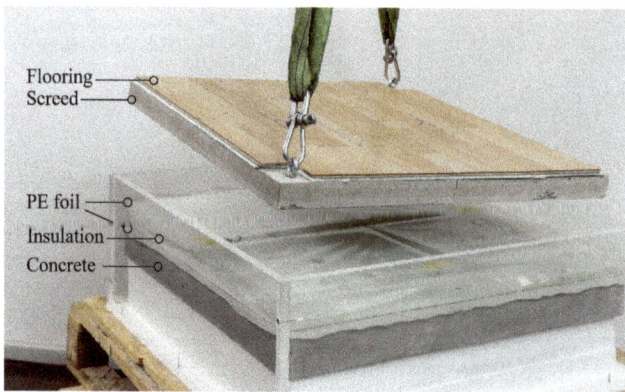

Figure 3. Modular test specimen with screed, insulation, and concrete base layer.

Table 1. Used materials and layer thicknesses for the screed (top) and insulation layer (bottom).

Material	Thickness D [cm]	Density [g·cm^{-3}]	Porosity * [%]
Cement screed (CT)	5, 6, 7	1.92	20.76
Anhydrite screed (CA)	5, 6,	2.05	27.18
Expanded polysterene (EP)	2, 5, 7, 10	0.027	-
Extruded polysterene (XP)	2, 5, 7, 10	0.037	-
Glass wool (GW)	2, 6, 10	0.061	-
Perlites (PS)	2, 6, 10	0.092	-

* Measured with mercury intrusion porosimetry.

2.2. Water Damage in Insulation Layer

To evaluate the resulting damage of added water, HIH-5030 humidity sensors were embedded in the insulation material, as shown in Figure 4. For EP, XP, and GW, this was accomplished with drilling holes of 3 cm diameter and depths varying from 50% to 75% of the respective insulation thickness. Top sealing was attained with waterproof tape. For the fine-grained PS, drilling holes were not necessary because the sensors could be placed easily.

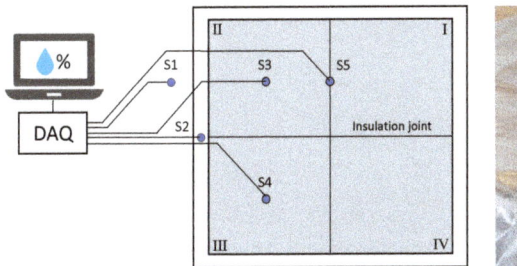

Figure 4. Evaluation of the resulting insulation damage through the use of embedded humidity sensors.

After adding equal amounts of water in all four sides of the setup, the moisture could spread for at least 12 hours to ensure stable conditions. In practical investigations, a threshold of 80% relative humidity is often considered as an adequate reason for renovations, since it provides optimum growth conditions for mould [22]. Following that, a setup was labeled as "damaged" only if all three sensors S3 to S5 exceeded this critical value. Thereafter, the measurement procedure, which is discussed in Section 2.4, was conducted for each of the six screeds.

2.3. Water Damage in Screed Layer

The quantification of screed moisture was carried out using the direct Darr method [23], which captures the loss of water by weighing samples before and after an oven-drying procedure. With the wet sample weight W_w and the dry sample weight W_d, the dry basis moisture content M_d is calculated as follows:

$$M_d = \frac{W_w - W_d}{W_d} \qquad (3)$$

Moisture content above 4 weight percent (wt%) and 0.5 wt% were valued as damage for cement and anhydrite screed, respectively. Due to preliminary investigations of the screed's hydration process, W_d was already known for each sample. Consequently, the sample's moisture content could be obtained by measuring W_w only. With 1.7 wt% to 2.3 wt% for CT and around 0.1 wt% for CA, these were rather low before simulating the damage. Therefore, we first flooded each sample by submersing them in water for 30 min (CT) or 10 min (CA). The moisture could then spread and evaporate for at least 2 days before the actual damage was induced. In consideration of practical screed damage that usually occurs after flooding from above, we then constantly poured water on top of the plates for 10 min. Besides continually weighing the samples, additional nuclear magnetic resonance (NMR) measurements were performed with the MOUSE [24,25] to obtain depth-resolved moisture distribution during the described saturation process. Figure 5 presents the exemplary NMR results with their respective water content measured on the 5 cm thick CT and CA screed.

Compared to CA, the CT screed shows an unbalanced water ingress for the sample's top and bottom side after the submersion. We explain this with a lower porosity of CT, not allowing the air in the bottom to be displaced towards the upper areas. The porosity also allows the water to spread more in CA after two days of rest. Nevertheless, sprinkling the samples resulted in quite a similar moisture distribution for both screed types with sufficiently high moisture content to be labeled as damage. After that, the screed was measured with all 14 insulation setups.

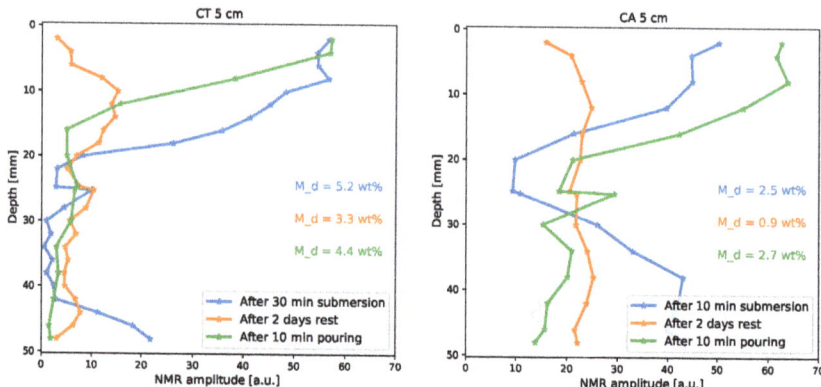

Figure 5. NMR measurements showing the depth-resolved moisture distribution during the saturation process of the 5 cm CT (**left**) and CA (**right**) screed.

2.4. Hardware and Measurement Procedure

The GPR measurements were carried out with the SIR 20 from GSSI and a 2 GHz antenna pair (bandwidth 1 GHz to 3 GHz) in common-offset configuration. As shown in Figure 6, the ground-coupled antenna pair is moved along two defined 40 cm survey lines that run from quadrant IV to I (1) and along the insulation joint (2). These joints were present for EP, XP, and GW, though not for the fine-grained PS. With 250 scans/meter, each survey line includes 100 A-scans to form one B-scan. An A-scan contains 512 samples covering a 11 ns time window.

Furthermore, each floor construction was investigated with a Troxler neutron probe placed in the setup center. To reduce the influence of individual deviations, 10 successive measurements with a respective time interval of 15 seconds were averaged.

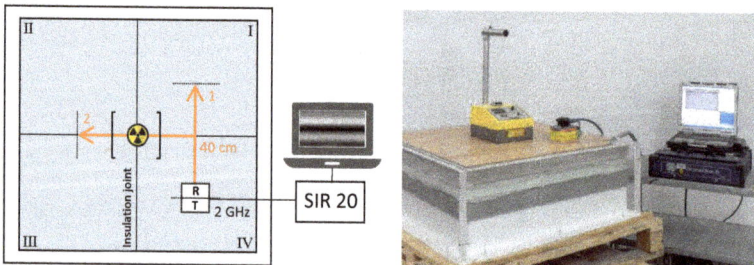

Figure 6. Measurement procedure with 40 cm long radar survey lines 1 and 2. The neutron probe is placed in the center of the construction.

2.5. Feature Extraction
2.5.1. A-Scan Features

As discussed in Section 1.1 previously, there are several signal features enabling the measurement of water with GPR. Before presenting the ones chosen in this work, it is important to understand the typical signal shapes that occur on layered floor constructions. Figure 7 gives an exemplary A-scan showing three prominent amplitude peaks and their respective origin in the setup. Since the direct wave (A_{DW}) partly travels through the superficial ground, it is influenced by the underlying nearest materials, here by the floor cover and the screed. The first reflection arises from the border between screed and insulation and is mostly recognized in the second dominant amplitude peak A_{RW1}. After that,

A_{RW2} shows the second reflection's amplitude, originating from the insulation-concrete interface below.

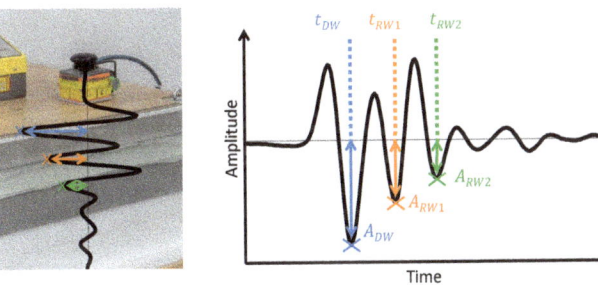

Figure 7. Exemplary A-scan with three prominent amplitudes influenced by present material interfaces.

All described reflection waves can interfere, especially for dry or thin layers, where resulting higher velocities or short traveling paths impede a clear separation in time. This is also why quantitative statements about actual water content cannot be reliable for such layered floor constructions. However, classifying the investigated scenarios is still possible and will be performed with the following signal features:

- Feature F_1: A_{DW} — Amplitude of direct wave [6–8]
- Feature F_2: A_{RW2} — Amplitude of second reflection [6–8]
- Feature F_3: f_{RW2} — Dominant frequency of second reflection (STFT) [26]
- Feature F_4: A_{RW1}/A_{DW} — Ratio between the amplitudes of first reflection and direct wave [9]
- Feature F_5: f_{RW1}/f_{DW} — Ratio between the dominant frequencies of first reflection and direct wave

The presented signal features cover all relevant signal parts, with insulation damage mostly influencing the second and first reflection and screed damage causing variations in the first reflection and the direct wave. However, the same features are also influenced by underlying layer thicknesses and different material types. Therefore, another preprocessing step is needed to overcome these construction-specific dependencies, which will be achieved by the B-scan features presented in the following section.

2.5.2. B-Scan Features

To achieve a damage classification independent of the underlying floor construction, we calculate the following scalar statistical values for each 1×100 A-scan feature vector $\vec{F}_1$ to $\vec{F}_5$, each including the respective feature elements F_1 to F_5 for all 100 A-Scans within one B-scan.

- Feature F_A: Standard deviation of $\vec{F}_1$
- Feature F_B: Standard deviation of $\vec{F}_2$
- Feature F_C: Span of $\vec{F}_3$
- Feature F_D: Standard deviation of $\vec{F}_4$
- Feature F_E: Span of $\vec{F}_5$

These measures for statistical distributions along a recorded survey line are motivated by the assumption that water damage often shows inhomogeneous deviations inside the respective B-scan. Such deviations can also be suitable to evaluate the spatial continuity of present damage. Both findings were generally recognized during our studies and will be discussed in the results Section 3. For the lower resolved frequency features, the span is expected to achieve better variance compared to the standard deviation, which works well on amplitude features with a higher resolution and range of values.

Figure 8 summarizes the discussed processing steps including the extraction of A-scan feature vectors out of B-scans and the following reduction to scalar B-scan features. The values shown for F_A to F_E are derived from the depicted A-scan feature plots (normalized by their means). They do not represent the actual values that were used for classification, since those were standardized with the *StandardScaler* function from the Scikit-Learn library. It does a mean removal for all features and scales them to unit variance, which is usually required by the classifiers discussed in Section 2.6. Regarding the magnitudes of amplitude, time, and frequency values, which are widely apart from each other, this step is necessary to avoid a baseless and unwanted dominance of certain features during the training process.

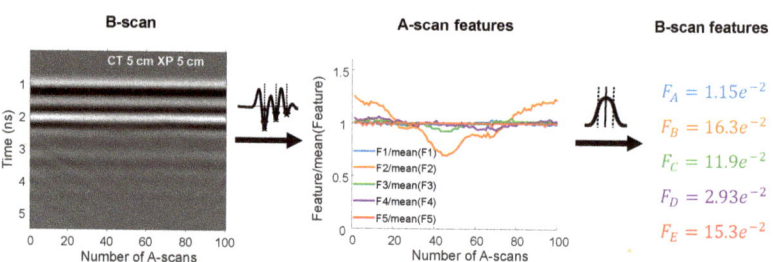

Figure 8. Processing steps to extract A- and B-scan features.

2.5.3. Feature Selection

The choice of this specific A- and B-scan feature set was made based on the achieved scores using the univariate feature selection method *SelectKBest* from the Scikit-Learn library in Python [27]. Using the *f_classif* scoring function, it estimates the degree of linear dependency between random variables (here, the features and damage scenario) by using the F-test. The five features presented before performed best in a set of 22 potential features including amplitude, time, and frequency values/ratios of each relevant reflection type in Figure 7. To avoid the use of insufficient input variables, which would impede efficient computation, only these five features were used to train the classifiers (next section), and all others were discarded. The respective scores of the chosen feature set will be shown in Section 3.2.

2.6. *Classification of Damage Scenarios*

With 84 different floor constructions for each of the three scenarios and two survey lines measured, a data set of 504 B-scans was produced, from which the features mentioned above were extracted. With this data, we trained the following four classifiers in standard configuration (default parameters only), which are all included in the Scikit-Learn library. The default parameters can be found in the respective documentation (e.g., default kernel of SVM: radial basis function):

- Multinomial logistic regression (MLR)
- Random forest (RF)
- Support vector machine (SVM)
- Artificial neural network (ANN)

The ANN consisted of two hidden layers with five neurons each (according to the number of features). To get a statistical comparability of the accuracies achieved, a $k = 20$-fold cross-validation was applied for all classifiers using the *cross_val_score* function from Scikit-Learn. Here, the parameter cv (cross-validation generator) was defined with *ShuffleSplit(n_splits = 20, test_size = 0.2, random_state = 0)* which produces 20 random splits of training and test data sets with a size of 80% and 20%, respectively. All classifiers were cross-validated with the same set of splits, which includes 20 consecutive training and test procedures for each classifier. The results were then statistically evaluated (mean and standard deviation) and are shown in Section 3.2. However, before discussing the classification, an impression of the collected data shall be given with exemplary measurement results from the experiments.

3. Results

3.1. Measurements at Modular Specimen

Figures 9 and 10 show the measurement results of all three scenarios at one respective floor construction. The first covers the configuration of a 7 cm CT screed combined with a 10 cm EPS insulation. The B-scans on top also contain text information about the underlying moisture states of screed and insulation, as well as the performed neutron probe measurement. All exemplary radar results were collected along survey line 1 (see Figure 6).

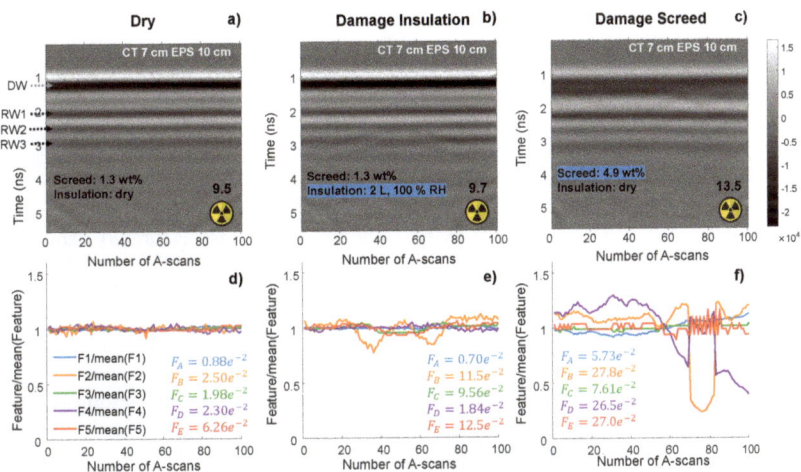

Figure 9. Measurements at a 7 cm CT and 10 cm EPS floor construction for the scenarios: (**a**) Dry, (**b**) damage insulation, and (**c**) damage screed. The bottom (**d**–**f**) shows the respective A-scan vector plots for each B-scan (top).

According to the general assumption mentioned in Section 2.5.2, the dry measurement in (a) has a homogeneous reflection pattern, whereas the two damage scenarios in (b) and (c) present clear deviations at specific time-spans. For the insulation damage in (b), we see amplitude changes in the second (RW2) and third (RW3) reflection around 2.5 ns to 3 ns, which come from the affected layer. Water added to EPS usually gathers inside the insulation joints, and from there, it slowly penetrates the material. Therefore, the areas of higher attenuation were located horizontally around the survey line's center, where the joint is crossed. These deviations become even clearer by considering the respective A-scan feature vector plots below each B-scan. Compared to the relatively flat lines for the dry measurement in (d), F_2 (A_{RW2}) particularly shows significant variations in (e), which is also captured by an increased standard deviation (F_B). These deviations are not immediately recognizable in the B-scan, since the third reflection RW3 shows a more significant variance. Insulation of 10 cm thickness usually developed two reflections, whereby the latter and therefore third reflection was not covered by the used feature set. However, due to their interference, RW2 also experienced a change in amplitude and is therefore suitable for recognizing damage. Since the neutron probe is more sensitive to moisture closer to its radiation source, a small amount of water inside the insulation is not sufficient to cause a significant increase.

In the case of screed damage in (c), the water induces deviations which appear in earlier time-spans, like in the direct wave DW or the first reflection RW1. As shown in Figure 5, all screed samples were poured from above, which is why the DW experiences a significant drop in its amplitude compared to other scenarios. A_{DW} is especially sensitive to superficial material properties and is therefore an appropriate feature to recognize flooding damage. In this case, F_4 being the ratio of A_{RW1} and A_{DW} shows a high dynamic in (f).

This is also because an unexpected reflection occurs at around 1.4 ns right after the direct wave, which was not present in other scenarios. The reason could be a steep water gradient providing a strong permittivity contrast and therefore a new reflector. This assumption is supported by the highest NMR amplitude measured for the 7 cm CT among all screeds, which was around 90 at the sample's surface. The other screeds had values of around 60 and did not show an extra reflection (compare Figure 10). Here, the new reflection at around 1.4 ns is interpreted as RW1, whereby the former RW1, originating from the screed bottom, is then seen as RW2. After a decrease of A_{RW1} between A-scans 50 and 65, it completely disappears between 70 and 80, causing a shift in reflection-counting. This leads to dominant jumps for F_2 and F_4, which cause an increased standard deviation and support the feature's sensitivity for water-induced deviations. The neutron probe is also capable of recognizing the increased moisture content with a difference of 4 digits compared to the dry measurement.

Figure 10 gives another example of a 5 cm CA screed and 6 cm GW floor construction. As before, the dry scenario in (a) shows a flat reflection pattern compared to notable deviations in the second reflection caused by a damaged insulation (b). Like with EPS, the water tended to accumulate in the joints between the GW plates and was slowly absorbed by the material. In this case, it formed a stronger permittivity contrast on the insulation's bottom, which resulted in an increased reflection amplitude in the survey line's center (see Equation (2)). This gets especially clear by considering Figure 11, in which parts of the GW insulation (measured by survey line 1) are shown. As all three plates were flipped by 90 degrees, the bottom edge belongs to the insulation joint between quadrants IV and I. The fact that only the first and lowest plate 1 shows marks of water ingress at this specific edge underlines the explanation of a strong permittivity contrast, which forms a thin reflector above the concrete plate. In this case, the neutron probe measures a slight increase due to an overall lower depth of the setup.

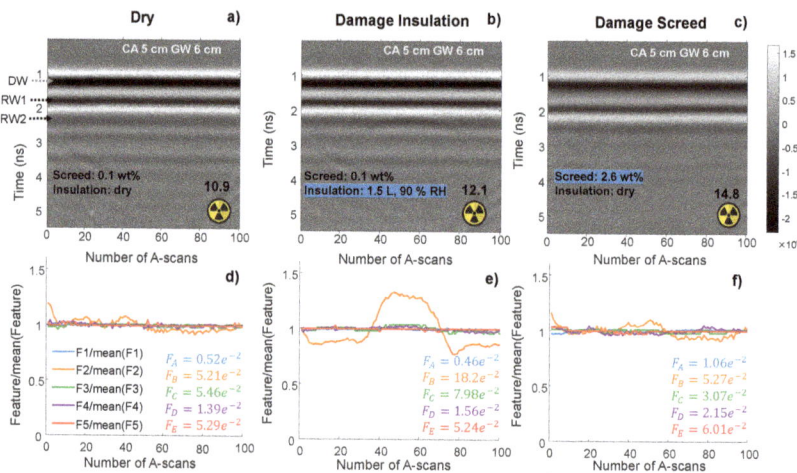

Figure 10. Measurements at an 7 cm CA and 6 cm GW floor construction for the scenarios: (**a**) dry, (**b**) damage insulation and (**c**) damage screed. The bottom (**d**–**f**) shows the respective A-scan vector plots for each B-scan (top).

Another interesting difference to the example before can be seen in the damaged screed scenario (c), which is even more representative for the whole measured data set. Like with all other screeds (except the 7 cm CT) the induced moisture damages appear comparatively homogeneous and do not show the expected deviations. This can be explained by an evenly distributed moisture gradient throughout the whole sample. The most dominant influence is the overall reduced amplitude for DW, RW1, and RW2, which becomes clear by comparing the dry scenario. However, it is not a clear indication for water without this prior knowledge. Nevertheless, by considering the values of F_A, F_D and F_E in (f), small increases can be registered, which might be sufficient to recognize the damage by trained classifiers. The validity of this statement shall be reviewed in the following section. Again, the screed damage is more visible for the neutron probe than moisture in the insulation layer.

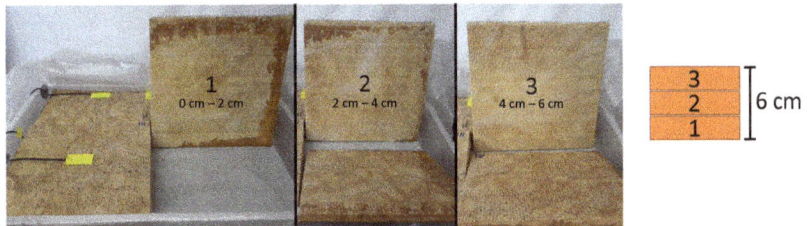

Figure 11. Water ingress (dark dyeing) in the 6 cm GW insulation. The pictures show the respective bottom of each used insulation plate (40 cm × 40 cm × 2 cm) in quadrant IV.

3.2. Damage Scenario Classification

Table 2 shows the achieved mean accuracies with the standard deviation of all trained classifiers mentioned in Section 2.6. By using the features presented in Section 2.5, all algorithms were capable of correctly recognizing 84.3% to 88.3% of the considered damage scenarios, without further knowledge about the underlying material or layer thickness. With regard to the broad variations considered in this data set, these accuracies are quite satisfying. To provide a better understanding of the presented results, Figure 12 shows confusion matrices containing each used classifier for the individual layer thicknesses of insulation and screed.

For a perfect classification with 100% accuracy, all confusion matrix cells (entries) except the main diagonal would be zero, which means that every scenario would have been classified correctly. Knowing that, the highest deviation of that perfect case gets immediately visible in Figure 12, which lies in the mid column of the top left matrix. It shows that more than half of the measured scenarios with a damaged insulation of 2 cm thickness were classified as dry. This can be explained by the low amount of water (around 0.5 L), that was necessary to cause relative humidities above 80%. Especially for GW and PS, the inserted water was absorbed by the outer edges and did not penetrate into measured areas. As a reference, Figure 13 again shows the flipped GW plate after the measurement with no signs of water ingress on the bottom edge (insulation joint between quadrant IV and I). Due to the significant number of unaffected B-scans, the classification results in Table 2 also show the accuracies for the excluded 2 cm insulation. All classifiers achieved a higher score and comparable standard deviations.

Table 2. Statistical comparison ($k = 20$-fold cross-validation) of the achieved accuracies for all trained classifiers.

Classifier	Accuracy (%)		Accuracy * (%)	
	Mean	Std	Mean	Std
MLR	86.4	3.0	89,7	3.1
RF	88.3	3.7	92.2	2.6
SVM	84.3	3.3	86.6	3.9
ANN	88.2	3.6	93.5	2.5

* Cases with insulation depth of $D = 2$ cm excluded.

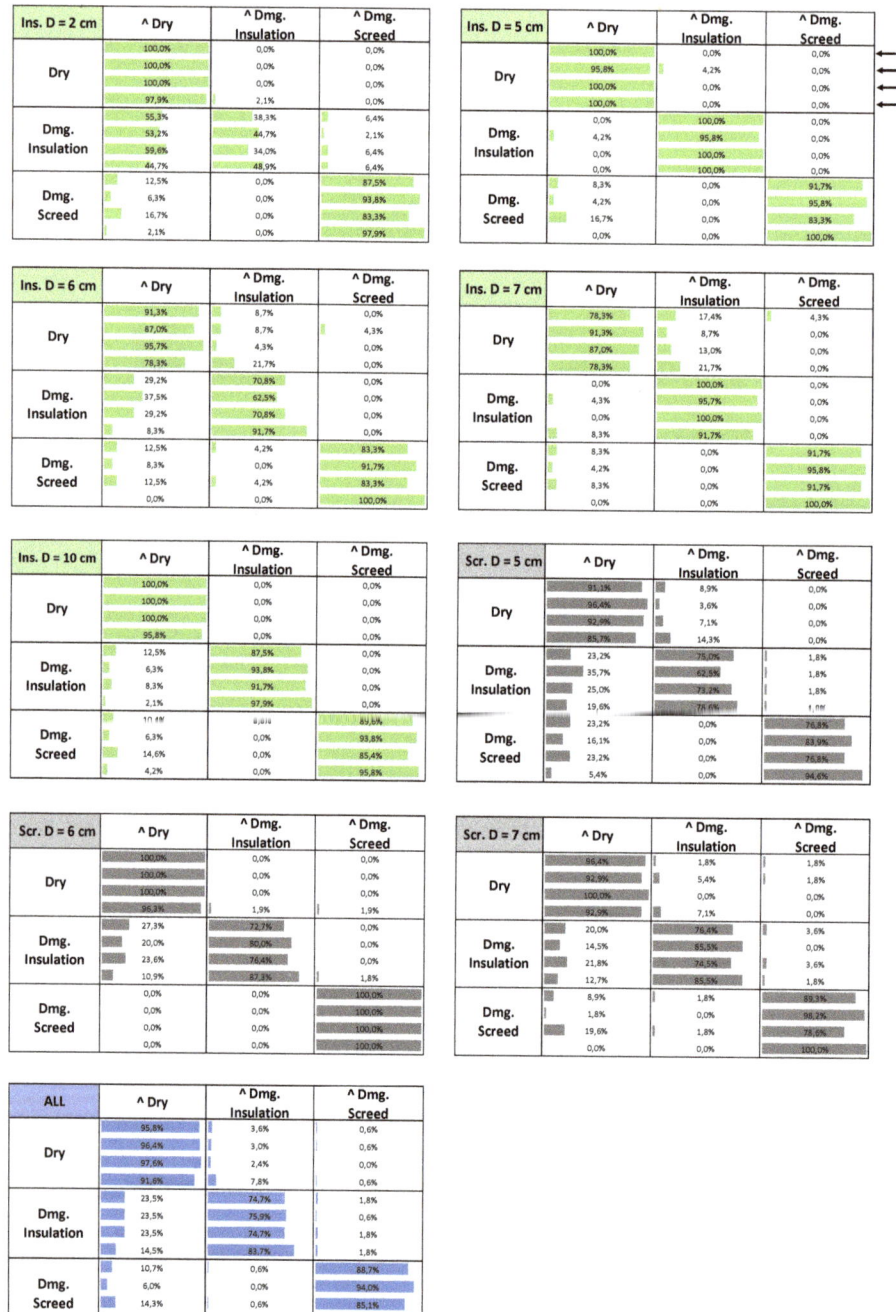

Figure 12. Combined confusion matrices for the individual insulation (green) and screed (gray) thicknesses considered in the experiment. The classifier's accuracies within one cell are presented in the same order as in Table 2. Rows and columns include the actual and the predicted (^) scenario, respectively. The blue confusion matrix summarizes the overall accuracies for each scenario.

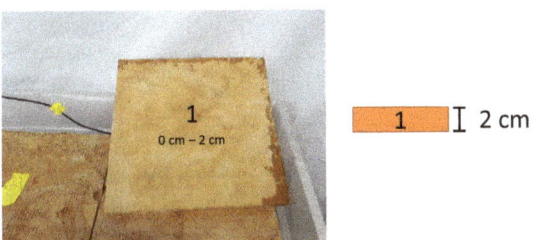

Figure 13. Water ingress (dark dyeing) in the 2 cm GW insulation. The picture shows the bottom of the used insulation plate (40 cm × 40 cm × 2 cm) in quadrant IV.

Additionally, for insulations of 6 cm thickness which only included GW and PS, the respective confusion matrix contains 8.3% to 37.5% false-negatives for damaged insulations. Since the GW of 6 cm already showed a measurable influence in Figures 10 and 11, the wrongly classified scenarios are located in the PS data set. In fact, survey line 1 for PS of 6 cm thickness presents a smooth reflection pattern, which is exemplarily shown in Figure 14b). Unfortunately, the structure of PS did not allow referencing pictures like for GW; however, the similarity between dry and damaged insulation suggests that no water penetrated in the measured area.

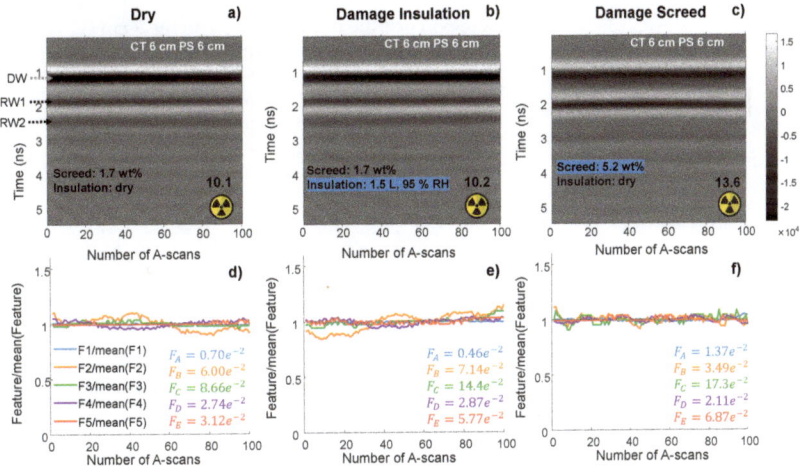

Figure 14. Measurements at a 6 cm CT and 6 cm PS floor construction for the scenarios: (**a**) dry, (**b**) damaged insulation, and (**c**) damaged screed. The bottom (**d**–**f**) shows the respective A-scan vector plots for each B-scan (top).

In general, most of the wrong classifications are false-negatives, which are represented by entries left of the main diagonals. Besides the mentioned reasons for damaged insulations, the damaged screed scenario also shows around 5% to 20% of measurements that were classified as dry in nearly every confusion matrix. With regard to the mostly homogeneous reflection patterns shown in Figures 10 and 14, these results are rather satisfying. It shows that even the slight deviations in F_A, F_D and F_E, as discussed in the previous section, are mostly sufficient to recognize the considered screed damages.

Overall, the four used classifiers achieved similar accuracies in all matrix entries. Only the damaged screed scenario reveals a more significant trend with a comparatively poorly-performing SVM, while ANN shows the best results.

Looking at the achieved scores of each extracted feature can give a better insight into the data's structure and their decisive components. Table 3 points out F_B as the best-performing feature, followed by F_C and F_A.

Table 3. Achieved scores of the applied B-scan features.

Feature	Origin in A-Scan	Score
F_A	A_{DW}	0.61
F_B	A_{RW2}	1.0
F_C	f_{RW2}	0.95
F_D	A_{RW1}/A_{DW}	0.48
F_E	f_{RW1}/f_{DW}	0.41

The reasons for these scores become clearer by considering the selected scatter plots in Figure 15 with standardized values. Combining the best-performing, RW2-related features F_B and F_C shows a good separation of the damaged insulation scenario with a broad distribution of possible values. However, due to the discussed appearance of smooth reflection patterns, the damaged screed is mostly indistinguishable from the dry scenario. In this case, features regarding DW and RW1 are obviously more decisive, which can be seen by a better separation in the middle and left scatter plot. However, the separation is not that clear as for the damaged insulation with F_B and F_C, which explains the comparatively lower scores. The blue outliers belong to the 7 cm CT screed shown in Figure 9, where the extra reflector caused an unusually strong deviation in F_D.

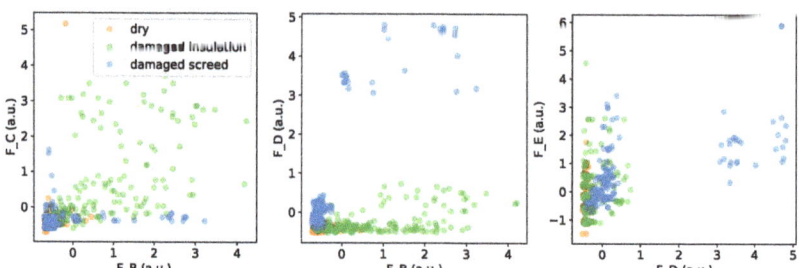

Figure 15. Scatter plots showing the feature combination of F_B & F_C (**left**), F_B & F_D (**middle**) and F_D & F_E (**right**).

4. Discussion

The results show that the proposed method regarding the horizontal distribution of specific A-scan features in one B-scan is suitable to classify moisture damages in unknown floor constructions. In a data set of 504 B-scans covering 252 different experimental setups, 84.3% to 88.3% of the scenario's dry, damaged insulation, and damaged screed were recognized correctly by the trained classifiers. A closer investigation of the produced false-negatives often revealed the measurement of undamaged areas which underlines the method's sensitivity and suggest even higher accuracies. In particular, the combination of amplitude and frequency features covering all relevant reflections in the GPR signal contributed to the successful results. Therefore, this study generally proposes an enhanced use of multivariate data analysis when performing moisture measurements with GPR.

The presented method also worked well as a supporting procedure for the neutron probe. In particular, moisture inside the insulation layer was mostly undetected by the sole use of the radiation measurement, whereas GPR achieved a satisfying sensitivity.

Since the data set only contained laboratory measurements under controlled conditions, the method still needs to be validated in practical on-site investigations of real damage cases. Here, unknown parameters like an unstable layer thickness or obstructive floor heating pipes could lead to misinterpretations which might produce an increased number of false-positive classifications. Upcoming works by the authors will address these questions. If satisfying accuracies can be achieved, the method will be capable of significantly reducing the need for destructive drilling cores to classify underlying damage scenarios, and therefore cut the costs of renovations.

Further, potential optimizations could be investigated regarding the classifiers' configurations, since only default parameters have been used so far. In addition, the use of deep learning (ANN) to automatically extract novel, relevant features out of radargrams (b-scans as input parameter) can be examined with the obtained dataset.

Author Contributions: Conceptualization, T.K., C.S., T.R. and S.K.; methodology, T.K., C.S., T.R. and S.K.; software, T.K.; validation, T.K.; formal analysis, T.K.; investigation, T.K.; resources, C.S., T.R. and S.K.; data curation, T.K.; writing—original draft preparation, T.K.; writing—review and editing, C.S. and S.K.; visualization, T.K.; supervision, C.S. and S.K.; project administration, C.S.; funding acquisition, C.S. and S.K. All authors have read and agreed to the published version of the manuscript.

Funding: This research received no external funding.

Acknowledgments: The authors would like to thank Thomas Kind and Christian Köpp for their helpful comments on the text. For their support in producing the screed samples, gratitude is owed to Hans-Carsten Kühne and Frank Haamkens.

Conflicts of Interest: The authors declare no conflict of interest.

Abbreviations

The following abbreviations are used in this manuscript:

ANN	Artificial neural network
CA	Anhydrite
CE	Civil engineering
CT	Cement
DAQ	Data acquisition
GPR	Ground penetrating radar
DW	Direct Wave
EM	Electromagnetic
EP	Expanded polysterene
GW	Glass wool
MLR	Multinomial logistic regression
NMR	Nuclear magnetic resonance
PS	Perlites
R	Receiver
RF	Random forest
RW	Reflection Wave
STFT	Short-time Fourier transform
SVM	Support vector machine
T	Transmitter
XP	Extruded polysterene

References

1. GDV. Annual of Gesamtverband der Deutschen Versicherungswirtschaft e.V. 2019. Available online: https://www.gdv.de/de/zahlen-und-fakten/versicherungsbereiche/wohngebaeude-24080 (accessed on 20 July 2021).
2. Chanasky, D.S.; Naeth, M.A. Field measurement of soil moisture using neutron probes. *Can. J. Soil Sci.* **1996**, *76*, 317–323. [CrossRef]
3. Huisman, J.A.; Hubbard, S.S.; Redman, J.D.; Annan, A.P. Measuring Soil Water Content with Ground Penetrating Radar: A Review. *Vadose Zone J.* **2003**, *2*, 476–491. [CrossRef]
4. Slater, L.; Comas, X. The Contribution of Ground Penetrating Radar to Water Resource Research. In *Ground Penetrating Radar: Theory and Applications*; Elsevier: Amsterdam, The Netherlands, 2009.
5. Saarenketo, T.; Scullion, T. Road evaluation with ground-penetrating radar. *J. Appl. Geophys.* **2000**, *43*, 119–138. [CrossRef]

6. Klysz, G.; Balayssac, J.P. Determination of volumetric water content of concrete using ground-penetrating radar. *Cem. Concr. Res.* **2007**, *37*, 1164–1171. [CrossRef]
7. Grote, K.; Hubbard, S.; Harvey, J.; Rubin, Y. Evaluation of infiltration in layered pavements using surface GPR reflection techniques. *J. Appl. Geophys.* **2005**, *57*, 129–153. [CrossRef]
8. Laurens, S.; Balayssac, J.P.; Rhazi, J.; Klysz, G.; Arliguie, G. Non-destructive evaluation of concrete moisture by GPR: Experimental study and direct modeling. *Mater. Struct.* **2005**, *38*, 827–832. [CrossRef]
9. Lai, W.L.; Kou, S.C.; Tsang, W.F.; Poon, C.S. Characterization of concrete properties from dielectric properties using ground-penetrating radar. *Cem. Concr. Res.* **2009**, *39*, 687–695. [CrossRef]
10. Kurz, F.; Sgarz, H. Measurement of Moisture Content in Building Materials using Radar Technology. In Proceedings of the International Symposium Non-Destructive Testing in Civil Engineering (NDT-CE), Berlin, Germany, 15–17 September 2015.
11. Klewe, T.; Strangfeld, C.; Kruschwitz, S. Review of moisture measurements in civil engineering with ground penetrating radar—Applied methods and signal features. *Constr. Build. Mater.* **2021**, *278*, 122250. [CrossRef]
12. Davis, J.L.; Annan, A.P. Electromagnetic Detection of Soil Moisture: Progress Report I. *Can. J. Remote Sens.* **1977**, *3*, 76–86. [CrossRef]
13. Soutsos, M.N.; Bungey, J.H.; Millard, S.G.; Shaw, M.R.; Patterson, A. Dielectric properties of concrete and their influence on radar testing. *NDT&E Int.* **2001**, *34*, 419–425. [CrossRef]
14. Daniels, D.J. *Ground Penetrating Radar*, 2nd ed.; The Institution of Engineering and Technology: London, UK, 2007.
15. Annan, A.P. Electromagnetic Principles of Ground Penetrating Radar. In *Ground Penetrating Radar: Theory and Applications*; Jol, H.M., Ed.; Elsevierr: Amsterdam, The Netherlands,2009; Chapter 1, pp. 4–38.
16. Bohren, C.; Huffman, D. *Absorption and Scattering of Light by Small Particles*; Wiley: Hoboken, NJ, USA, 1983.
17. Debye, P.J.W. Polar molecules. *J. Soc. Chem. Ind.* **1929**, *48*, 1036–1037. [CrossRef]
18. Cole, K.S.; Cole, R.H. Dispersion and Absorption in Dielectrics I. Alternating Current Characteristics. *J. Chem. Phys.* **1941**, *9*, 341–351. [CrossRef]
19. Kaatze, U. Complex permittivity of water as a function of frequency and temperature. *J. Chem. Eng. Data* **1989**, *34*, 371–374. [CrossRef]
20. Balanis, C.A. *Advanced Engineering Electromagnetics*, 2nd ed.; John Wiley & Sons: Somerset, NJ, USA, 2012.
21. Yelf, R. Where is true time zero? In Proceedings of the Tenth International Conference on Ground Penetrating Radar, Delft, The Netherlands, 21–24 June 2004. [CrossRef]
22. Coppock, J.B.M.; Cookson, E.D. The effect of humidity on mould growth on constructional materials. *J. Sci. Food Agric.* **1951**, *2*, 534–537. [CrossRef]
23. ASTM D2216-19. Standard Test Methods for Laboratory Determination of Water (Moisture) Content of Soil and Rock by Mass. *ASTM* **2019**, *4*, 8.
24. Blümich, B.; Blümler, P.; Eidmann, G.; Guthausen, A.; Haken, R.; Schmitz, U.; Saito, K.; Zimmer, G. The NMR-mouse: Construction, excitation, and applications. *Magn. Reson. Imaging* **1998**, *16*, 479–484. [CrossRef]
25. Blümich, B.; Perlo, J.; Casanova, F. Mobile single-sided NMR. *Prog. Nucl. Magn. Reson. Spectrosc.* **2008**, *52*, 197–269. [CrossRef]
26. Lai, W.L.; Kind, T.; Kruschwitz, S.; Wöstmann, J.; Wiggenhauser, H. Spectral absorption of spatial and temporal ground-penetrating radar signals by water in construction materials. *NDT&E Int.* **2014**, *67*, 55–63. [CrossRef]
27. Pedregosa, F.; Varoquaux, G.; Gramfort, A.; Michel, V.; Thirion, B.; Grisel, O.; Blondel, M.; Prettenhofer, P.; Weiss, R.; Dubourg, V.; et al. Scikit-learn: Machine Learning in Python. *J. Mach. Learn. Res.* **2011**, *12*, 2825–2830.

Article

Longitudinal Monostatic Acoustic Effective Bulk Modulus and Effective Density Evaluation of Underground Soil Quality: A Numerical Approach

Yuqi Jin [1,2], Tae-Youl Choi [2] and Arup Neogi [1,3,*]

1. Department of Physics, University of North Texas, P.O. Box 311427, Denton, TX 76203, USA; yuqijin@my.unt.edu
2. Department of Mechanical and Energy Engineering, University of North Texas, 3940 North Elm Suite F101, Denton, TX 76207, USA; tae-youl.choi@unt.edu
3. Advanced Materials and Manufacturing Processes Institute, University of North Texas, 3940 North Elm Street, Box Q, Discovery Park Annex, Denton, TX 76207, USA
* Correspondence: arup@unt.edu

Abstract: In this study, we introduce a novel method using longitudinal sound to detect underground soil voids to inspect underwater bed property in terms of effective bulk modulus and density of the material properties. The model was simulated in terms of layered material within a monostatic detection configuration. The numerical model demonstrates the feasibility of detecting an underground air void with a spatial resolution of about 0.5 λ and can differentiate a soil firmness of about 5%. The proposed technique can overcome limitations imposed by conventional techniques that use spacing-consuming sonar devices and suffer from low penetration depth and leakage of the transverse sound wave propagating in an underground fluid environment.

Keywords: ultrasonic elastography; underground detection; soil inspection; underwater acoustics

1. Introduction

According to the US Geological Survey in 2014, the average cost of karst collapses in the United States over the past 15 years is more than $300 million per year. The subsidence from sinkhole collapse is especially highest in Florida, Texas, Alabama, Missouri, Kentucky, Tennessee, and Pennsylvania. It is impossible to know when a catastrophic sinkhole collapse occurs. However, it is possible to predict the occurrence of such likely events. Sinkholes in karst terrain occur naturally and from anthropogenic activity, e.g., groundwater development, oil and gas drilling, surface loading, and urban expansion into previously undeveloped sinkhole-prone areas and drought or precipitation extremes [1,2]. Most states with substantial damage attributed to karst sinkholes have public resources documenting sinkholes and sinkhole density locations, except for Texas [3].

Hence, the appropriate geophysical methods to provide subsurface information are crucial for the migration of catastrophic disasters due to subsidence or sinkholes. Non-instructive tests or non-destructive (NDT) such as ground-penetrating radar (GPR) [4,5], spectral analysis of surface waves (SASW) [6,7], multi-channel analysis of surface waves (MASW) [8,9], and micro-tremor array measurement (MAM) [10,11] are useful methodologies to detect underground voids. They can provide 2D or 3D subsurface stiffness profiles from the measurements at the ground surface. Each method has advantages and limitations. For example, GPR can identify layering sites, but it cannot resolve material properties. However, seismic methods SASW and MASW can resolve layer thickness and stiffness of materials. These methods can be inserted into boreholes and can be used to measure subsurface characteristics from the inside of a borehole. Accurate voids detection can be appreciable. However, the softening process that occurs before the air voids formation was not usually involved in the checklist for inspection. The density and mechanical properties

Citation: Jin, Y.; Choi, T.-Y.; Neogi, A. Longitudinal Monostatic Acoustic Effective Bulk Modulus and Effective Density Evaluation of Underground Soil Quality: A Numerical Approach. *Appl. Sci.* **2021**, *11*, 146. https://dx.doi.org/10.3390/app11010146

Received: 10 December 2020
Accepted: 23 December 2020
Published: 25 December 2020

Publisher's Note: MDPI stays neutral with regard to jurisdictional claims in published maps and institutional affiliations.

Copyright: © 2020 by the authors. Licensee MDPI, Basel, Switzerland. This article is an open access article distributed under the terms and conditions of the Creative Commons Attribution (CC BY) license (https://creativecommons.org/licenses/by/4.0/).

undergo a clear decrease when the soil became soft. Unfortunately, those methods are limited to provide more in-depth engineering information, such as soil type, strength, stability, and so on [12].

Furthermore, these techniques do not resolve subsurface layering in the presence of certain anomalies. Impedance contrasts, moisture, and cavities can affect different tests in different ways. Ultrasonic techniques have been recently used to study the underwater distribution in soil [13] and soil properties. Recently pulsed velocity ultrasound has been used to detect hard objects in farmland [14]. However, most of these techniques require close contact of the transducer with the soil.

Therefore, it is necessary to find an appropriate non-contact method for mapping subsurface voids and monitoring the soil's healthy in terms of mechanical properties and density while providing material properties through the evaluation of geophysical methods. Electromagnetic and seismic monitoring systems are the most commonly used techniques to detect voids on land. To overcome the limitations in these methods, such as compactness, low penetrate depth, and leakage of the transverse sound wave propagating in underground fluid, we introduce a recently developed elastographic mapping technique. The effective bulk modulus and effective density detection (EBME) [15] have been applied to underground soil health monitoring and void detection in a compact monostatic setup.

2. Numerical Experiment Design

As Figure 1 shows, the typical underground is formatted in layers structure modeled as ambient air, soil layer with and without voids, and an underlying rock layer. The basic principle of the model employed in this work involves using low-frequency acoustic waves to detect the soil's effective density. It is based on the amplitude ratio of the reflected wave between the soil layer and the underlying layer due to acoustic impedance mismatch. The effective density can be presented in terms of an absolute value or a relative scale estimated from the recently invented non-invasive imaging technique: effective bulk modulus elastography (EBME) [15,16]. The previous studies used this technique to distinguish different materials such as hard and soft materials and similar tissue phantoms in terms of effective bulk modulus and effective density [15]. The application of EBME showed that the unique technique could differentiate the various regions of 3D printed plastic differing in density due to the air porosity introduced during the printing process under various conditions of printing. One of the fabricated samples had five density zones varying from 100% to 60%, which is similar to the varying packing density of porosity in soil due to environmental conditions. The effective density imaging technique remotely evaluated the absolute elastic values of the various regions with a maximum 6% error in absolute density values [17]. The technique can be applied to the underlying layers using acoustic radiation force to estimate the effective density in both lateral and axial directions [17]. This work motivated us to use this technique to study the void formation and the packing density in soil using a remote and rapid scanning technique applied to characterize other material systems.

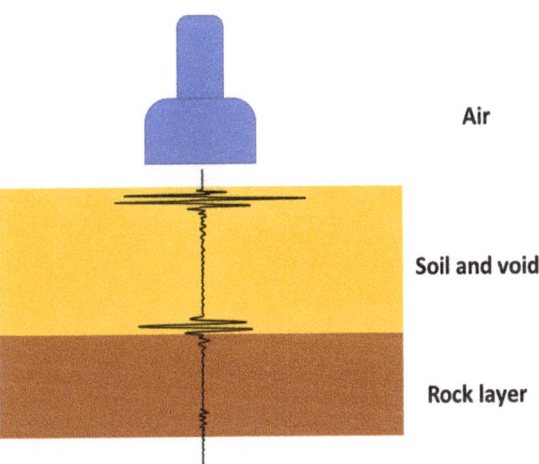

Figure 1. Material model of the layers used for simulations.

2.1. Effective Bulk Modulus and Effective Density Calculation

Effective bulk modulus and Effective density in scanned imaging were calculated as [15]:

$$\rho = c^{-1} Z_0 \left(\frac{-1 - \frac{p_1}{p_e - p_0} - \sqrt{4\frac{p_1}{p_e - p_0} + 1}}{\frac{p_1}{p_e - p_0} - 2} \right), \frac{Z}{Z_0} > 1,$$

$$\rho = c^{-1} Z_0 \left(\frac{\left(1 - \frac{p_1}{p_e - p_0}\right) + \sqrt{1 - 4\frac{p_1}{p_e - p_0}}}{\frac{p_1}{p_e - p_0} + 2} \right), \frac{1}{3} < \frac{Z_1}{Z_0} < 1. \quad (1)$$

$$\rho = c^{-1} Z_0 \left(\frac{\left(1 - \frac{p_1}{p_e - p_0}\right) - \sqrt{1 - 4\frac{p_1}{p_e - p_0}}}{\frac{p_1}{p_e - p_0} + 2} \right), 0 < \frac{Z_1}{Z_0} \leq \frac{1}{3}.$$

$$K = c Z_0 \left(\frac{-1 - \frac{p_1}{p_e - p_0} - \sqrt{4\frac{p_1}{p_e - p_0} + 1}}{\frac{p_1}{p_e - p_0} - 2} \right), \frac{Z}{Z_0} > 1,$$

$$K = c Z_0 \left(\frac{\left(1 - \frac{p_1}{p_e - p_0}\right) + \sqrt{1 - 4\frac{p_1}{p_e - p_0}}}{\frac{p_1}{p_e - p_0} + 2} \right), \frac{1}{3} < \frac{Z_1}{Z_0} < 1. \quad (2)$$

$$K = c Z_0 \left(\frac{\left(1 - \frac{p_1}{p_e - p_0}\right) - \sqrt{1 - 4\frac{p_1}{p_e - p_0}}}{\frac{p_1}{p_e - p_0} + 2} \right), 0 < \frac{Z_1}{Z_0} \leq \frac{1}{3}.$$

In the above equations, p_e, p_0, and p_1 are, respectively, the highest amplitude of the source pulse from the probe, the reflection of the wavefront back from the front interface of the sample layer, and the second echo back from the interface between the target sample layer and next material layer separately. p_e was the maximum amplitude value of the emission source. It was set in the software to 1 µPa on the absolute scale. p_0 and p_1 were the maximum absolute values of the detected reflection signal amplitudes obtained from the probe's upper surface. The values were averaged from the ten linear distributed arrays on the probe line laterally. c is indicated the sound velocity from the time of flight in the sample layer at the measured location, described as $c = 2d/(t_1 - t_0)$, where t_1 and t_0, are the first peak of the first and second echoes. d is the thickness of the target layer. Effective density ρ is calculated from $\rho = Z/c$, where Z is the acoustic impedance of the sample at the scanned location. The baseline impedance value was referred to in the previous layer is $Z_0 = \rho_0 c_0$.

For the multiple layers of various materials:

$$p_k = \frac{Z_n}{Z_0}(p_e - \text{sig}(Z_1 - Z_0)|p_0|) \cdot \left(\prod_{i=2}^{k} t_{i-1,i}\right) r_{k-1,k} \left(\prod_{i=1}^{k} t_{i,i-1}\right), k = 1, 2, \cdots, n. \quad (3)$$

where the transmission and the reflection coefficients are $t_{i-1,i} = (2Z_n)/(Z_{n-1} + Z_n)$ and $r_{k-1,k} = (Z_n - Z_{n-1})/(Z_n + Z_{n-1})$ the reflection coefficient at the interface between layer $(n-1)$ and n. The reflection coefficient of the interface between the last layer and ambient material is expressed as $r_{k-1,0} = (Z_0 - Z_{n-1})/(Z_{n-1} + Z_0)$. n numbers of Z_n values are obtained by solving n numbers of Equation (2) for the n numbers of layers in the samples. The effective density values are expressed as $\rho_n = Z_n c_n^{-1}$.

2.2. Numerical Modeling

The numerical simulations were performed using COMSOL Multiphysics. The geometry was designed in two-dimension to reduce computational time. The whole detected region was eight meters in length and 4 m wide, including a 1.5 m layer of air thickness between the probe and soil. We also consider 2 m of the rocky layer under the thick layer of soil. The physical properties of the regular soil layer were defined as $c = 800$ m/s and $\rho = 2000$ kg/m^3, the rock layer was presumed to have $c = 2000$ m/s and $\rho = 3000$ kg/m^3. The room temperature speed of sound in the air was considered to be $c = 342$ m/s and density as $\rho = 1.225$ kg/m^3. The physical properties of air and rock layers were provided by the built-in materials library in COMSOL Multiphysics software. The parameters related to soil properties were used from the literature [18,19]. The soft-soil was defined to exhibit a 5% reduction in the speed of sound and its density compared to regular soil. We also considered that the softer soil had a 5% decrease in the speed of sound and density from the soft soil. The time-dependent wave equation was simulated with a general pulse form with its pulse function expressed as $\sin(\omega_0 t)e^{-f_0(t-3T_0)^2}$, where ω_0 was the angular frequency of the pulse at the operating frequency f_0 (2000 Hz), and $T_0 = 1/f_0$ is the time period. t was the time interval over which the event was simulated. The time window used for the estimation of the wave propagation was $480 T_0$.

Each of the simulated model illustrated in Figure 2 shows the geometrical configuration used in this study, which were considered to be 2.05 m tall (in the vertical direction) and 1 m wide (in the horizontal direction). The top probe was 0.05 m-thick and 0.5 m-wide. The rest of the 2 m region was generally separated into three major zones. From top to bottom, the air ambient layer was 0.15 m thick. The major soil zone thickness was 1.6 m, which had a 0.25 m rock layer under it. For the cases of soft soil, softer soil, and air void existing in Figure 2B–D, the center of the anomalous regions located at the center of the soil layer. Figure 2B,C, show that the soft soil and softer soil regions were 0.8 m wide and 0.5 m thick. In Figure 2D, the air void was a circle with 0.05 m in diameter.

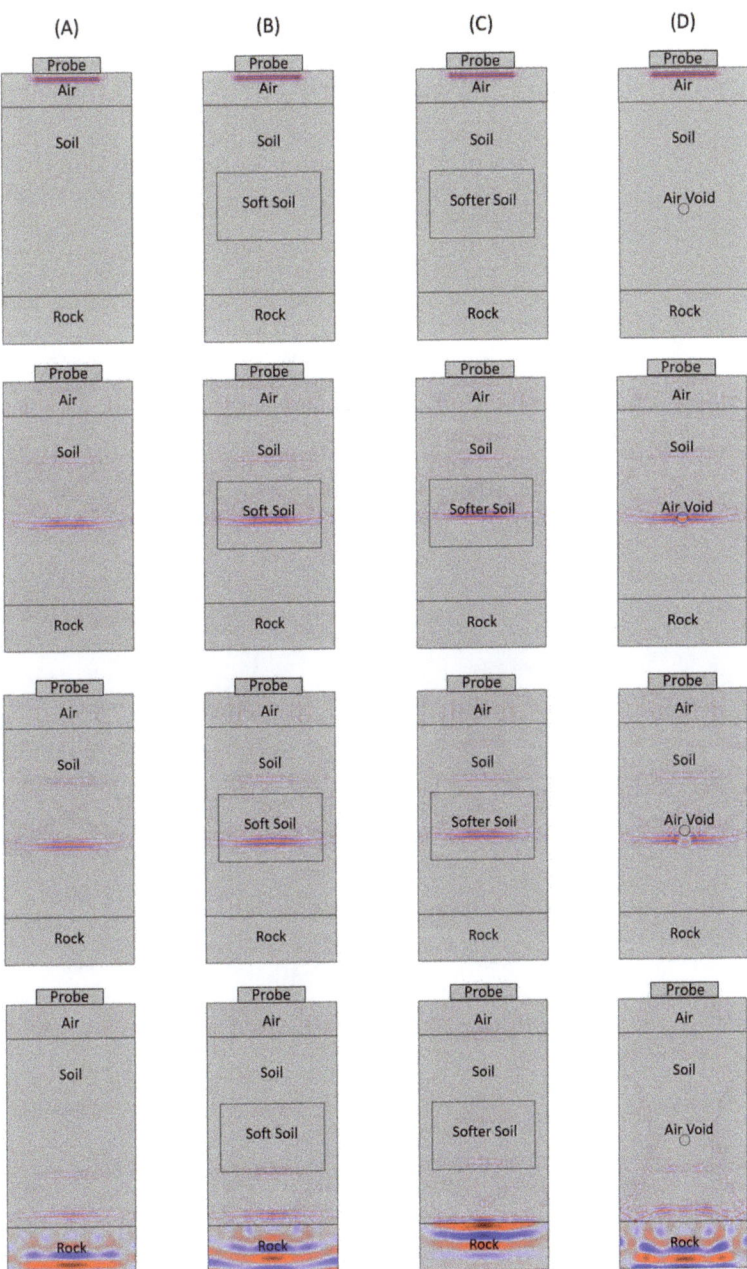

Figure 2. Sound wave propagation through air, soil and rocky layers in four different cases. The red and blue color scale indicated the positive and negative amplitude of the sound wave pressure. Case (**A**): healthy soil. Case (**B**): healthy soil with 5% properties reduction for the soft soil region. Case (**C**): healthy soil with a softer soil region with 5% lower values compared to the soft soil in (**B**). (**D**): healthy soil with a small internal air void. The size of the air void is smaller than the sound wavelength.

3. Results and Discussion

A finite element analysis-based numerical simulation was performed to simulate the feasibility of using effective density detection in determining soil voids and the overall density of the porous soil. The simulation was used to visualize the transient sound wave propagation in the four soil with different conditions. As Figure 2 shows, the models of the initial study were categorized into healthy soil (A), soil with a region of soft soil Figure 2B, soil with a region of softer soil Figure 2C, and healthy soil with an internal small air gap (of about 0.5 λ in size) Figure 2D. In the axial propagation of the wave in the detection setup, the low-frequency sound wave pulse has a small amplitude reflection at the interface between Air and soil layers. In the soil layers, the sound wave propagation was delayed in the case of Figure 2B,C comparing with healthy soil (A). Moreover, in the case of Figure 2D, the small air void caused a scattering effect on the propagating wave without any clear temporal delay. The sound wave was reflected back into the transducer with a larger amplitude at the interface between the soil layer and the rock layer under it. The reflected wave propagated in the opposite direction of the wave emission direction. The backward trip of the wave undergoes another temporal delay in the case of Figure 2B,C, and once scattering effect in case (D). By measuring the reflected signal pressure over a roundtrip propagation of the wave, a significant difference was found between the four cases, as shown in Figure 3.

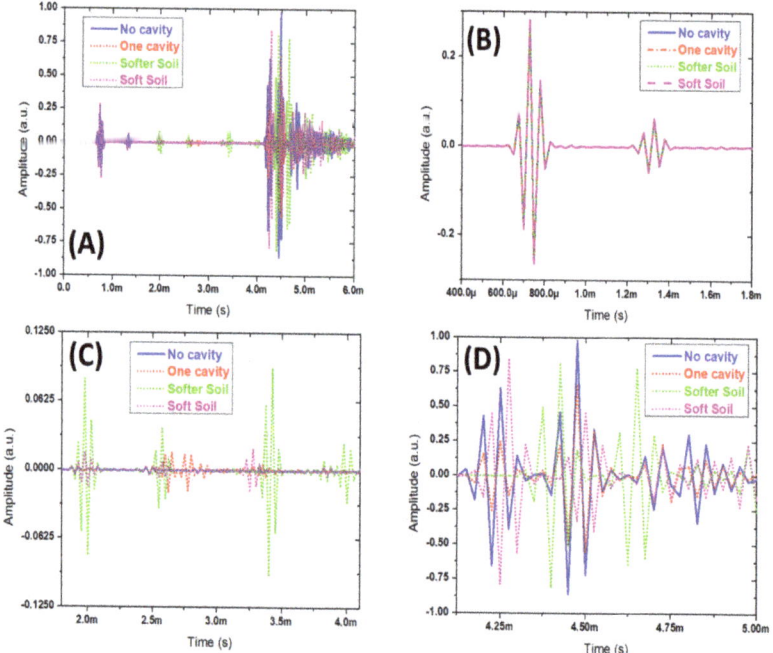

Figure 3. (**A**) Temporal reflected signal collected at the surface of the probe in all four cases mentioned in Figure 2. "No cavity" was the healthy soil condition in case (**A**). "One cavity" indicated the soil with a small air void in case (**D**). "Soft soil" was the case (**B**), which has a soft soil zone in the soil layer, which has similar properties. "Softer soil" was the case (**C**), which has a soft soil zone in the soil layer which has dissimilar properties. The time window width was 6 ms. The wave completed a roundtrip in the simulated model in the calculated time length. (**B**) was the zoomed-in view between 0.4 ms and 1.8 ms. (**C**) was the zoomed-in view between 2 ms and 4 ms. (**D**) was the zoomed-in view between 4.125 ms and 5 ms.

In Figure 3, Figure 3A shows the entire range of the temporal response of the propagating wave. Figure 3B–D shows the zoomed-in response from the various temporal regions plotted in Figure 3A. Figure 3B shows that the echoes from the two reflections can be expected from the interface between the air and soil and its second roundtrip envelope. Figure 3C depicts the zoomed-in time window, and the propagation of the wave occurs within the soil layer's internal region. No reflection pulse can be observed from the healthy soil line. The first and third echo on the green line occurred from the front and back interface between softer soil and healthy soil. The second and fourth pulses were the second roundtrip of the first and third reflection that occurred inside the softer soil cavity. The only two visible reflections on the soft soil were the echo from the two mismatch interface around the anomalous zone and occurred along the direction of the propagating wave. The red line has two lower amplitude echoes occurring from the air cavity's front and back interface on the wave propagation direction. In Figure 3D, there were three major reflection envelopes on each line. The first echo was reflected from the rocky layer's front interface, and the second echo was from the back boundary of the rock layer. The third reflection envelope occurred from the back surface and underwent a roundtrip internally within the rocky layer.

Figure 3A showed the temporal measurement of the reflected sound wave at the probe surface regarding the normalized sound pressure for all the four proposed cases. Before 2 ms, the sound wave has experienced the same media, and the first two echo were reflected at the interface between air and soil layers due to acoustic impedance mismatch. Between 2 ms and 4 ms, the temporal signal's zoomed-in view shows the reflections from the soft soil, softer soil, and air void in case Figure 3B–D compared to the normal soil ("No cavity"). The larger echo amplitude in case Figure 3C was due to the larger difference between the density and speed of sound properties in softer soil and healthy soil. The difference between the density and speed of sound in soft soil in case (Figure 3B) and healthy soil is smaller comparing with case Figure 3C (softer soil), which led to a smaller amplitude reflected echoes. The red dotted trace showed the two small symmetric reflection envelopes from the air void, as shown in Figure 3D. The air void's axial size could be estimated from the temporal distance between the highest peak of the two small symmetric reflection envelopes.

Figure 3D shows the signal between 4.125 ms and 5 ms, which indicated the reflected wave from the interface between the soil and rock layers. As the sound wave in the four cases experiences different soil conditions during the two roundtrip propagation, the reflection from the interface between the soil and rock layers shows a clear time difference between the cases. Comparing Figure 3A, which is for healthy soil, and Figure 3D with one small air void, the reflected wave does not have any time delay for the arrival pulse. The attenuated amplitude of the signal in case Figure 3D was affected by the scattering due to the small air void. Due to slightly lower density and speed of sound in the soil's temperate region, the reflected wave in the case, Figure 3B, had a small amount of time delay compared to the healthy soil. When the decreased density and sound velocity values occurred more in the soft soil zone, the time delay between the soil with and without the soft region was larger, as illustrated in Figure 3C.

From the temporal waveform in terms of pressure, as shown in Figure 3, the calculated relative density and effective bulk modulus values in healthy soil were normalized to 1. In the case of Figure 3B, the effective bulk modulus decreased to 0.923, and the effective density value of the soil layer reduced to 0.984. In the case of C, the effective bulk modulus decreased to 0.907, and the effective density value of the soil layer was lowered to 0.962. In the case of D, the effective bulk modulus decreased to 0.985, and the effective density value of the soil went down to 0.985. As the listed equations indicated, the amount of reflection time delay back from the rock layer provides a larger effect on the estimated effective bulk modulus and density values than the echo amplitude difference since sound speed is a square term in the equation. However, in Figure 3D, since the rock layer interface echo does not have a clear time delay in the measurement, the effective bulk modulus and effective density value decreased linearly by a small amount based on a decrease of the echo amplitude due to the scattering effect. In the numerical simulation, the defined media was not dispersive. In reality, the dispersion [20,21] may be introduced by the nonuniform soil status. In a highly dispersive medium, the frequency components in a broadband acoustic pulse have different speed of sound (phase velocity). This dispersion would elongate the pulse width and modify the velocity of the acoustic pulses. For an accurate estimation of the effective bulk modulus and effective density values, the single frequency component phase velocity should be considered and used in place of the speed of sound value from the pulse envelope.

Unlike the existing seismic methods of soil void detecting techniques, EBME uses a monostatic low-frequency sound wave to estimate the effective bulk modulus and effective density over the entire soil layer's depth in the effective area as large as the sound probe beyond the long-wavelength limit. The technique can monitor the soil's health in terms of relative content of air, water (effective elasticity), and foreign objects (Appendix A). The significant advantage of the EBME void detection is the requirement of compact equipment set up on the ground. Almost all the existing sonic soil void detecting techniques use a bistatic setup. To place the emission source and receiver on the ground required more space. Some techniques require either sound wave emission source array or detector array on the ground, which the preparation is time-consuming. In addition, the alignment of the array is important to obtain accurate results. The non-flat ground condition would introduce non-negligible uncertainty to the detection. Since the EBME technique uses a high penetrating low-frequency sound wave, the probe is not necessary to contact the ground. The air layer between the probe surface and soil does not decrease the signal-to-noise ratio a lot. In this way, the limitation from the non-flat ground surface condition could be overcome. A sound wave array system would be preferred as it provides a 2D surface scan. Using EBME, the 2D surface scan could be carried out through a raster scan since it is independent of the ground surface condition and does not require any contact with the ground surface. Most seismic methods apply transversal mode sound wave or radiational stress, which approaches a limitation of wave propagation in the fluid such as underground water. Since most fluids do not have a shear modulus to transmit transverse mode vibration, the existing seismic methods have difficulty determining the underground structure and properties once the underground water layer exists. The EBME technique uses longitudinal mode wave, which can propagate in both solid and fluid media to overcome the limitation of the underground fluid propagation issue compared to existing seismic methods. Ground-penetrating radar is another non-destructive method in soil void detection. The radar emitted an MHz microwave into the soil and collecting reflection. Since the electromagnetic wave has a shorter wavelength, the resolution of ground penetrating radar is usually appreciable. However, the increased resolution by small wavelength introduces large noise as a tradeoff. The wavelength of the low-frequency EBME technique can minimize noise since it is employing a long-wavelength sound. Instead of using a long-wavelength sound wave to detect the comparable size of soil void, the EBME technique uses a relatively scaled effective bulk modulus and effective density to estimate the volume fraction of the void inside the soil by using a rapid non-contact raster scan. Once the porous soil's target area was determined, a higher frequency sound wave detector [22], acoustic lens/collimator [23–26], or electromagnetic radar could be applied in the region to find detailed information of the specific voids for further interest. This work showed the initial feasibility of underground acoustic detection. Real experiments will be further performed as future works in the lab scaled-down condition and further in real condition with the influence factors studies such as vegetation on the ground surface and unknown solid objects in the soil layer.

4. Conclusions

This study proposed a novel method to detect underground soil voids and monitor the soil healthy in terms of effective bulk modulus and effective density demonstrated by numerical simulation. The technique can detect about 0.5 λ size air void in soil and

5% reduction in soil density due to the decrease in the measured effective bulk modulus and density compared to a healthy reference soil. The proposed technique would provide better penetration depth than electromagnetic methods, more compactness than the multi-detectors array systems, and better resolution than conventional sonic techniques. Compared to surface wave and shear wave techniques, this study's novel method can overcome the limitation of non-guided propagation of the transverse wave in underground water or other fluid.

Author Contributions: Conceptualization, A.N. and T.-Y.C.; methodology, T.-Y.C. and Y.J.; software, Y.J.; validation, Y.J., T.-Y.C. and A.N.; formal analysis, T.-Y.C.; investigation, Y.J.; resources, A.N.; writing—original draft preparation, Y.J., T.-Y.C. and A.N.; writing—review and editing, A.N.; visualization, A.N.; supervision, A.N.; project administration, A.N.; funding acquisition, A.N. All authors have read and agreed to the published version of the manuscript.

Funding: This work is supported by an Emerging Frontiers in Research and Innovation (EFRI) grant from the National Science Foundation (NSF) Grant No. 1741677. The support from the infrastructure and support of the Center for Agile & Adaptive and Additive Manufacturing (CAAAM) funded through State of Texas Appropriation #190405-105-805008-220 is also acknowledged.

Institutional Review Board Statement: Not applicable.

Informed Consent Statement: Not applicable.

Data Availability Statement: Data availability from corresponding author.

Conflicts of Interest: The authors declare no conflict of interest.

Appendix A

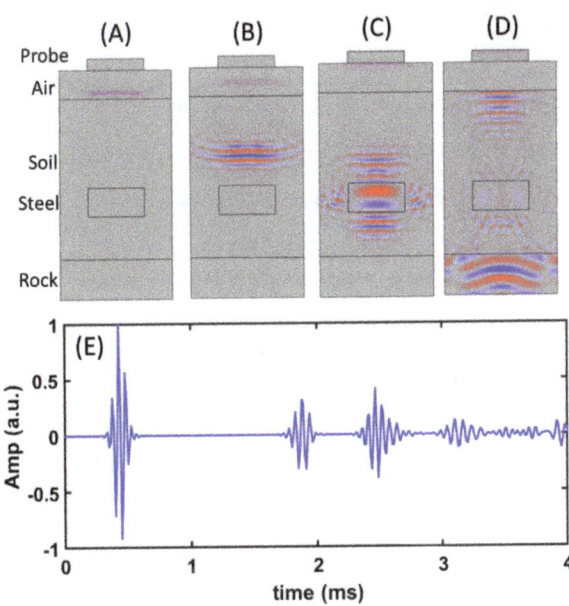

Figure A1. (**A–D**) Various time points of an additional case with steel block embedded into the soil layer, which could be clearly recognized. (**E**) Time of flight signal of the case.

References

1. Wilson, W.L.; Shock, E.J. New sinkhole data spreadsheet manual (v1.1). In *Subsurface Evaluations*; Winter Springs: Winter Springs, FL, USA, 1996.
2. Tihansky, A.B. Sinkholes, west-central Florida. In *Land Subsidence in the United States: US Geological Survey Circular*; USGS: Reston, VA, USA, 1999; pp. 121–140.

3. Kuniansky, E.L.; Weary, D.J.; Kaufmann, J.E. The current status of mapping karst areas and availability of public sinkhole-risk resources in karst terrains of the United States. *Hydrogeol. J.* **2016**, *24*, 613–624. [CrossRef]
4. Daniels, D.J. Ground penetrating radar. In *Encyclopedia of RF and Microwave Engineering*; Wiley: Hoboken, NJ, USA, 2005.
5. Huisman, J.A.; Hubbard, S.S.; Redman, J.D.; Annan, A.P. Measuring soil water content with ground penetrating radar. *Vadose Zone J.* **2003**, *4*, 476–491. [CrossRef]
6. Satoh, T.; Yamagata, K.; Poran, C.J.; Rodriguez, J.A. *Soil Profiling by Spectral Analysis of Surface Waves*; University of Missouri: St. Louis, MI, USA, 1991.
7. Murillo, C.A.; Thorel, L.; Caicedo, B. Spectral analysis of surface waves method to assess shear wave velocity within centrifuge models. *J. Appl. Geophys.* **2009**, *68*, 135–145. [CrossRef]
8. Miller, R.D.; Xia, J.; Park, C.B.; Ivanov, J.M. Multichannel analysis of surface waves to map bedrock. *Leading Edge* **1999**, *18*, 1392–1396. [CrossRef]
9. Park, C.B.; Miller, R.D.; Xia, J.; Ivanov, J. Multichannel analysis of surface waves (MASW)—Active and passive methods. *Leading Edge* **2007**, *26*, 60–64. [CrossRef]
10. Tokeshi, J.C.; Karkee, M.B.; Sugimura, Y. Reliability of Rayleigh wave dispersion curve obtained from f–k spectral analysis of microtremor array measurement. *Soil Dyn. Earthq. Eng.* **2006**, *26*, 163–174. [CrossRef]
11. Roberts, J.; Asten, M. Estimating the shear velocity profile of Quaternary silts using microtremor array (SPAC) measurements. *Explor. Geophys.* **2005**, *36*, 34–40. [CrossRef]
12. Kim, Y.; Nam, B.H.; Youn, H. Development of a Probabilistic Spatio-Magnitude Sinkhole Hazard Model. In Proceedings of the Geo-Congress 2019: Engineering Geology; Site Characterization, and Geophysics, Philadelphia, PA, USA, 24–27 March 2019; pp. 81–90.
13. Tanaka, K.; Suda, T.; Hirai, K.; Sako, K.; Fukagawa, R. Monitoring of soil moisture and groundwater level using ultrasonic waves to predict slope failures. *Jpn J. Appl. Phys.* **2009**, *48*, 09KD12. [CrossRef]
14. Zhao, Z.; Ma, K.; Luo, Y. Application of Ultrasonic Pulse Velocity Test in the Detection of Hard Foreign Object in Farmland Soil. In *2019 IEEE 4th Advanced Information Technology, Electronic and Automation Control Conference (IAEAC)*; IEEE: Piscataway, NJ, USA, 2019; Volume 1, pp. 1311–1315.
15. Jin, Y.; Walker, E.; Krokhin, A.; Heo, H.; Choi, T.-Y.; Neogi, A. Enhanced Instantaneous Elastography in Tissues and Hard Materials Using Bulk Modulus and Density Determined without Externally Applied Material Deformation. *IEEE Trans. Ultrason. Ferroelectr. Freq. Control* **2019**, *67*, 624–634. [CrossRef]
16. Jin, Y.; Yang, T.; Heo, H.; Krokhin, A.; Shi, S.Q.; Dahotre, N.; Choi, T.-Y.; Neogi, A. Novel 2D Dynamic Elasticity Maps for Inspection of Anisotropic Properties in Fused Deposition Modeling Objects. *Polymers* **2020**, *12*, 1966. [CrossRef]
17. Jin, Y.; Walker, E.; Heo, H.; Krokhin, A.; Choi, T.Y.; Neogi, A. Nondestructive ultrasonic evaluation of fused deposition modeling based additively manufactured 3D-printed structures. *Smart Mater. Struct.* **2020**, *29*, 045020. [CrossRef]
18. Oelze, M.; Darmody, R.; O'Brien, W. Measurement of attenuation and speed of sound in soils for the purposes of imaging buried objects. *J. Acoust. Soc. Am.* **2001**, *109*, 2287. [CrossRef]
19. Oelze, M.L.; O'Brien, W.D.; Darmody, R.G. Measurement of attenuation and speed of sound in soils. *Soil Sci. Soc. Am. J.* **2002**, *66*, 788–796. [CrossRef]
20. Jin, Y.; Heo, H.; Walker, E.; Krokhin, A.; Choi, T.Y.; Neogi, A. The effects of temperature and frequency dispersion on sound speed in bulk poly (Vinyl Alcohol) poly (N-isopropylacrylamide) hydrogels caused by the phase transition. *Ultrasonics* **2020**, *104*, 105931. [CrossRef] [PubMed]
21. Jin, Y.; Yang, T.; Ju, S.; Zhang, H.; Choi, T.-Y.; Neogi, A. Thermally Tunable Dynamic and Static Elastic Properties of Hydrogel Due to Volumetric Phase Transition. *Polymers* **2020**, *12*, 1462. [CrossRef] [PubMed]
22. Lockwood, G.R.; Turnball, D.H.; Christopher, D.A.; Foster, F.S. Beyond 30 MHz. *IEEE Eng. Med. Biol. Mag.* **1996**, *15*, 60–71. [CrossRef]
23. Walker, E.L.; Reyes-Contreras, D.; Jin, Y.; Neogi, A. Tunable Hybrid Phononic Crystal Lens Using Thermo-Acoustic Polymers. *ACS Omega* **2019**, *4*, 16585–16590. [CrossRef] [PubMed]
24. Zubov, Y.; Djafari-Rouhani, B.; Jin, Y.; Sofield, M.; Walker, E.; Neogi, A.; Krokhin, A. Long-range nonspreading propagation of sound beam through periodic layered structure. *Commun. Phys.* **2020**, *3*, 1–8. [CrossRef]
25. Yang, T.; Jin, Y.; Choi, T.-y.; Dahotre, N.B.; Neogi, A. Mechanically tunable ultrasonic metamaterial lens with a subwavelength resolution at long working distances for bioimaging. *Smart Mater. Struct.* **2021**, *30*, 015022. [CrossRef]
26. Walker, E.L.; Jin, Y.; Reyes, D.; Neogi, A. Sub-wavelength lateral detection of tissue-approximating masses using an ultrasonic metamaterial lens. *Nat. Commun.* **2020**, *11*, 1–13. [CrossRef] [PubMed]

Article

Verification of a Nondestructive Method for Assessing the Humidity of Saline Brick Walls in Historical Buildings

Anna Hoła and Łukasz Sadowski *

Department of Building Engineering, Wroclaw University of Science and Technology, Wybrzeże Wyspiańskiego 27, 50-370 Wroclaw, Poland; anna.hola@pwr.edu.pl
* Correspondence: lukasz.sadowski@pwr.edu.pl

Received: 28 August 2020; Accepted: 29 September 2020; Published: 2 October 2020

Abstract: The paper presents the results of the verification of the neural method for assessing the humidity of saline brick walls. The method was previously developed by the authors and can be useful for the nondestructive assessment of the humidity of walls in historic buildings when destructive intervention during testing is not possible due to conservation restrictions. However, before being implemented in construction practice, this method requires validation by verification on other historic buildings, which to date has not been done. The paper presents the results of such verification, which has never been carried out before, and thus extends the scope of knowledge related to the issue. For experimental verification of the artificial neural network (ANN), the results of moisture tests of two selected historic buildings, other than those used for ANN learning and testing processes, were used. An artificial unidirectional multilayer neural network with backward error propagation and the algorithm for learning conjugate gradient (CG) was found to be useful for this purpose. The obtained satisfactory value of the linear correlation coefficient R of 0.807 and low average absolute error $|\Delta f|$ of 1.16% confirms this statement. The values of average relative error $|RE|$ of 19.02%, which were obtained in this research, were not very high for an in-situ study. Moreover, the relative error values $|RE|$ were mostly in the range of 15% to 25%.

Keywords: historic buildings; brick walls; nondestructive testing; artificial neural networks

---

1. Introduction

The problem of excessive moisture in brick walls usually concerns old buildings that have been used for several dozen years or more [1–6]. A special group of such old buildings concerns those that are included in the register of monuments and those that are subjected to special conservation protection. The building owners' duties include their proper maintenance and preservation; however, all works interfering with the historic tissue require prior permission from conservation authorities [7].

In the objects in question, the reason for excessive moisture is most often the lack of moisture insulation, the effect of which is the direct contact of the wall with the ground. Due to this, the water molecules contained in the ground—along with the salts dissolved in them—gradually penetrate into the wall's components, in turn moisturizing and salting it. The level of dampness and salinity of walls in such buildings usually significantly exceeds the permissible levels determined on the basis of classifications commonly accepted in technical literature [2–6], which are presented in Figure 1.

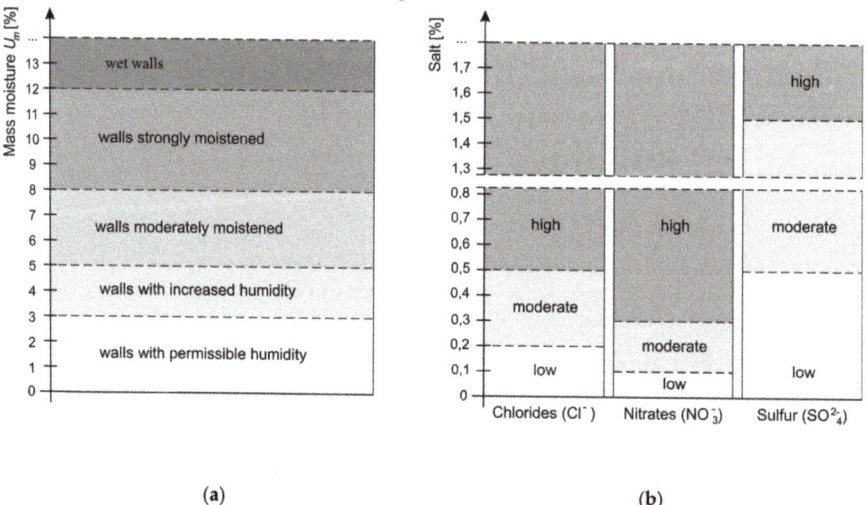

Figure 1. Classification of the moisture (**a**) and salinity (**b**) of brick walls (based on [8–14]).

The effect of long-term excessive moisture and salinity in a wall is its susceptibility to frost destruction over time and the falling off of plaster. This is followed by the chipping of brick and mortar fragments, which gradually reduces the wall's cross-section and decreases its strength.

In order to stop the above-mentioned destructive processes, it is first of all necessary to perform tests aimed at determining the amount and causes of the moisture. Of all the known methods for testing the humidity of brick walls, the most reliable results are obtained when using the destructive direct gravimetric method [11]. It allows the moisture value to be determined on the surface of the tested partition, as well as in its thickness. When using it, however, it is necessary to collect in-situ samples of material for laboratory tests. In the case of historic buildings, masonry sampling for moisture testing is only possible to a very limited extent. Paper [12] highlights the fact that destructive methods are not suitable for historical buildings of great architectural significance and also that nondestructive testing methods play an important role in the diagnostics of these buildings.

Of the nondestructive methods, electrical methods such as the dielectric, resistance, and microwave methods are considered to be very useful and are commonly used for testing brick wall moisture [15–18]. When conducting tests using these methods, it is not the humidity that is directly measured but instead a different physical feature of the wall. The value of this feature is affected by the water that is contained in the wall. Therefore, in the dielectric method the phenomenon of changing the dielectric constant of the tested material due to moisture is used; in the resistance method the change in material resistance depends on the amount of water it contains; while in the microwave method the attenuation of microwaves passing through moist material is measured.

The measurement result when using these methods is the unnamed parameter X. Therefore, in order to determine the moisture value, it is necessary to scale the apparatus on the tested object and determine the correlation relationship U_m-X, in which U_m usually means mass humidity. For this purpose, a specified number of moist masonry samples should be taken, and their mass humidity should be determined using the gravimetric method. The obtained humidity values should then be correlated with the corresponding readings of the meter that was used for the testing. A lack of consent of conservator services for taking samples often results in the skipping of the scaling procedure, and instead the results of scaling that were conducted by the apparatus manufacturer are being used. It is usually made on the basis of testing moist masonry samples that contain water without soluble salts, whereas in moist brick masonry, walls of existing objects such salts are present. As was shown in [19], their content significantly affects the results of tests that were obtained using electrical methods and also causes the overstatement of the estimated humidity value when compared to the actual value. In order to eliminate troublesome scaling of the apparatus on the examined object, it was proposed to use the possibilities offered by artificial intelligence. Nowadays, artificial intelligence is extensively used when predicting humidity values in various parts of buildings made of brick. The authors of works [20–22] recently determined the compressive strength of brick–mortar masonry based on neural networks, neuro-fuzzy inference systems, and nondestructive tests. The authors of work [23] employed support vector machines, neural networks, and Gaussian Naïve Bayes techniques for the evaluation of damage in a turn-of-the-century, six-story building with timber frames and masonry walls. The authors of work [24] used mobile deep learning for damage detection of historic masonry buildings. Another attempt to classify the damage of masonry historic structures based on convolutional neural networks and still images was presented in [25]. Convolutional neural networks were also used in [26,27] for the prediction and metamodeling of the hygrothermal behavior and performance of building components. In [28], a method based on numerical experiments for the identification of the thermal resistance of exterior walls of buildings was presented. However, there is still a lack of research that aims to use artificial neural networks and nondestructive methods to assess the humidity of saline brick walls in historic buildings.

This is why the authors of this publication, based on experimental research and numerical analysis, have developed a nondestructive neural method of assessing the humidity of saline brick walls in historic buildings, which is described in detail in [29]. This method is based on the use of artificial neural networks that are learned and tested on a data set built for this purpose. However, this method requires validation by verification on other historic buildings. This has not been done before and can be seen to be necessary before implementing it in construction practice. This issue is currently the subject of research, and the purpose of this work is to present the first results of such verification carried out on two historic buildings selected for this purpose. As mentioned before, such verification has not been done to date and is therefore a new contribution to the knowledge related to the issue.

2. Description of the Previously Developed Nondestructive Method for Assessing the Humidity of Saline Brick Walls in Historical Buildings

As described in [29], data collection was obtained on the basis of research of in-situ brick walls of several selected historic buildings from various historical periods from the 14th to 19th centuries. The tests were carried out in several hundred places. The data set, which is fully available in [29], therefore included several hundred sets of test results, each of which consists of six parameters. Two basic dimensionless parameters, X_D and X_M, describing the moisture of the wall, were determined using nondestructive, dielectric (Gann Hydromette Uni 2 meter with an active ball probe) and microwave (Trotec T 600 meter) methods. The use of the resistance method for testing was abandoned because its use in historic buildings is practically impossible due to the need to drill small holes in a wall for the proper application of the measuring head, which conservation services often do not agree to. The three auxiliary parameters A, C, and S, describing the molar concentration of nitrate, chloride, and sulphate salts harmful to the wall, were determined using the semi-quantitative method. However, the sixth parameter, U_m, describing the actual mass moisture of the wall, and which is needed to teach artificial neural networks, was determined using the gravimetric method with the use of a laboratory dryer. As a result of numerical analysis, the type and structure of neural networks were selected, as shown in Figure 2.

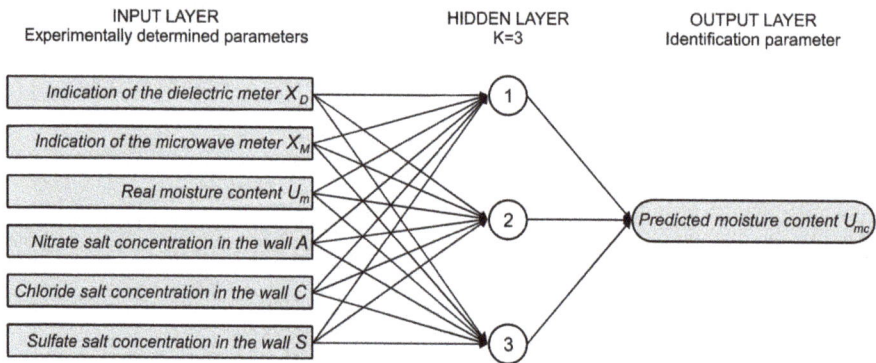

Figure 2. Structure of the artificial neural network used in [29].

Figure 3 shows the relationship between the real mass humidity U_m obtained by the gravimetric method and the U_{mc} humidity that was identified using an artificial neural network with the structure shown in Figure 2. Figure 3a shows this relationship in the learning process, while Figure 3b shows this relationship in the testing process. Figure 3a shows that the network correctly maps learning data and accurately identifies test data. This is demonstrated by the location of the points along the regression line, which corresponds to the ideal mapping, as well as the high values of the linear correlation coefficient R, which are 0.919 and 0.928, respectively, for the learning and testing process.

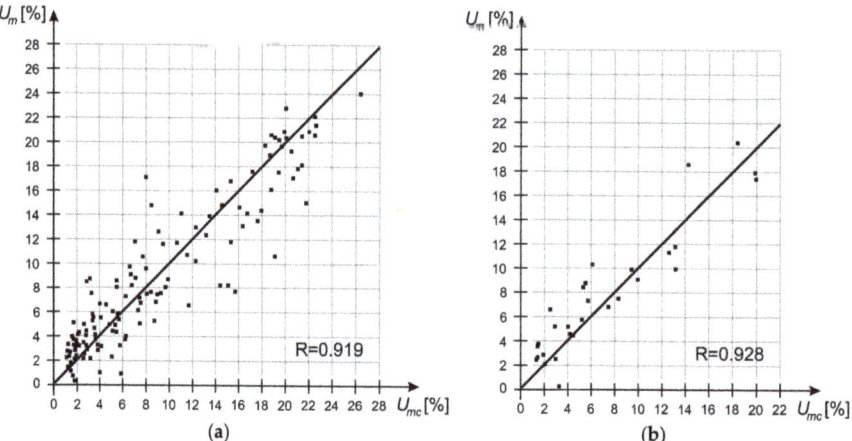

Figure 3. The relationship between the mass humidity, obtained using the gravimetric method, and humidity, which was identified by the network for the process of: (**a**) learning and (**b**) testing (based on the results presented in [29]).

The obtained satisfactory learning and testing results of the above-mentioned network indicate the possibility of a wider use of the developed method of the noninvasive assessment of the humidity of saline brick walls. This method seems to be very useful, especially in the case of historical buildings, where destructive interference during humidity and salinity tests should be kept to a minimum due to conservation restrictions [30–33].

3. Test Results Used to Verify the ANN Model

In order to carry out the experimental verification of the aforementioned artificial neural network, in-situ brick wall moisture tests were carried out on two historic buildings erected at the end of the 19th century, for which full ceramic brick and lime mortar were used. The research was carried out in autumn during cloudy days with no wind, sun, or rain. The outside air humidity was around 75% and the temperature was about 16 °C. The indoor relative humidity was 85% ± 5% and the air temperature was 18 °C ± 3 °C.

Figures 4 and 5 present photos from the examined objects. In turn, the characteristics of the objects are given in Table 1. In the case of both buildings, traces of moisture are clearly visible on their walls, and the height of the capillary water fringe reaches over 100 cm above the ground level.

Figure 4. Photographs of selected fragments of a former farm building in Katy Wroclawskie.

Figure 5. Pictures of selected fragments of the Redemptorist monastery in Wroclaw.

Table 1. Characteristics of the facilities where the tests were conducted for the purpose of verifying the artificial neural network.

Description of the Object	
OBJECT NO. 1	
Name, location	post-farm building, Katy Wroclawskie, Poland
Date of construction	end of the nineteenth century
Description of the object	■ erected on a rectangular plan with dimensions of 55.0 m × 10.0 m ■ the building has no basement, has two floors above ground, and a barren attic covered by a steep gable roof ■ served as an economic function for farm buildings; currently has a residential and service function
Place of measurement	■ brick basement walls ■ wall thickness in the range of 51 cm to 64 cm
OBJECT NO. 2	
Name, location	Redemptorist monastery, Wroclaw, Poland
Date of construction	end of the nineteenth century
Description of the object	■ erected on a rectangular plan with dimensions of 27.0 m × ~17.0 m ■ basement, two floors above ground, and a usable attic covered with a steep gable roof
Place of measurement	■ brick walls of the basement—external and internal ■ wall thickness in the range of 55 cm to 75 cm

Table 2 contains a set of data consisting of 15 sets of results, which were obtained on the basis of the tests of the historic buildings described above. Sets 1 to 7 were obtained from the post-farm building in Katy Wroclawskie and sets 8 to 15 from the monastery building in Wroclaw. As with previous studies [29], each set of results included two dimensionless parameters X_D and X_M, determined nondestructively using the dielectric (with the use of the Gann Hydromette Uni 2 meter with an active ball probe) and microwave (with the use of the Trotec T 600 meter) methods, respectively; three parameters A, C, and S, describing the molar concentrations of nitrate, chloride, and sulfate salts in walls, determined using the semi-quantitative method; as well as the U_m parameter obtained using the gravimetric method (with the use of a laboratory dryer), describing the actual mass humidity of the wall (in%). The samples were dried at a temperature of + 105 °C ± 1 °C.

Table 2. Test results of sets prepared for experimental verification of the artificial neural network.

Data set no. [-]	Height of the Measuring Point above Ground/Floor Level [cm]	Indication of the Dielectric Meter X_D [-]	Indication of the Microwave Meter X_M [-]	Salt Concentration in the Wall A [%]	Salt Concentration in the Wall C [%]	Salt Concentration in the Wall S [%]	Real Mass Humidity of the Wall U_m [%]
1	100	135.00	58.40	0.04	0.20	0.10	8.32
2	50	137.10	68.30	0.45	0.50	0.01	9.75
3	15	130.00	72.50	0.50	0.20	0.30	6.87
4	100	89.70	47.20	0.35	1.20	0.25	4.57
5	100	89.30	35.80	0.02	0.20	0.00	5.50
6	15	39.80	30.60	0.03	0.40	0.00	3.00
7	10	133.60	73.40	0.50	0.45	0.50	7.21
8	10	134.00	23.10	0.29	0.35	0.50	6.18
9	50	104.00	35.70	0.22	0.45	0.05	4.00
10	100	78.00	39.60	0.42	0.30	0.00	4.95
11	50	87.60	31.00	0.03	0.30	0.01	4.17
12	100	129.90	49.60	0.03	0.20	0.03	7.14
13	100	34.60	24.10	0.10	0.25	0.03	2.01
14	50	100.24	39.50	0.23	0.33	0.16	5.09
15	10	111.06	51.16	0.27	0.44	0.21	6.43

Table 3 summarizes selected statistical characteristics of the obtained test results.

Table 3. List of selected statistical characteristics of parameters.

Name and Symbol of the Characteristic	Parameter Symbol					
	X_D [-]	X_M [-]	A [%]	C [%]	S [%]	U_m [%]
Average value $\bar{x}_i$	34.60	23.10	0.02	0.20	0.00	2.01
Maximum value x_{max}	137.10	73.40	0.50	1.20	0.50	9.75
Minimum value x_{min}	102.26	45.33	0.23	0.38	0.14	5.68
Standard deviation S_x	32.15	16.12	0.18	0.24	0.17	1.97

4. Verification of the Artificial Neural Network

The results of the experimental verification of an unidirectional multilayer neural network (with back error propagation, the conjugate gradient algorithm (CG), and the number of hidden layer neurons equal to 3), which was previously taught and tested based on the results of the research presented in [29], are presented below. A total of 15 result sets were accepted for verification. This network generated a mass humidity value U_{mc} at each point. These values were compared with the actual U_m values obtained experimentally using the gravimetric method.

Figure 6 shows the relationship between U_m humidity obtained on the basis of gravimetric tests and U_{mc} humidity identified by the network for the experimental verification process. The obtained results show that the artificial neural network correctly mapped the verification data. This is evidenced by the location of points along the regression line, which correspond to the ideal mapping, and also the satisfactory value of the linear correlation coefficient R equal to 0.807.

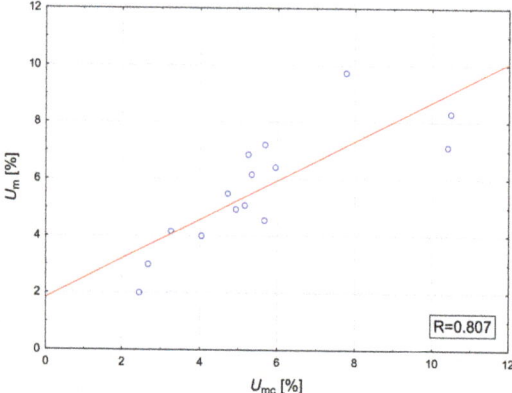

Figure 6. Relationship between U_m mass humidity obtained on the basis of gravimetric tests and U_{mc} humidity identified by the neural network for experimental verification.

Table 4 presents a comparative comparison of the U_m and U_{mc} humidity values determined using the gravimetric method and by means of the artificial neural network, respectively.

The results of the experimental verification presented in Table 3 indicate the correct identification of validation data. This is evidenced by the low average absolute error $|\Delta f|$ value of 1.16% and the average relative error $|RE|$ of 19.02% (not very high for an in-situ study). It is also worth noting that the average U_{mc} humidity value of 5.59% identified by the ANN is close to the U_m humidity value of 5.68% obtained by means of a test using the gravimetric method. According to the authors, the obtained results can be considered satisfactory.

Table 4. Comparative list of selected U_m and U_{mc} humidity values determined using the gravimetric method and artificial neural network, respectively.

Designation of Measuring Points	Moisture Content Obtained during Tests Using the Gravimetric Method	Moisture Content Identified by the ANN	Absolute Error	Relative Error
	U_m	U_{mc}	$|\Delta f|$	$|RE|$
	%	%	%	%
1	8.32	10.49	2.17	20.71
2	9.75	7.77	1.98	25.46
3	6.87	5.25	1.62	30.82
4	4.57	5.67	1.10	19.40
5	5.50	4.72	0.78	16.42
6	3.00	2.67	0.33	12.53
7	7.21	5.68	1.53	26.93
8	6.18	5.34	0.84	15.80
9	4.00	4.04	0.04	0.87
10	4.95	4.93	0.02	0.31
11	4.17	3.25	0.92	28.25
12	7.14	10.41	3.27	31.43
13	2.01	2.45	0.44	18.03
14	5.09	5.16	1.01	17.38
15	6.43	5.95	1.29	21.01
Mean value	**5.68**	**5.59**	**1.16**	**19.02**

Figure 7 presents the relative error histogram $|RE|$ of U_m mass humidity obtained using the gravimetric method, as well as the U_{mc} humidity identified by means of an artificial neural network for experimental verification. The figure also shows that the relative error values $|RE|$ are mostly in the range of 15% to 25%, which indicates a satisfactory distribution of the obtained results.

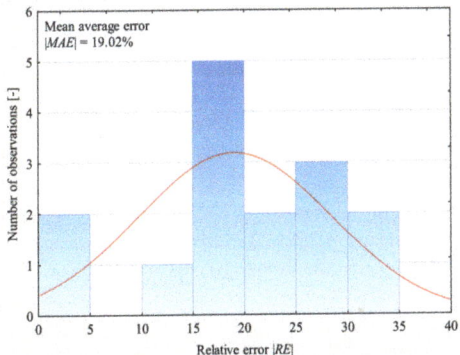

Figure 7. Relative error histogram $|RE|$ of U_m mass humidity obtained using the gravimetric method, as well as U_{mc} humidity identified by means of an artificial neural network for experimental verification (red curve shows the perfect fit for the normal distribution of the relative error).

5. Conclusions

Based on the research and analysis carried out to verify the neural method of assessing the humidity of saline brick walls, which was previously developed by the authors, the following conclusions can be drawn:

- It was shown that it is possible to reliably assess the humidity of a saline brick wall using an artificial neural network with a properly selected structure and learning algorithm on the basis of parameters assessed using the nondestructive dielectric and microwave methods.
- An artificial unidirectional multilayer neural network with backward error propagation and the algorithm for learning conjugate gradient (CG) is predisposed for this purpose. The obtained satisfactory value of the linear correlation coefficient R of 0.807 confirms this statement.
- The correct identification of validation data was evidenced by the low average absolute error $|\Delta f|$ value of 1.16% and the average relative error $|RE|$ of 19.02% (not very high for an in-situ study). Moreover, the relative error values $|RE|$ were mostly in the range of 15% to 25%.
- The average U_{mc} humidity value of 5.59% identified by the ANN was very close to the U_m humidity value of 5.68% obtained by means of a test using the gravimetric method.

Further verification work is currently underway. This paper presents the results obtained to date, which are based on the results of testing two buildings. Ultimately, the group of buildings that will be used for verification purposes will be more numerous.

Author Contributions: Conceptualization, A.H. and Ł.S.; methodology, Ł.S.; software, Ł.S.; validation, A.H. and Ł.S.; resources, A.H.; data curation, A.H. and Ł.S.; writing-original draft preparation, Ł.S.; writing—review and editing, A.H.; visualization, A.H.; All authors have read and agreed to the published version of the manuscript.

Funding: This research received no external funding.

Conflicts of Interest: The authors declare no conflict of interest.

References

1. Bajno, D.; Bednarz, L.; Matkowski, Z.; Raszczuk, K. Monitoring of thermal and moisture processes in various types of external historical walls. *Materials* **2020**, *13*, 505. [CrossRef] [PubMed]
2. Cabeza, A.P.; Camino, M.S.O.; Rodríguez, M.A.E.; Llorente, A.Á.; Pérez, M.P.S. Moisture influence on the thermal operation of the late 19th century brick facade, in a historic building in the city of Zamora. *Energies* **2020**, *13*, 1307. [CrossRef]
3. Guimarães, A.S.; Delgado, J.M.P.Q.; de Freitas, V.P.; Azevedo, A.C. Influence of different joints on moisture transport in building walls—A brief review. In *Diffusion Foundations*; Trans Tech Publications Ltd.: Bäch, Switzerland, 2019; Volume 22, pp. 19–23.
4. Falchi, L.; Slanzi, D.; Balliana, E.; Driussi, G.; Zendri, E. Rising damp in historical buildings: A venetian perspective. *Build. Environ.* **2018**, *131*, 117–127. [CrossRef]
5. Glavaš, H.; Hadzima-Nyarko, M.; Haničar, I.B.; Barić, T. Locating hidden elements in walls of cultural heritage buildings by using infrared thermography. *Buildings* **2019**, *9*, 32. [CrossRef]
6. Suchocki, C.; Damięcka, M.S.; Katzer, J.; Janicka, J.; Rapiński, J.; Stałowska, P. Remote detection of moisture and bio-deterioration of building walls by time-of-flight and phase-shift terrestrial laser scanners. *Remote Sens.* **2020**, *12*, 1708. [CrossRef]
7. Hoła, J. Degradacja budynków zabytkowych wskutek nadmiernego zawilgocenia–wybrane problemy. *Bud. Archit.* **2018**, *17*, 133–148. [CrossRef]
8. Goetzke, A.P.; Hoła, J. Influence of burnt clay brick salinity on moisture content evaluated by non-destructive electric methods. *Arch. Civ. Mech. Eng.* **2016**, *16*, 101–111. [CrossRef]
9. Czech Technical Standards. *Waterproofing of Buildings—The Rehabilitation of Damp Masonry and Additional Protection of Buildings Against Ground Moisture and Against Atmospheric Water—The Basic Provision*; CSN P 73 0610; Czech Technical Standards: Prague, Czech Republic, 2000.
10. WTA. *WTA 2-6-99-D, Erganzungen Zum Merkblatt 2-2-91-D Sanierputzsysteme*; WTA Publications: Zürich, Switzerland, 2001.
11. Hoła, A. Measuring of the moisture content in brick walls of historical buildings the overview of methods. In Proceedings of the 3rd International Conference on Innovative Materials, Structures and Technologies (IMST 2017), Riga, Latvia, 27–29 September 2017.
12. Rymarczyk, T.; Kłosowski, G.; Kozłowski, E. A non-destructive system based on electrical tomography and machine learning to analyze the moisture of buildings. *Sensors* **2018**, *18*, 2285. [CrossRef]
13. Concu, G.; Trulli, N.; Valdés, M. Knowledge acquisition of existing buildings by means of diagnostic surveying. Case studies. *Int. J. Struct. Glass Adv. Mater. Res.* **2018**, *2*, 22–29. [CrossRef]
14. Ruiz, L.V.; Flores, V.S.; Prieto, E.V. In situ assessment of superficial moisture condition in facades of historic building using non-destructive techniques. *Case Stud. Constr. Mater.* **2019**, *10*, e00228.
15. Freimanis, R.; Vaiskunaite, R.; Bezrucko, T.; Blumberga, A. In situ moisture assessment in external walls of historic building using non-destructive methods. *Environ. Clim. Technol.* **2019**, *23*, 122–134. [CrossRef]
16. Chastre, C.; Ludovico, M.M. Nondestructive testing methodology to assess the conservation of historic stone buildings and monuments. In *Handbook of Materials Failure Analysis with Case Studies from the Construction Industries*; Hamdy, A.S.M., Aliofkhazraei, M., Eds.; Elsevier: Alpharetta, GA, USA, 2018; pp. 255–294. ISBN 978-0-08-101928-3. [CrossRef]
17. Ince, I.; Bozdag, A.; Tosunlar, M.B.; Hatir, M.E.; Korkanc, M. Determination of deterioration of the main façade of the ferit pasa cistern by non-destructive techniques (Konya, Turkey). *Environ. Earth Sci.* **2018**, *77*, 420. [CrossRef]
18. Goetzke, A.P.; Hoła, A.; Sadowski, Ł. A non-destructive method of the evaluation of the moisture in saline brick walls using artificial neural networks. *Arch. Civ. Mech. Eng.* **2018**, *18*, 1729–1742. [CrossRef]
19. Goetzke, A.P. Identyfikacja Wilgotności Murów Ceglanych na Podstawie Badań Nieniszczących z Wykorzystaniem Sztucznych Sieci Neuronowych. Ph.D. Thesis, Wrocław University of Science and Technology, Wrocław, Poland, 2016.
20. Mishra, M.; Bhatia, A.S.; Maity, D. A comparative study of regression, neural network and neuro-fuzzy inference system for determining the compressive strength of brick–Mortar masonry by fusing nondestructive testing data. *Eng. Comput.* **2019**, 1–15. [CrossRef]

21. Mishra, M.; Bhatia, A.S.; Maity, D. Support vector machine for determining the compressive strength of brick-mortar masonry using NDT data fusion (case study: Kharagpur, India). *SN Appl. Sci.* **2019**, *1*, 564. [CrossRef]
22. Mishra, M.; Bhatia, A.S.; Maity, D. Predicting the compressive strength of unreinforced brick masonry using machine learning techniques validated on a case study of a museum through nondestructive testing. *J. Civ. Struct. Health Monit.* **2020**, *10*, 389–403. [CrossRef]
23. Nazarian, E.; Taylor, T.; Weifeng, T.; Ansari, F. Machine-learning-based approach for post event assessment of damage in a turn-of-the-century building structure. *J. Civ. Struct. Health Monit.* **2018**, *8*, 237–251. [CrossRef]
24. Wang, N.; Zhao, X.; Zhao, P.; Zhang, Y.; Zou, Z.; Ou, J. Automatic damage detection of historic masonry buildings based on mobile deep learning. *Autom. Constr.* **2019**, *103*, 53–66. [CrossRef]
25. Wang, N.; Zhao, Q.; Li, S.; Zhao, X.; Zhao, P. Damage classification for masonry historic structures using convolutional neural networks based on still images. *Comput. Aided Civil Infrastruct. Eng.* **2018**, *33*, 1073–1089. [CrossRef]
26. Tijskens, A.; Janssen, H.; Roels, S. Optimising convolutional neural networks to predict the hygrothermal performance of building components. *Energies* **2019**, *12*, 3966. [CrossRef]
27. Tijskens, A.; Roels, S.; Janssen, H. Neural networks for metamodelling the hygrothermal behaviour of building components. *Build. Environ.* **2019**, *162*, 106282. [CrossRef]
28. Chen, L.; Zhan, C.; Li, G.; Zhang, A. An artificial neural network identification method for thermal resistance of exterior walls of buildings based on numerical experiments. In *Building Simulation*, 3rd ed.; Tsinghua University Press: Beijing, China, 2019; Volume 12, pp. 425–440.
29. Hoła, A.; Sadowski, Ł. A method of the neural identification of the moisture content in brick walls of historic buildings on the basis of non-destructive tests. *Autom. Constr.* **2019**, *106*, 102850. [CrossRef]
30. Lourenço, P.B.; Luso, E.; Almeida, M.G. Defects and moisture problems in buildings from historical city centres: A case study in Portugal. *Build. Environ.* **2006**, *41*, 223–234. [CrossRef]
31. Brown, J.P.; Rose, W.B. Humidity and moisture in historic buildings: The origins of building and object conservation. *APT Bull. J. Preserv. Technol.* **1996**, *27*, 12–23. [CrossRef]
32. Sandrolini, F.; Franzoni, E. An operative protocol for reliable measurements of moisture in porous materials of ancient buildings. *Build. Environ.* **2006**, *41*, 1372–1380. [CrossRef]
33. D'Agostino, D. Moisture dynamics in an historical masonry structure: The Cathedral of Lecce (South Italy). *Build. Environ.* **2013**, *63*, 122–133. [CrossRef]

© 2020 by the authors. Licensee MDPI, Basel, Switzerland. This article is an open access article distributed under the terms and conditions of the Creative Commons Attribution (CC BY) license (http://creativecommons.org/licenses/by/4.0/).

Article

Atlas of Defects within a Global Building Inspection System

Clara Pereira, Jorge de Brito *, José D. Silvestre and Inês Flores-Colen

CERIS, Instituto Superior Técnico, Universidade de Lisboa, Av. Rovisco Pais, 1, 1049-001 Lisboa, Portugal; clareira@sapo.pt (C.P.); jose.silvestre@tecnico.ulisboa.pt (J.D.S.); ines.flores.colen@tecnico.ulisboa.pt (I.F.-C.)
* Correspondence: jb@civil.ist.utl.pt; Tel.: +351-21-841-8118

Received: 30 July 2020; Accepted: 21 August 2020; Published: 25 August 2020

Abstract: Building inspection systems are essential to optimise building maintenance. In the context of developing a global building inspection system, the lack of an expeditious tool to identify defects and their urgency of repair was detected. This study intends to propose an atlas of defects applicable to several types of building elements/materials, simplifying issues associated with the diagnosis of building pathology. A database was devised using previously developed components of the global inspection system: the classification list of defects and the urgency of repair parameters. Such a database was structured using several pages, each one with tables organised according to types of defects, building elements/materials and levels of urgency of repair (five-level scale—0–4). The atlas of defects has 38 pages in total, each for a different type of defect. The levels of urgency of repair are illustrated with photographs and described with concise classification criteria. Not all levels of urgency of repair apply to all defect–building element/material combinations; levels 1, 2 and 3 are those most often considered. The proposed atlas of defects is an innovative approach, useful to assist surveyors during technical inspections of buildings, whose concept may be adapted to other inspection systems.

Keywords: building defects; building diagnosis; building envelope; building inspection system; urgency of repair

---

1. Introduction

1.1. Background, Problem and Purpose

The built environment has a central position in the pathway to achieve a sustainable development [1]. Several issues may be key in this context, such as urban and spatial planning, energy use, greening, material selection, design strategies, thermal comfort and indoor air quality. Additionally, building maintenance has a relevant role. Maintenance strategies, if well planned, prolong the service life of buildings, hence contributing to their constant reuse, reducing the need for new buildings and decreasing construction and demolition waste. A global reduction of consumption of resources is thus possible through effective building maintenance.

Building maintenance plans should include a combination of proactive (preventive or predictive) and reactive strategies [2]. Nevertheless, proactive strategies are preferred, as reactive maintenance tends to result from more severe damage to buildings, thus involving higher costs. Although predictive proactive strategies are those that depend the most on building inspection, to assess the real state of building elements, all strategies benefit from building inspections in some way [3,4]. In this context, inspection results are paramount, as they are used to determine the type and boundaries of repair actions, thus influencing the efficiency of those actions and global costs. To make inspection procedures as unbiased as possible, they have to be systematised in order to increase objectivity and improve the diagnosis results. That is the role of building inspection systems. If such systems are adopted, procedures are standardised, the collection of information is organised, the technical language is homogenised, communication is improved, inspection activities become more agile and the whole process becomes less dependent on the surveyors' knowledge and experience [5]. First and foremost, the inspection results will be more reliable.

The architecture of building inspection systems should include methods to quantify the severity of degradation and the urgency of repair of building elements, as they are associated with maintenance optimisation [6–8]. The quantification of the degree of deterioration of building elements allows prioritising interventions [9] to comply with functional requirements at a minimal cost [10].

While systematising the knowledge on building pathology, a research team from Instituto Superior Técnico (IST), University of Lisbon (UL), developed a set of expert inspection systems, each one referring to a specific type of building element/material. These systems include classification lists of defects, their causes, diagnosis methods and repair techniques, as well as correlation matrices between defects and the other items. However, building inspections are not usually focused on a single type of building element/material, considering the building as a whole instead. So, surveyors would have to use different inspection systems to assess the real condition of different types of building elements/materials. To pragmatically tackle this issue, a global building inspection system is under development at IST–UL based on the individual inspection systems (Table 1). The harmonised classification lists of the global system, referring to defects, their causes, diagnosis methods and repair techniques, have already been published [11–14]. Additionally, each type of defect, in each type of building element/material, is given specific parameters to determine its urgency of repair [11]. Within this context, the main research question emerged: how can an expeditious tool to identify defects and their urgency of repair be achieved?

This study intends to propose an atlas of defects applicable to several types of non-structural building elements/materials, trying to clarify complex issues associated with the analysis of building pathology. The elements of the global building inspection system already developed are the basis of this research—the harmonised classification list of defects and the criteria to determine the urgency of repair [11]. The proposed atlas is expected to assist building surveyors during technical inspections, whether occasional or periodic.

A brief review of quick tools to assess building pathology is carried out. Then, the materials and methods to develop the atlas of defects are described. Next, the results are presented, showing excerpts of the atlas of defects, which are then discussed in Section 4.

1.2. Expeditious Tools to Assess Building Pathology

Investigating building pathology is a complex process involving the observation of anomalous occurrences, deciding whether they are defects and determining their causes and origin. It may comprise elaborate discovery procedures, involving several in situ non-destructive and destructive diagnosis methods or even laboratory tests on collected samples.

Focusing on the observation of building elements to detect defects, the inspection may be aided by some simple tools. Some of these, like binoculars or the zoom of a camera, help to get a closer look at building elements and decide whether a phenomenon is worth recording or just a temporary observation hindrance (e.g., an obstacle or the reflection from another building). Other tools provide basic data on building elements and detected defects, like a tape measure, which adds dimensional information to observations, or a spirit level, which immediately confirms the orthogonality of edges and surfaces. Additionally, portable and light comparison tools developed explicitly for inspection procedures are swift diagnosis methods that can be easily carried out in every inspection site and do not require advanced knowledge or a sophisticated apparatus [15]. Crack width rulers [16–18] and colour systems' samples [19–21] are two examples of such tools widely used, both using comparison as an operation principle.

Table 1. Expert inspection systems developed at Instituto Superior Técnico (IST), University of Lisbon (UL), which were the basis of the global inspection system.

References	Building Elements/Materials		Validation Sample
[22,23]	roofs	external claddings of pitched roofs (ECPR)	207 surfaces 164 buildings
[24–26]		flat roofs (FL)	105 surfaces 105 buildings
[27,28]	façade elements	door and window frames (DWF)	295 frames 96 buildings
[29,30]	façade claddings	wall renders (WR)	150 surfaces 55 buildings
[31,32]		external thermal insulation composite systems (ETICS)	146 façades 14 buildings
[33,34]		painted façades (PF)	105 façades 41 buildings
[35,36]		architectural concrete surfaces (ACS)	110 surfaces 53 buildings
[37–39]	façade claddings and floorings	adhesive ceramic tiling (ACT)	88 surfaces 46 buildings
[40,41]		natural stone claddings (NSC)	128 surfaces 59 buildings
[42,43]	floorings	wood floorings (WF)	98 floorings 35 buildings
[44]		epoxy resin industrial floor coatings (ERIFC)	29 floorings 23 buildings
[45,46]		vinyl and linoleum floorings (VLF)	101 floorings 6 buildings

Using crack width rulers, the surveyor is provided with a small transparent rectangle with a set of printed lines ordered from thinnest to thickest [16]. Each line is identified by its thickness, allowing the easy determination of the width of a crack throughout its length. In other words, graphical data are complemented with quantitative data, providing a user-friendly instrument [17,18].

Colour system's samples are a set of cards manufactured in a resistant paper and printed with solid colours in order to be compared with those of building elements observed on-site. Each solid colour is identified with a code according to the predetermined colour system, such as the natural colour system [19] or the Munsell colour system [20,21]. The use of colour samples allows comparing (i) the observed colour with that of the beginning of the service life of the building element and (ii) the observed colour in different areas of a building element or building. Once more, graphical information, complemented with coded information, allows a more accurate diagnosis methodology.

Additionally, standards EN ISO 4628-4:2016 [47] and EN ISO 4628-2:2016 [48] propose comparison methods to evaluate the degradation of paint and varnish coatings, namely the assessment of the degree of cracking and blistering, respectively. These standards provide pictorial criteria, allowing surveyors to match the visual characteristics of detected defects with those of the standard, referring to the size, quantity and density of cracking and blistering. In these situations, the assessment of coatings is also enhanced by the use of graphical information.

The advantages of the mentioned easy-to-use, graphical and informative tools to aid inspection procedures raise interest in an analogous tool to help to observe defects and systematise their diagnosis. Such a tool, proposed in this study as an "atlas of defects", would be an image-based scale to ascertain the type of detected defect and assess its urgency of repair, complemented with essential written criteria difficult to express through photographs.

Several scales are widely used in various fields of science, for instance, to measure [9] wind force (Beaufort scale), the intensity of tornadoes (Fujita–Pearson scale), hurricane winds (Saffir–Simpson hurricane wind scale), earthquakes (modified Mercalli intensity scale), the magnitude of earthquakes (Richter scale), mineral hardness (Mohs scale of mineral hardness), the likelihood of developing pressure ulcers (Norton scale) and the state of a person's consciousness (Glasgow coma scale). Psychometric scales are used in questionnaires, like the visual analogue scale (VAS) or the Likert scale. One of the main differences between scales is the ability to use measurable data (e.g., Richter scale) or only to consider features of phenomena (Mohs scale), sometimes with considerable levels of subjectivity.

Likewise, in the context of building inspection, several scales have been proposed to measure the degradation of building elements. Each scale is associated with a specific study or inspection system. Some of the most relevant are described next, in a non-exhaustive perspective.

The study of a Markov decision model for rationalising building maintenance [49] makes use of a six-point scale to determine the state of a building component during an inspection. This ordinal scale goes from 1 to 6, whose condition states are described as excellent (1), good (2), reasonable (3), moderate (4), bad (5) and very bad (6). Each condition is further described with some general characteristics, which are used for direct assignment of a condition state. The set of conditions is used to associate actions with the transition between condition states to estimate and optimise maintenance costs.

The "Méthode d'Évaluation de scenarios de Dégradation probables d'Investissements Correspondants" (MEDIC) was developed for the "Energy Performance, Indoor environment Quality, Retrofit" (EPIQR) system [50–52]. The MEDIC consists of using four codes (a, b, c and d) to describe the deterioration state of building elements. Code a corresponds to an element in good condition and code d to an element that needs replacement. This method is based on curves of probability of deterioration determined for each type of building element. The surveyor directly assigns a code to the building element, and then, using the predetermined curves, the probability of transition to another deterioration state may be better estimated [50].

Allehaux and Tessier [53] developed a methodology to evaluate the obsolescence of office buildings. This method was prepared to audit all current building services expected in office buildings. For each predetermined obsolescence criterion and object, three ratings may be attributed according to descriptive parameters. Ratings A, B and C correspond to good, medium and poor or not sufficient condition state, respectively.

The National Aeronautics and Space Administration (NASA) developed a maintenance guide for facilities [54]. The proposed maintenance model includes three sets of metrics: the system condition index (SCI), the facility condition index (FCI) and the deferred maintenance cost estimate. Condition assessment ratings support the definition of SCI and FCI. A five-tier condition rating system is used, namely excellent (5), good (4), fair (3), poor (2), and bad (1). When a system does not exist, rating 0 is attributed.

Rodrigues et al. [55] developed a visual survey methodology to assess the defects of the building envelope of social housing. Within this methodology, an evaluation scale was proposed, composed of eight degradation levels with specific parameters for each level according to the building element, taking the intensity, extent and location of damages into account. Level 10 is the most favourable degradation level, referring to an exceptional condition, not requiring any intervention, just planned maintenance actions for conservation purposes. Level 3 is the most severe degradation level, referring to an unacceptable condition, which is unsuitable for rehabilitation and requiring replacement. The survey methodology includes a step of aggregation of results for each type of building element. That aggregation considers the same scale, attributing a global degradation level according to the relative frequencies of degradation levels of each inspected building element.

Elhakeem and Hegazy [56] proposed using the severity of various deficiencies, defined by the surveyor on a scale from 0 to 100, to compute the overall deterioration index. That index also considers a weight attributed to each type of deficiency. The deterioration index is used to generate repair scenarios to optimise maintenance decisions, taking costs into account.

Morgado et al. [57,58] proposed methodologies to implement maintenance planning in pitched and flat roofs. Those methodologies encompass the establishment of intervention priorities for maintenance actions. In these studies, priorities are determined by the combination of four criteria: environmental aggressiveness of the surroundings, the extent of the defect, the level of deterioration of the building elements, and the severity of the defect. The last two criteria use two different scales, both for direct assignment. The degradation condition of the building elements is rated according to a four-level scale, ranging from 0 to 3, with the following definitions: no visible degradation (0), superficial degradation (1), moderate degradation (2) and generalised degradation (3). To determine the severity of the defects, a five-levels scale is used, from grades A to E, where A refers to "negative influence on the aesthetical aspect" and E to "risk to the safety of users". The combination of the different criteria is computed using weights and multiplying factors to determine a quantifiable intervention priority.

The condition assessment method developed by the US Army Corps of Engineers [59] establishes two main types of assessment: detailed distress surveys and direct condition ratings. These methods are intended to be used in different scenarios, the latter being faster to use and the former more meticulous. While in detailed distress surveys the surveyor identifies each type of defect detected in a building component, when using direct condition ratings, the surveyor evaluates each component in more general terms. To measure the degradation in detailed distress surveys, each detected defect is classified according to severity levels (low, medium and high), whose criteria are defined for each type of defect in each type of building component. With direct condition ratings, the surveyor assesses each component according to a set of general rating criteria. Three main ratings are defined: green, amber and red, from little to serious serviceability loss due to degradation. Each rating is further subdivided into three categories: high (+), low (−) and middle. Both methodologies are used to compute condition indexes at different levels: component-section, building component, system and building.

Severity ratings of defects are also used in research about a decision-making process to select façade materials minimising potential defects at the design stage [60]. In this situation, severity ratings are used to determine the criticality of defects according to a five-point scale, where the seriousness of defects is assessed. These severity ratings are as follows: extremely minor (1), minor (2), moderate (3), major (4) and extremely major (5). Another scale determines the frequency rating of defects.

Bortolini and Forcada [61] developed a building inspection system for evaluating the technical performance of existing buildings. The third step of that system includes the assessment of the severity of defects using a severity rating associated with the effects of defects and with their urgency of repair. Three levels of severity are determined according to general criteria: severity 1 refers to low impact, severity 2 to moderate impact and severity 3 to severe impact. The assessment results are used to support maintenance recommendations.

The set of individual expert inspection systems developed by the authors' research team at IST–UL includes, in each detailed file of defects, the classification parameters to determine the severity/repair urgency level of each detected defect. These parameters are used to rate defects from level 0 to level 2. Level 0 corresponds to the need for immediate intervention, level 1 to the need for intervention in the medium term, and level 2 to the need for monitoring the progression of the defect. When the set of individual inspection systems was used to develop the global inspection system, where the atlas of defects was included, the three levels were adapted to a five-level scale, considering implicit gravity differences between different building elements/materials, i.e., the effects of defects may be aggravated by the characteristics of some materials. Furthermore, the global inspection system was developed in conjunction with service life prediction models for the same scope of building elements/materials [62]. Those models already included a methodology to determine the severity of degradation of building elements. For this reason, in the global building inspection system, the rating of defects was restricted to the urgency of repair [11].

At IST-UL, a research team has been developing service life prediction methodologies for elements/materials of the building envelope [62]. Those methodologies start by determining the following: the weighted severity of degradation of building elements (%) according to the area of the building element affected by each defect; that defect's multiplying factor as a function of its condition (from 0 to 4); and the type of defect's weighting coefficient, which corresponds to the relative importance of each defect. The obtained severity of degradation is then placed on a discrete scale that qualitatively defines the degradation level. So, two different scales are used: (i) one to determine the detected defects' multiplying factor according to its condition, varying from 0 to 4, where 0 correspond to very slight degradation and 4 to severe degradation; and (ii) another to define the building elements' degradation level, varying from no visible degradation (level A) to generalised degradation (level E), matching percentages of severity of degradation.

Ruiz et al. [9] studied the optimal metric for condition rating scales. This study tested a representative group of experts on how they would classify 33 different cases of building pathology in building elements through direct assignment, according to a proposed scale with 11 levels of severity. The statistical analysis of the answers allowed determining that the proposed scale needed improvement since only a 32% probability of correct classification of the phenomena was achieved. Using a clustering algorithm, a five-level scale was determined to provide the lowest standard deviation of the global error. IST–UL's global inspection system also proposes five levels of urgency of repair. Additionally, Ruiz et al. [9] conclude that a catalogue of images of building elements with reference values of severity would be a valuable contribution to classify building pathology, increasing the accuracy of assigning levels of severity of degradation. The proposed atlas of defects is the realisation of such a catalogue.

Table 2 summarises the mentioned scales used to measure the degradation of building elements. The scales' measure, levels and type of use are highlighted.

Table 2. Examples of scales used to measure the degradation of building elements.

Reference	Measure	Levels	Use
[49]	state of a building component	1–6 (excellent to very bad)	direct assignment
[50]	deterioration state of elements	a, b, c and d (from the least to the most severe)	direct assignment
[53]	obsolescence criteria ratings for elements of office buildings	A, B and C (from good to poor or not sufficient)	algorithm
[54]	condition assessment ratings of building systems	5, 4, 3, 2 and 1 (from excellent to bad)	direct assignment
[55]	level of severity of degradation of building elements	3–10 (8 levels, from the most to the least severe)	algorithm
[56]	inspected severity of deficiencies	0–100 (from the least to the most severe)	direct assignment
[57,58]	degradation condition of building elements	0, 1, 2 and 3 (from not visible to generalised degradation)	direct assignment
[57,58]	severity of defects	A, B, C, D and E (from the least to the most severe)	direct assignment
[59]	distress severity (detailed distress survey)	low, medium or high	algorithm
[59]	direct rating of serviceability problems due to degradation	from Green (+) to Red (−) (9 levels, from the least to the most severe)	direct assignment
[60]	severity rating (seriousness) of defects	1, 2, 3, 4 and 5 (from extremely minor to extremely major)	direct assignment
[61]	severity rating of defects (effects and urgency of repair)	1, 2 and 3 (from low to severe impact)	direct assignment
[8,11,14,16,18, 19,21,25,27,29– 31]	severity/repair urgency level of defects	0, 1 and 2 (from the most to the least severe)	algorithm
[62]	defects' condition	0, 1, 2, 3 and 4 (from the least to the most severe)	algorithm
[62]	severity of degradation of building elements	A, B, C, D and E (from the least to the most severe)	algorithm
[9]	severity of degradation of building elements	0, 1, 2, 3 and 4 (from the least to the most severe)	direct assignment

2. Materials and Methods

Considering the objective of creating a catalogue to ease matching building degradation phenomena with defects in a classification list and with a level of urgency of repair, a database was devised for reference during fieldwork. Previously developed components of the global building inspection system were used, namely the classification list of defects [11] and the detailed files of defects, where the urgency of repair was characterised.

First, the structure of such a database was designed. The atlas of defects has several pages, each corresponding to a defect listed in the classification of defects of the global building inspection system. Each page is arranged as a table with columns corresponding to types of building elements/materials and rows to levels of urgency of repair. The number of columns varies according to the field of application specified in the detailed file of the defect (Figure 1). The levels of urgency of repair are also established in the files of defects (Figure 1), varying from 0 to 4, where level 0 is the most urgent and level 4 the least urgent. So, the table has five rows. In each cell of each page's table, an exemplifying photograph and a concise description are available, establishing the criteria to identify a defect in a type of building element/material with a specific level of urgency of repair.

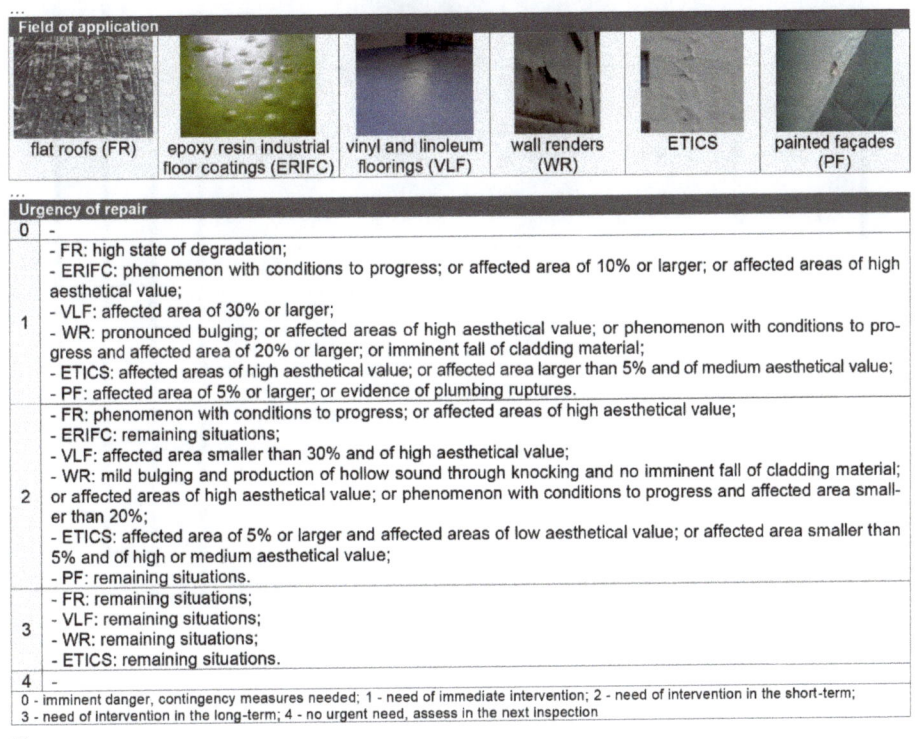

Figure 1. Excerpt of the detailed file of defect "A-B4 Blistering/bulging" showing the fields "field of application" and "urgency of repair".

The levels of urgency of repair are defined as follows [11]:

- 0: imminent danger, contingency measures needed;

- 1: need for immediate intervention;
- 2: need for intervention in the short-term;
- 3: need for intervention in the long-term;
- 4: no urgent need, assess in the next inspection.

During fieldwork, the surveyor attributes a level of urgency of repair to each detected defect based on the application of algorithms. In other words, each level of urgency of repair corresponds to a set of conditions determined for each type of building defect according to the type of building element/material.

With the structure of the atlas of defects laid down (Figure 2), all the pages were filled with graphical and descriptive contents. The photographic content of the atlas of defects was collected from the authors' research team, mainly from inspection campaigns, or taken specifically to fill the atlas of defects.

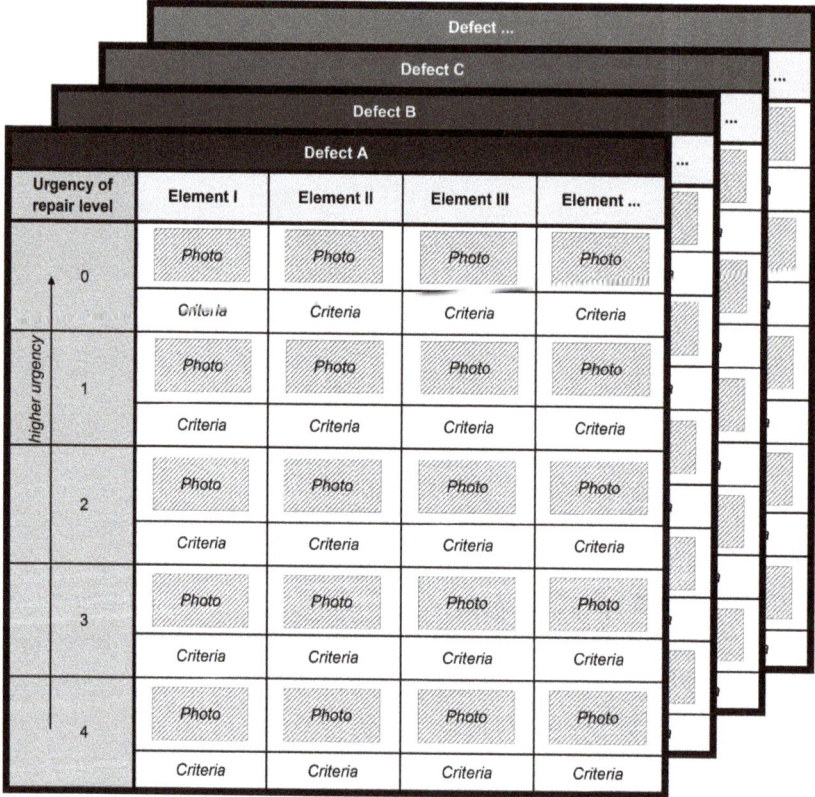

Figure 2. Structure of the atlas of defects.

3. Results

Figures 3 and 4 show excerpts of the atlas of defects of the global building inspection system. Figure 3 partially shows the page of defect "A-A3 Dirt and accumulation of debris" with columns corresponding to external claddings of pitched roofs, door and window frames, and wall renders. Figure 4 shows a part of the page of defect "A-C2 Oriented cracking on the current surface", presenting the columns for adhesive ceramic tiling, natural stone claddings and architectural concrete surfaces.

Urgency of repair level	A-A3 Dirt and accumulation of debris		
	External claddings of pitched roofs	Door and window frames	Wall renders
0			
1			 - Affected areas of high aesthetical value; - Or high probability of repetition of the phenomenon and affected area larger than 65%.
2	 - Leakage occurs; - Or phenomenon with conditions to progress.	 - High chemical aggressiveness of debris; - Or affected areas of high aesthetical value; - Or incorrect operation of moveable leaves; - Or obstruction of drainage holes.	 Affected areas of medium aesthetical value and larger than 20%.
3	 Affected areas of high aesthetical value.	 Remaining situations.	Remaining situations.
4	 Remaining situations.		

Figure 3. Excerpt of the atlas of defects: page of defect "A-A3 Dirt and accumulation of debris" with columns corresponding to three different types of building elements/materials.

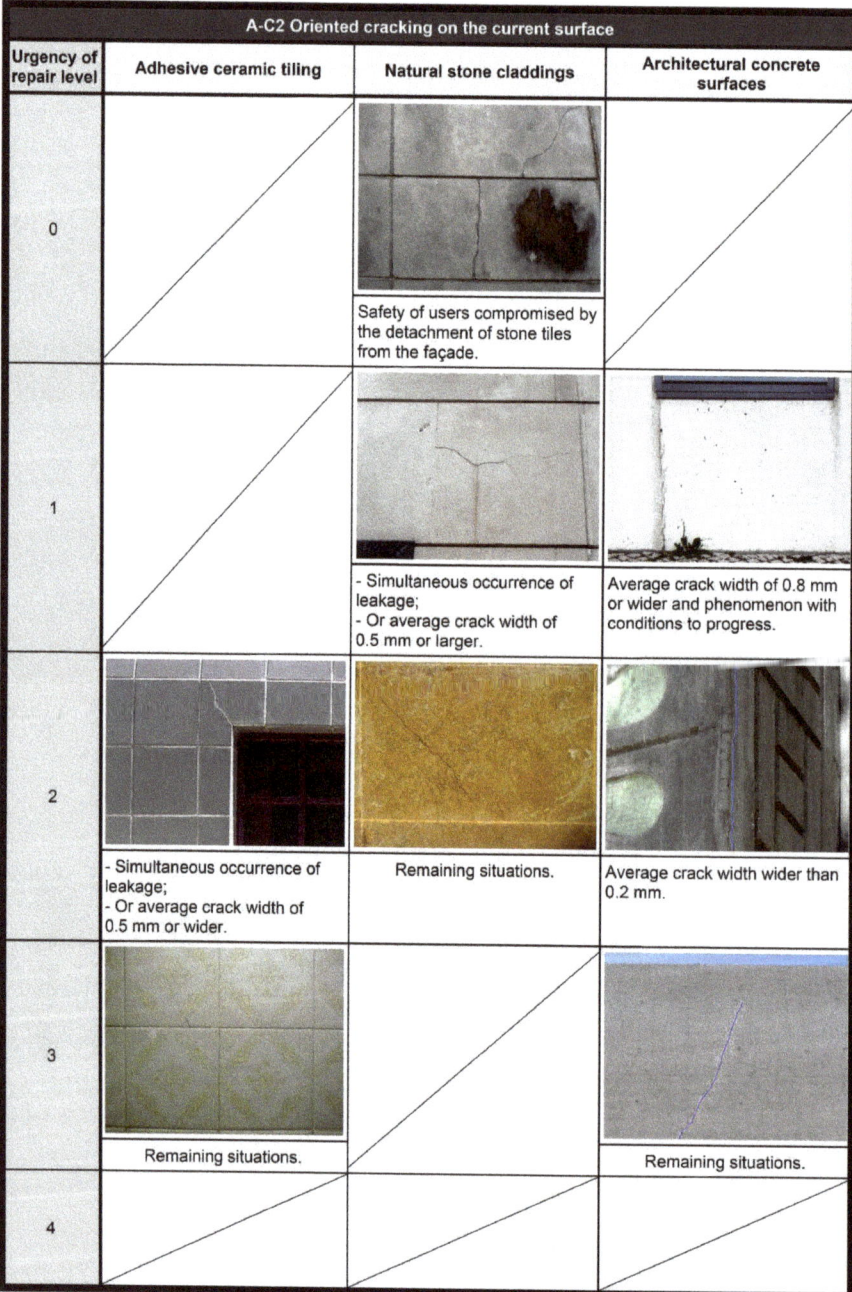

Figure 4. Excerpt of the atlas of defects: page of defect "A-C2 Oriented cracking on the current surface" with columns corresponding to three different types of building elements/materials.

Figures 3 and 4 show that not all levels of urgency of repair apply to a type of defect in every type of building element/material. For instance, in Figure 3, for external claddings of pitched roofs, only levels 2, 3 and 4 are defined. In the same figure, for door and window frames, only levels 2 and 3 are determined. This is associated with the effects of a defect in different building elements/materials. In these cases, "dirt and accumulation of debris" (A-A3) in claddings of pitched roofs or in door and window frames is considered to only result in interventions in the short term in the most urgent cases, as this defect does not usually raise safety concerns. In pitched roofs, lighter cases of dirt and accumulation of debris are even considered to only require monitoring in subsequent inspections (level 4).

However, in the case of "Oriented cracking on the current surface" (A-C2) on natural stone claddings, defects may go from levels 0 to 2 of urgency of repair (Figure 4). The occurrence of cracks on natural stone claddings is considered to endanger users and passers-by in severe situations, requiring contingency measures (level 0).

Considering the number of types of defects (38) included in the corresponding classification list and the (varying) number of building elements/materials each one applies to, the atlas of defects has a total of 179 columns (combination defect–building element/material). Of the latter, 94% are filled in the row corresponding to level 2 of urgency of repair, while only 1% are filled in the row corresponding to level 4. The row of level 0 is also sparsely filled, with only 13% of columns filled. Levels 1, 2 and 3 are those more often defined for defect–building element/material combinations, with rows filled in more than 50% of columns.

Additionally, out of the 23 defect–building element/material combinations that are defined in level 0, 14 (61%) refer to defects occurring in natural stone claddings. The possibility of detachment of a stone slab from a façade is regarded as a menace to the safety of passers-by due to its weight, thus increasing the prospective level of urgency of repair of defects on natural stone claddings.

Observing the descriptive contents of Figures 3 and 4, different types of criteria may be distinguished. Some refer to the simultaneous occurrence of other defects (e.g., leakage), some to the context of the defect (e.g., the aesthetical value of affected areas), some to the characteristics of the defect (e.g., extent) and others to the effects of the defect (e.g., safety issues). Many criteria are not identifiable through exemplificative photographs. It is the case of the phenomenon's conditions to progress, aesthetical value of the affected areas, safety risks for passers-by and percentage quantification of the affected area.

A phenomenon's conditions to progress can only be conclusively analysed on-site. Data like temperature, relative humidity, aggressiveness of the environment and maintenance means, among others, should be taken into account. They may be recorded for off-site assessment, but at least one visit to the site under inspection is required. Some signs of this type of data may be identified in photographs, but they generally go beyond the visual information (e.g., a cloudy photograph may not correspond to low temperatures). Therefore, non-visual information should be taken into account while reading the atlas of defects (e.g., Figure 3, level 2 of "A-A3 Dirt and accumulation of debris" in external claddings of pitched roofs).

The aesthetical value of the affected areas is generally determined by the importance of the façade, roof or flooring in the context of the building. Typically, in current buildings, the front façade has high aesthetical value due to being adjacent to a public street and encompassing the main entrance to the building, while side and rear façades normally have medium and low aesthetical value, respectively. Since the photographs that illustrate the atlas of defects tend to focus on the pathological phenomena, the aesthetical value of the building elements where they are detected is not obvious. This is the case in the pictures illustrating the column of wall renders in Figure 3. If the case of "Dirt and accumulation of debris" (A-A3), illustrating level 3 of urgency of repair in wall renders, occurred on the front façade of the building, it would then be considered of level 1, thus requiring immediate intervention, instead of intervention in the long-term.

As for safety risks for passers-by, the criteria to determine the urgency of repair of some defects in some building elements/materials may refer to slippery floorings, risk of stumbling or the possibility of elements detaching and falling from façade claddings (as proposed by the system of evaluation of façades of Ruiz et al. [63]). These occurrences are not usually evident in photographs, being identified through a tactile assessment or depending on the height of the phenomenon. For instance, the possibility of a ceramic tile falling from a façade is associated with different safety risks according to the height of the defect: risks are lower immediately above the floor than at 10 m high. Still, that height is not usually understandable in photographs centred on defects. So, the unclear expression of safety risks needs to be considered while referring to the atlas of defects. This is the case in the picture illustrating the occurrence of "Oriented cracking on the current surface" (A-C2) on a natural stone cladding with level 0 of urgency of repair (Figure 4).

The percentage quantification of the affected area is a characteristic of the defect that depends not only on its extent but also on the extent of the whole surface. Although it is possible to estimate the absolute area affected by the defect, whether using on-site measurements or reference elements in the photograph (like tiles or windows with known size), the whole surface where the defect occurs is not usually visible in defect-centred photographs. For that reason, pictures like those illustrating the column of wall renders in Figure 3 do not provide all the information needed to determine the urgency of repair of situations of "Dirt and accumulation of debris" (A-A3).

In short, the usefulness of the atlas of defects depends on the inseparable combination of exemplifying photographs with descriptive criteria.

Moreover, it should be stressed that the same defect may manifest itself in different ways, even in the same type of building element/material (e.g., different fungi). In its current state, the atlas of defects is not an exhaustive catalogue of all possible forms of defects.

4. Discussion

Considering the excerpts of the atlas of defects presented in Figures 3 and 4, cases of defects "A-A3 Dirt and accumulation of debris" and "A-C2 Oriented cracking on the current surface" may be analysed and classified in terms of urgency of repair.

Figure 5 shows three different window frames affected by "Dirt and accumulation of debris" (A-A3). The window frames in Figure 5a,b are of anodised aluminium and the one in Figure 5c of reinforced concrete. To determine the urgency of repair of each window frame, the criteria determined in the atlas of defects should be analysed, starting at the most urgent level of repair, which is level 2, for door and window frames (Figure 3). In Figure 5, none of the observed debris has high chemical aggressiveness. In terms of aesthetical value, the window in Figure 5b is located on a front façade, thus having high aesthetical value. The observance of this criterion alone makes this a case of level 2 of urgency of repair, even though only a white spot of paint is observed in the window frame. As for the incorrect operation of moveable leaves, Figure 5b,c show fixed windows, but Figure 5a shows a sliding window, where the accumulation of debris next to the sliding guides probably hinders the movement of the leaves and prevents them from shutting completely. So, the case in Figure 5a is also of level 2 of urgency of repair. This case also shows obstruction of the drainage holes of the window frame, which reinforces the need for intervention in the short-term (level 2). The case in Figure 5c does not match any of the criteria stated for level 2 and thus is considered a situation with level 3 of urgency of repair. It is interesting to conclude that both cases in Figure 5a,b have the same level of urgency of repair while being so different, with the debris of Figure 5a appearing to be much more worrying. If considered in the classification parameters, the aesthetical value of the affected areas may be very influential.

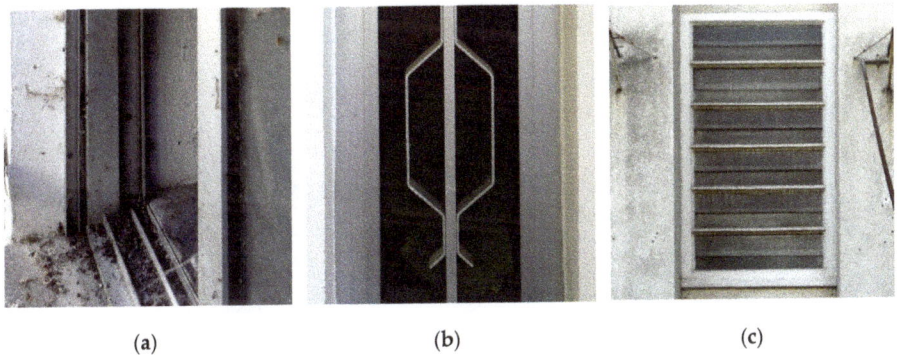

Figure 5. Cases of defect "A-A3 Dirt and accumulation of debris" in window frames: (**a**) accumulation of debris affecting the operation of moveable leaves and obstructing drainage holes; (**b**) dirt in a window frame on a front façade; (**c**) dirt in a window frame on a rear façade.

Accumulation of debris is also a factor considered by Fernandes et al. [64] to determine the degradation level of window frames. The accumulation of debris is considered to have a low impact on the overall degradation condition of window frames, thus being associated with a weighting coefficient of 0.1. Furthermore, whether this defect affects sealings, fittings or framework materials and coatings, it is always rated as level 1 of defect condition (from 0 to 4, from the least to the most severe), even if the accumulation of debris affects more than 20% of the component. An assessment of the defect's effects is taken into account, considering that corrective cleaning operations can eliminate this defect and regular cleaning may prevent its reoccurrence. Still, Fernandes et al.'s [64] approach does not invalidate the urgency of repair approach, as the low impact of a defect's effects in terms of degradation does not contradict the importance of eliminating such defect, even more so when the defect may cause other damages in the future.

Figure 6 presents six cases (a–f) of "A-A3 Dirt and accumulation of debris" on wall renders. Two cases per building are shown: Figure 6a,b occur in the same building, Figure 6d,e occur in another building and Figure 6c,d occur in yet another building. For wall renders, defect A-A3 is classified from levels 1 to 3 of repair urgency (Figure 3). As for window frames, to assign a level of urgency of repair to each case in Figure 6, they must be analysed in light of the criteria determined for each level, starting from the most urgent one. So, if the affected areas are of high aesthetical value, level 1 applies, which is the case in Figure 6a,f. Then, situations with an affected area larger than 65% of the surface and high probability of repetition of the phenomenon are also considered as level 1 of urgency of repair. Figure 6d observes these criteria, with about 77% of the surface affected by the accumulation of dirt associated with a thermophoresis phenomenon and no measures implemented to stop the defect from progressing. If the affected areas are of medium aesthetical value and larger than 20% of the rendered surface, they should be classified within level 2 of urgency of repair. Figure 6c,e occur on side façades, thus having medium aesthetical value, but only the affected area of Figure 6e represents more than 20% of the rendered surface (about 58%). So, Figure 6e requires intervention in the short-term (level 2). The remaining situations, i.e., Figure 6b,c, are of level 3 of urgency of repair, as their characteristics do not match any of the parameters defined for levels 1 and 2.

Summing up, the urgency of repair of cases in Figure 6 is as follows:

- Figure 6a: level 1;
- Figure 6b: level 3;
- Figure 6c: level 3;
- Figure 6d: level 1;
- Figure 6e: level 2;

- Figure 6f: level 1.

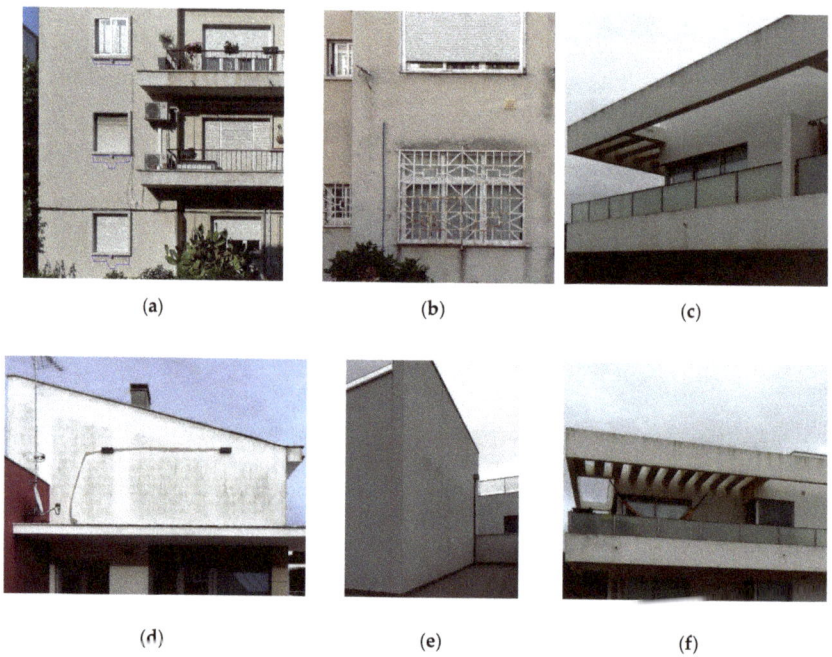

Figure 6. Cases of defect "A-A3 Dirt and accumulation of debris" on wall renders: (**a**) dirt below windowsills on a front façade, (**b**) dirt on a rear façade, (**c**) dirt on a side façade, (**d**) dirt on a rear façade, (**e**) dirt on a side façade, (**f**) dirt on a front façade.

Once again, the criterion associated with the aesthetical value of the affected areas is decisive. For instance, Figure 6c,f have identical characteristics, but the importance of Figure 6f, showing the front façade, which is adjacent to the public street and forms a uniform set of façades that frames that street, determines that, when prioritising the repairs, this case should be repaired first.

The analysis of van Hees et al. [65] of the damage of old mortars systematises the factors affecting the mortars' degradation processes. Of those, some may be highlighted for affecting the development of "Dirt and accumulation of debris" (A-A3) in the cases shown in Figure 6. In terms of environmental factors, the moisture supply and the air pollution impact all the mentioned cases. Furthermore, the moisture supply, in these cases, is associated with rain runoff and surface water. In Figure 6c,f, the effect of rainwater runoff is more evident as no coping/capping stone/flashing is observed on the top of the rendered surfaces. As wet–dry cycles occur successively, different types of dirt particles tend to accumulate on the wet surface of the rendered façade [66]. So, a design factor is also associated with these cases of degradation (detailing of the building) [65]. On the other hand, in Figure 6a, the differential washing effect of rainwater affects the façade below windowsills. When rainwater reaches a façade, the water flow over the surface carries dirt particles, washing them off. Some areas of the façade are not affected by this phenomenon and keep accumulating dirt [66]. Moreover, in Figure 6d,e, dirt accumulates on areas where water condensation occurs. This is called a thermophoresis phenomenon, which is associated with poor thermal insulation conditions [67]. Additionally, the lack of maintenance and the properties of the mortar system may affect the degradation processes in Figure 6. The porosity of materials may also affect the accumulation of dirt on the façades [65].

Looking now at instances of defect "A-C2 Oriented cracking on the current surface", Figure 7 shows four situations of oriented cracking on architectural concrete surfaces. Analysing the criteria to define the urgency of repair, in Figure 4, the main classification parameter is the average crack width. Although a crack width ruler was not used in the photographs in Figure 7, due to the elevated position and consequent lack of access to the cracks within the façades, other elements may be considered as size references, such as bug holes, fastening marks or panel/casting joints. Two criteria have to be met to include an oriented cracking phenomenon in architectural concrete surfaces in the level 1 set (need for immediate intervention): (i) the average crack width must be 0.8 mm or wider, and (ii) the defect must have conditions to progress. Considering that the exposure conditions of all cases in Figure 7 remain unchanged, all cases meet the second criterion. As for the crack width, Figure 7a,b show clearly wider cracks than Figure 7c,d. Therefore, the former should be accounted for as level 1 of urgency of repair. A-C2 occurrences in architectural concrete surfaces belong to level 2 of urgency of repair when the average crack width is greater than 0.2 mm (and thinner than 0.8 mm). Careful analysis of Figure 7c,d allows discovering some differences between them, the most important being that the crack in Figure 7c is better defined throughout its length, while the crack in Figure 7d almost disappears in some places. Although the analysis of the threshold is not very precise and is more comparative in nature, given the provided data, Figure 7c should be rated as level 2 and Figure 7d as level 3 of urgency of repair.

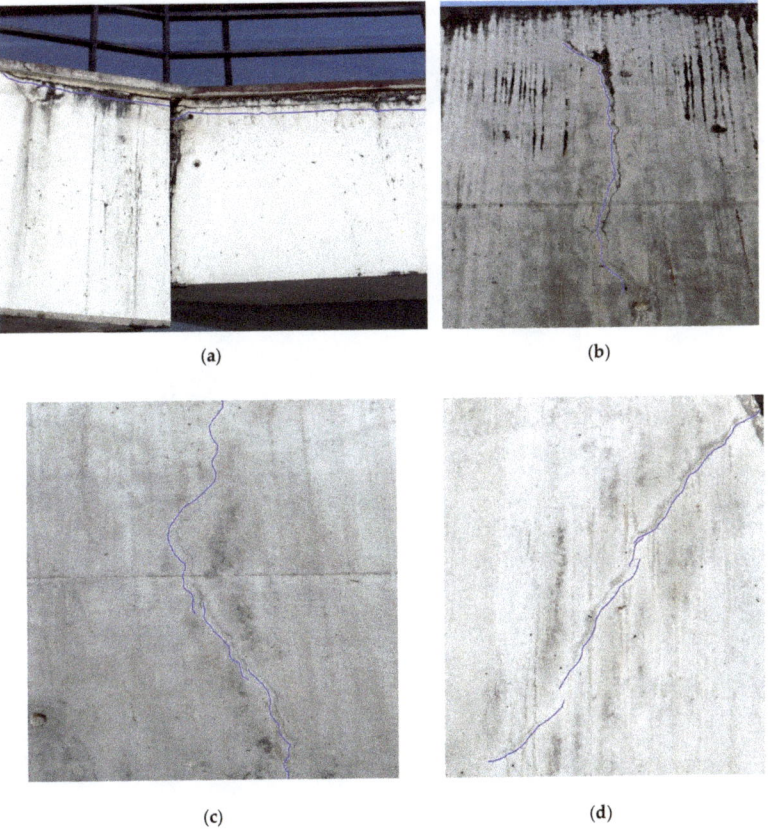

Figure 7. Cases of defect "A-C2 Oriented cracking on the current surface" in architectural concrete surfaces.

Figure 7a–d all show different cases of defect "A-C2 Oriented cracking on the current surface" in architectural concrete surfaces. As the explanation would be the same for all, only a general explanation was provided.

The National Precast Concrete Association [68] also suggests thresholds to classify the width of cracks on architectural concrete. Three types of cracks are proposed—fine, medium and wide—with the following width ranges (original values are provided in inches): fine cracks are less than 0.25 mm wide, medium cracks are between 0.25 mm and 1.01 mm wide and wide cracks are more than 1.01 mm wide. This classification is only slightly different from that proposed in Figure 4, not changing significantly the assessment of cracking phenomena in Figure 7.

In Figure 8, three different cases of "Oriented cracking on the current surface" (A-C2) on adhesive ceramic tiling are observed. These cases may be rated according to the levels of urgency of repair defined in the atlas of defects (Figure 4); for adhesive ceramic tiling, only levels 2 and 3 are set for defect A-C2. Oriented cracking occurrences may be of level 2 of urgency of repair if one out of two conditions is observed: (i) simultaneous occurrence of leakage; or (ii) average crack width of 0.5 mm or greater. All the other occurrences are accounted for as level 3. Again, no crack width ruler was used to measure the cracks in Figure 8, but the size of movement joints and other elements in the pictures may be used as a reference. Figure 8c shows almost imperceptible and very thin cracks. This occurrence does not meet any level 2 criteria, thus being classified as level 3 of urgency of repair. As for Figure 8a,b, Figure 8a appears to show wider cracking, but both cases present cracks wider than 0.5 mm. Additionally, Figure 8a also shows signs of leakage through the crack.

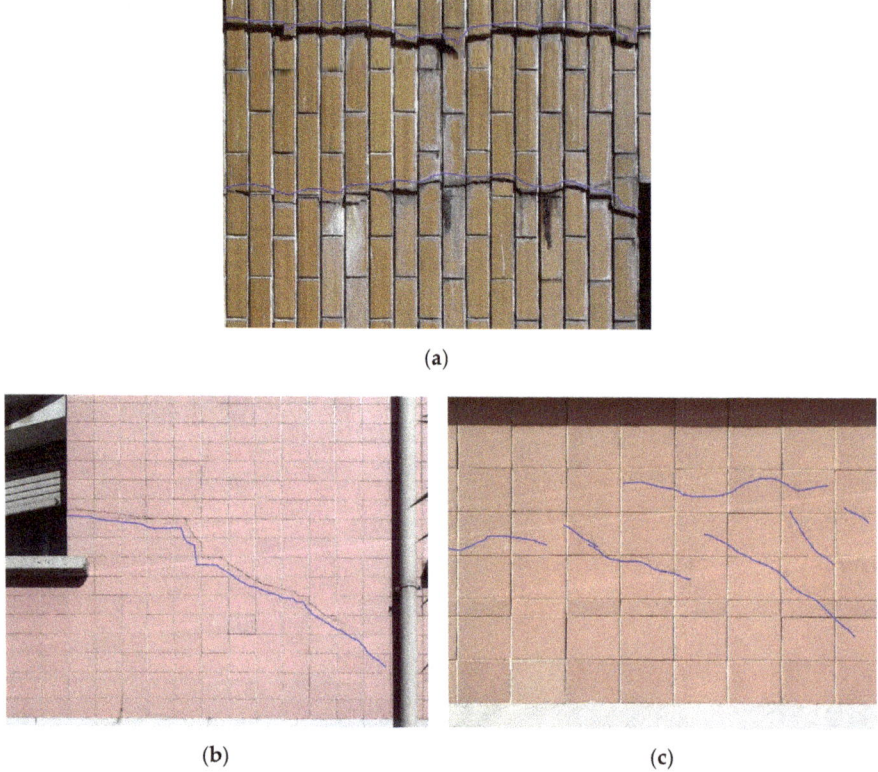

Figure 8. Cases of defect "A-C2 Oriented cracking on the current surface" in adhesive ceramic tiling.

Figure 8a–c all show different cases of defect "A-C2 Oriented cracking on the current surface" in adhesive ceramic tiling. As the explanation would be the same for all, only a general explanation was provided.

Cracking phenomena in adhesive ceramic tiling are analysed by Gaspar [69]. They may be caused by expansion and shrinkage phenomena affecting the bedding material and, consequently, the tile–mortar interface. The different coefficients of expansion of materials used in the substrate cause shear and tensile stresses that may cause failure. This may be the case in Figure 8a. Additionally, the significant width of the crack may cause water seepage, resulting in subsequent damage.

The exercise of attributing an urgency-of-repair level to each case in Figures 5–8 may be extended to using other scales for comparison purposes. The methodological approach of Ruiz et al. [9], optimising the metric of a condition rating scale, justifies adopting that scale in this comparative analysis. The scale proposed by Ruiz et al. [9] establishes five grades of severity of degradation of building elements for direct assignment, namely null (0), mild (1), moderate (2), high (3) and extreme (4). Transposing the definition of each severity grade of building elements to building defects, the results in Table 3 are obtained. To compare the grades attributed using the scale of the atlas of defects and the one proposed by Ruiz et al. [9], a third scale is necessary, unifying the descendant direction of urgency of repair/severity of degradation. In the unified scale, the most urgent/severe level is level I, as opposed to level V, which is the least urgent/severe. The unified grades equivalent to the urgency-of-repair grade and the severity grade are shown between brackets in each column of Table 3.

Table 3. Comparison of urgency-of-repair levels with grades of severity.

Cases of Defects		Urgency of Repair (Atlas of Defects) [1]	Severity Grade (Ruiz et al. [9]) [2]	Comparison Using a Unified Scale [3]
Figure 5	a	2 (III)	3 (II)	one level difference
	b	2 (III)	1 (IV)	one level difference
	c	3 (IV)	2 (III)	one level difference
Figure 6	a	1 (II)	1 (IV)	two levels difference
	b	3 (IV)	2 (III)	one level difference
	c	3 (IV)	1 (IV)	match
	d	1 (II)	2 (III)	one level difference
	e	2 (III)	2 (III)	match
	f	1 (II)	1 (IV)	two levels difference
Figure 7	a	1 (II)	3 (II)	match
	b	1 (II)	2 (III)	one level difference
	c	2 (III)	1 (IV)	one level difference
	d	3 (IV)	1 (IV)	match
Figure 8	a	2 (III)	3 (II)	one level difference
	b	2 (III)	2 (III)	match
	c	3 (IV)	1 (IV)	match

[1] 0: imminent danger, contingency measures needed; 1: need for immediate intervention; 2: need for intervention in the short term; 3: need for intervention in the long term; 4: no urgent need, assess in the next inspection. [2] 0: null severity; 1: mild severity; 2: moderate severity; 3: high severity; 4: extreme severity [9]. [3] A unified scale was used for comparison purposes, ranging from I to V, where level I is the most urgent/severe and level V is the least.

The last column of Table 3 analyses the difference and correspondence between grades. Levels I and V are not identified in either scales. Only four grades (out of 16) are a match in both scales, with no common denominator, as two are in level IV, one in level III and one in level II. Two cases show more marked differences between grades according to different scales: Figure 6a,f. In both cases, the aesthetical value of the façades determined the classification of defects in level 1 of urgency of repair. So, when confronted with a scale that only assesses the degradation conditions, disregarding other contextual limitations, the severity of these cases is relieved, as only mild signs of "Dirt and accumulation of debris" (A-A3) are observed. This comparison highlights the conceptual differences between urgency of repair and severity of degradation [11].

5. Conclusions

Considering the question that resulted in the presented research (how can an expeditious tool to identify defects and their urgency of repair be achieved?), this paper has demonstrated that it is possible to develop a catalogue of defects for the non-structural building envelope, taking into account (i) a well-defined classification of defects; (ii) a predetermined scope of building elements/materials; and (iii) a scale of urgency-of-repair levels, whose classification is determined by specific criteria. The atlas of defects combines visual and descriptive contents, aiming at a simple identification approach, easy to use during fieldwork.

The proposed atlas of defects may be used by surveyors during technical inspections. Furthermore, occasional surveyors, like architects and engineers, may also benefit from a tool like the atlas of defects when performing building assessments. Technical building inspections are used in several situations, like within maintenance plans, before deciding on retrofitting or rebuilding options, at the design stage (retrofitting) and to assess insurance claims. Additionally, researchers may use the proposed tool.

The use of such an atlas of defects is expected to help achieve a more reliable diagnosis of building defects, with a more accurate and intuitive identification of the type of defects and their level of urgency of repair. In turn, that improved reliability should improve the effectiveness of maintenance and repair actions, thus prolonging the service life of building elements at lower costs.

The atlas of defects that is proposed was developed within a global building inspection system, but, with the presented methodology, other atlases may be developed within different systems, given the basic materials and, most importantly, a well-defined scope and classification systems (of defects and urgency of repair or others). Although the proposed atlas is focused on non-structural building elements, its layout may be extrapolated to structural building elements.

To the best of the authors' knowledge, there is no such catalogue of building defects and respective urgency of repair in the literature. Furthermore, the need for the proposed atlas of defects has been identified by previous research [9].

In the future, the atlas of defects may be improved with further photographs representing a wider variety of significant situations within the same type of defect and building element/material.

Furthermore, the whole building inspection system will be computerised, including the atlas of defects. That step will provide an advantage in terms of developing a building pathology database, which may be useful in the context of maintenance plans and research. The visual aid of the atlas of defects represents an advantage in terms of swiftness of identification and characterisation. Moreover, the criteria to determine the urgency of repair may be included in algorithms, and the user of the computerised inspection system may only need to answer some simple questions.

Author Contributions: Conceptualization, J.d.B.; methodology, C.P., J.d.B. and J.D.S.; software, C.P.; validation, C.P.; formal analysis, C.P.; investigation, C.P.; resources, C.P., J.B., J.D.S. and I.F.-C.; data curation, C.P.; writing—original draft preparation, C.P.; writing—review and editing, J.B., J.D.S. and I.F.-C.; visualization, C.P.; supervision, J.d.B., J.D.S. and I.F.C.; project administration, J.B.; funding acquisition, J.B. All authors have read and agreed to the published version of the manuscript.

Funding: This research was funded by Fundação para a Ciência e a Tecnologia (FCT), grant number PTDC/ECI-CON/29286/2017, and also by FCT, grant number SFRH/BD/131113/2017.

Acknowledgments: The authors gratefully acknowledge the support of CERIS, from IST-UL.

Conflicts of Interest: The authors declare no conflict of interest. The funders had no role in the design of the study; in the collection, analyses, or interpretation of data; in the writing of the manuscript, or in the decision to publish the results.

References

1. Kohler, N.; Yang, W. Long-Term management of building stocks. *Build. Res. Inf.* **2007**, *35*, 351–362. [CrossRef]
2. Palmer, R.D. *Maintenance Planning and Scheduling Handbook*, 2nd ed.; McGraw Hill: New York, NY, USA, 2006; Volume 111, ISBN 007150155X.
3. Flores-Colen, I.; de Brito, J. A systematic approach for maintenance budgeting of buildings façades based on predictive and preventive strategies. *Constr. Build. Mater.* **2010**, *24*, 1718–1729. [CrossRef]
4. Flores-Colen, I.; de Brito, J. Discussion of proactive maintenance strategies in façades' coatings of social housing. *J. Build. Apprais.* **2010**, *5*, 223–240. [CrossRef]
5. de Brito, J.; Pereira, C.; Silvestre, J.D.; Flores-Colen, I. *Expert Knowledge-Based Inspection Systems. Inspection, Diagnosis and Repair of the Building Envelope*; Springer: Cham, Switzerland, 2020; ISBN 9783030424459.
6. Robertsen, E. Design for durability–A practical approach. In *Durability of Building Materials and Components 8*; Lacasse, M.A., Vanier, D.J., Eds.; Institute for Research in Construction: Ottawa, ON, Canada, 1999; pp. 2107–2117.
7. Khan, F.I.; Haddara, M.M. Risk-Based maintenance (RBM): A quantitative approach for maintenance/inspection scheduling and planning. *J. Loss Prev. Process. Ind.* **2003**, *16*, 561–573. [CrossRef]
8. Teo, E.A.-L.; Harikrishna, N. A quantitative model for efficient maintenance of plastered and painted façades. *Constr. Manag. Econ.* **2006**, *24*, 1283–1293. [CrossRef]
9. Ruiz, F.; Aguado, A.; Serrat, C.; Casas, J.R. Optimal metric for condition rating of existing buildings: Is five the right number? *Struct. Infrastruct. Eng.* **2019**, *15*, 740–753. [CrossRef]
10. Goyet, J.; Straub, D.; Faber, M.H. Risk-Based inspection planning of offshore installations. *Struct. Eng. Int.* **2002**, *12*, 200–208. [CrossRef]
11. Pereira, C.; Silva, A.; de Brito, J.; Silvestre, J.D. Urgency of repair of building elements: Prediction and influencing factors in façade renders. *Constr. Build. Mater.* **2020**, *249*, 118743. [CrossRef]
12. Pereira, C.; de Brito, J.; Silvestre, J.D. Harmonising the classification of the causes of defects in a global building inspection system: Proposed methodology and analysis of fieldwork data. *Sustainability* **2020**, *12*, 5564. [CrossRef]
13. Pereira, C.; de Brito, J.; Silvestre, J.D. Harmonising the classification of diagnosis methods within a global building inspection system: Proposed methodology and analysis of fieldwork data. *Eng. Fail. Anal.* **2020**, *115*, 104627. [CrossRef]
14. Pereira, C.; de Brito, J.; Silvestre, J.D. Harmonised classification of repair techniques in a global inspection system: Proposed methodology and analysis of fieldwork data. *J. Perform. Constr. Facil.* **2020**, in press.
15. Flores-Colen, I.; de Brito, J.; de Freitas, V.P. Expedient in situ test techniques for predictive maintenance of rendered façades. *J. Build. Apprais.* **2006**, *2*, 142–156. [CrossRef]
16. Elcometer Inspection Equipment: Elcometer 143—Crack Width Ruler. Available online: https://www.elcometer.com/images/stories/PDFs/Datasheets/English/143.pdf (accessed on 18 November 2019).
17. Johnson, R.W. The significance of cracks in low-rise buildings. *Struct. Surv.* **2002**, *20*, 155–161. [CrossRef]
18. Akbari, R. Crack survey in unreinforced concrete or masonry abutments in short- and medium-span bridges. *J. Perform. Constr. Facil.* **2013**, *27*, 203–208. [CrossRef]
19. Hård, A.; Sivik, L.; Tonnquist, G. NCS, natural color system—From concept to research and applications. Part I *Color Res. Appl.* **1996**, *21*, 180–205. [CrossRef]
20. ASTM International. *ASTM D1535-14(2018) Standard Practice for Specifying Color by the Munsell System*; ASTM International: West Conshohocken, PA, USA, 2018.
21. Tyler, J.E.; Hardy, A.C. An analysis of the original Munsell color system. *J. Opt. Soc. Am.* **1940**, *30*, 587–590. [CrossRef]
22. Garcez, N.; Lopes, N.; de Brito, J.; Silvestre, J. System of inspection, diagnosis and repair of external claddings of pitched roofs. *Constr. Build. Mater.* **2012**, *35*, 1034–1044. [CrossRef]
23. Garcez, N.; Lopes, N.; de Brito, J.; Sá, G. Pathology, diagnosis and repair of pitched roofs with ceramic tiles: Statistical characterisation and lessons learned from inspections. *Constr. Build. Mater.* **2012**, *36*, 807–819. [CrossRef]
24. Walter, A.; de Brito, J.; Lopes, J.G. Current flat roof bituminous membranes waterproofing systems–Inspection, diagnosis and pathology classification. *Constr. Build. Mater.* **2005**, *19*, 233–242. [CrossRef]

25. Conceição, J.; Poça, B.; de Brito, J.; Flores-Colen, I.; Castelo, A. Inspection, diagnosis, and rehabilitation system for flat roofs. *J. Perform. Constr. Facil.* **2017**, *31*, 04017100. [CrossRef]
26. Conceição, J.; Poça, B.; de Brito, J.; Flores-Colen, I.; Castelo, A. Data analysis of inspection, diagnosis, and rehabilitation of flat roofs. *J. Perform. Constr. Facil.* **2019**, *33*, 04018100. [CrossRef]
27. Santos, A.; Vicente, M.; de Brito, J.; Flores-Colen, I.; Castelo, A. Analysis of the inspection, diagnosis, and repair of external door and window frames. *J. Perform. Constr. Facil.* **2017**, *31*, 04017098. [CrossRef]
28. Santos, A.; Vicente, M.; de Brito, J.; Flores-Colen, I.; Castelo, A. Inspection, diagnosis, and rehabilitation system of door and window frames. *J. Perform. Constr. Facil.* **2017**, *31*, 04016118. [CrossRef]
29. Sá, G.; Sá, J.; de Brito, J.; Amaro, B. Statistical survey on inspection, diagnosis and repair of wall renderings. *J. Civ. Eng. Manag.* **2015**, *21*, 623–636. [CrossRef]
30. Sá, G.; Sá, J.; de Brito, J.; Amaro, B. Inspection and diagnosis system for rendered walls. *Int. J. Civ. Eng.* **2014**, *12*, 279–290.
31. Amaro, B.; Saraiva, D.; de Brito, J.; Flores-Colen, I. Statistical survey of the pathology, diagnosis and rehabilitation of ETICS in walls. *J. Civ. Eng. Manag.* **2014**, *20*, 511–526. [CrossRef]
32. Amaro, B.; Saraiva, D.; de Brito, J.; Flores-Colen, I. Inspection and diagnosis system of ETICS on walls. *Constr. Build. Mater.* **2013**, *47*, 1257–1267. [CrossRef]
33. Pires, R.; de Brito, J.; Amaro, B. Inspection, diagnosis, and rehabilitation system of painted rendered façades. *J. Perform. Constr. Facil.* **2015**, *29*, 04014062. [CrossRef]
34. Pires, R.; de Brito, J.; Amaro, B. Statistical survey of the inspection, diagnosis and repair of painted rendered façades. *Struct. Infrastruct. Eng.* **2015**, *11*, 605–618. [CrossRef]
35. da Silva, C.; Coelho, F.; de Brito, J.; Silvestre, J.; Pereira, C. Inspection, diagnosis, and repair system for architectural concrete surfaces. *J. Perform. Constr. Facil.* **2017**, *31*, 04017035. [CrossRef]
36. da Silva, C.; Coelho, F.; de Brito, J.; Silvestre, J.; Pereira, C. Statistical survey on inspection, diagnosis and repair of architectural concrete surfaces. *J. Perform. Constr. Facil.* **2017**, *31*, 04017097. [CrossRef]
37. Silvestre, J.D.; de Brito, J. Ceramic tiling inspection system. *Constr. Build. Mater.* **2009**, *23*, 653–668. [CrossRef]
38. Silvestre, J.D.; de Brito, J. Ceramic tiling in building façades: Inspection and pathological characterization using an expert system. *Constr. Build. Mater.* **2011**, *25*, 1560–1571. [CrossRef]
39. Silvestre, J.D.; de Brito, J. Inspection and repair of ceramic tiling within a building management system. *J. Mater. Civ. Eng.* **2010**, *22*, 39–48. [CrossRef]
40. Neto, N.; de Brito, J. Validation of an inspection and diagnosis system for anomalies in natural stone cladding (NSC). *Constr. Build. Mater.* **2012**, *30*, 224–236. [CrossRef]
41. Neto, N.; de Brito, J. Inspection and defect diagnosis system for natural stone cladding. *J. Mater. Civ. Eng.* **2011**, *23*, 1433–1443. [CrossRef]
42. Delgado, A.; Pereira, C.; de Brito, J.; Silvestre, J.D. Defect characterization, diagnosis and repair of wood flooring based on a field survey. *Mater. Constr.* **2018**, *68*, 1–13. [CrossRef]
43. Delgado, A.; de Brito, J.; Silvestre, J.D. Inspection and diagnosis system for wood flooring. *J. Perform. Constr. Facil.* **2013**, *27*, 564–574. [CrossRef]
44. Garcia, J.; de Brito, J. Inspection and diagnosis of epoxy resin industrial floor coatings. *J. Mater. Civ. Eng.* **2008**, *20*, 128–136. [CrossRef]
45. Carvalho, C.; de Brito, J.; Flores-Colen, I.; Pereira, C. Inspection, diagnosis, and rehabilitation system for vinyl and linoleum floorings in health infrastructures. *J. Perform. Constr. Facil.* **2018**, *32*, 04018078. [CrossRef]
46. Carvalho, C.; de Brito, J.; Flores-Colen, I.; Pereira, C. Pathology and rehabilitation of vinyl and linoleum floorings in health infrastructures: Statistical survey. *Buildings* **2019**, *9*, 116. [CrossRef]
47. European Committee for Standardization. *EN ISO 4628-4:2016 Paints and Varnishes–Evaluation of Degradation of Coatings–Designation of Quantity and Size of Defects, and of Intensity of Uniform Changes in Appearance–Part. 4: Assessment of Degree of Cracking (ISO 4628-4:2016)*; European Committee for Standardization: Brussels, Belgium, 2016.
48. European Committee for Standardization. *EN ISO 4628-2:2016 Paints and varnishes–Evaluation of Degradation of Coatings–Designation of Quantity and Size of Defects, and of Intensity of Uniform Changes in Appearance–Part. 2: Assessment of Degree of Blistering (ISO 4628-2:2016)*; European Committee for Standardization: Brussels, Belgium, 2016.
49. van Winden, C.; Dekker, R. Rationalisation of building maintenance by Markov decision models: A pilot case study. *J. Oper. Res. Soc.* **1998**, *49*, 928–935. [CrossRef]

50. Flourentzou, F.; Brandt, E.; Wetzel, C. MEDIC—A method for predicting residual service life and refurbishment investment budgets. *Energy Build.* **2000**, *31*, 167–170. [CrossRef]
51. Brandt, E.; Wittchen, K.B.; Faist, A.; Genre, J.L. EPIQR—A new surveying tool for maintenance and refurbishment. In *Durability of Building Materials and Componentes 8*; Lacasse, M.A., Vanier, D.J., Eds.; National Research Council Canada: Ottawa, ON, Canada, 1999; pp. 1576–1584.
52. Flourentzos, F.; Droutsa, K.; Wittchen, K.B. EPIQR software. *Energy Build.* **2000**, *31*, 129–136. [CrossRef]
53. Allehaux, D.; Tessier, P. Evaluation of the functional obsolescence of building services in European office buildings. *Energy Build.* **2002**, *34*, 127–133. [CrossRef]
54. National Aeronautics and Space Administration. *Reliability-Centered Maintenance Guide for Facilities and Collateral Equipment*; National Aeronautics and Space Administration: Washington, DC, USA, 2008.
55. Rodrigues, M.F.S.; Teixeira, J.M.C.; Cardoso, J.C.P. Buildings envelope anomalies: A visual survey methodology. *Constr. Build. Mater.* **2011**, *25*, 2741–2750. [CrossRef]
56. Elhakeem, A.; Hegazy, T. Building asset management with deficiency tracking and integrated life cycle optimisation. *Struct. Infrastruct. Eng.* **2012**, *8*, 729–738. [CrossRef]
57. Morgado, J.; Flores-Colen, I.; de Brito, J.; Silva, A. Maintenance planning of pitched roofs in current buildings. *J. Constr. Eng. Manag.* **2017**, *143*, 05017010. [CrossRef]
58. Morgado, J.; Flores-Colen, I.; de Brito, J.; Silva, A. Maintenance programs for flat roofs in existing buildings. *Prop. Manag.* **2017**, *35*, 339–362.
59. Uzarski, D.R.; Grussing, M.N.; Mehnert, B.B. *ERDC/CERL SR-18-7 Knowledge-Based Condition Assessment Reference Manual for Building Component-Sections. For. Use with BUILDER™ and BuilderRED™ (v. 3 Series)*; Construction Engineering Research Laboratory, US Army Engineer Research and Development Center: Champaign, IL, USA, 2018.
60. Lee, J.-S. Value engineering for defect prevention on building façade. *J. Constr. Eng. Manag.* **2018**, *144*, 04018069. [CrossRef]
61. Bortolini, R.; Forcada, N. Building inspection system for evaluating the technical performance of existing buildings. *J. Perform. Constr. Facil.* **2018**, *32*, 04018073. [CrossRef]
62. Silva, A.; de Brito, J.; Gaspar, P.L. Green Energy and Technology. In *Methodologies for Service Life Prediction of Buildings. With a Focus on Façade Claddings*; Springer: Cham, Switzerland, 2016; ISBN 978-3-319-33288-8.
63. Ruiz, F.; Aguado, A.; Serrat, C.; Casas, J.R. Condition assessment of building façades based on hazard to people. *Struct. Infrastruct. Eng.* **2019**, *15*, 1346–1365. [CrossRef]
64. Fernandes, D.; de Brito, J.; Silva, A. Methodology for service life prediction of window frames. *Can. J. Civ. Eng.* **2019**, *46*, 1010–1020. [CrossRef]
65. van Hees, R.P.J.; Binda, L.; Papayianni, I.; Toumbakari, E. Characterisation and damage analysis of old mortars. *Mater. Struct.* **2004**, *37*, 644–648. [CrossRef]
66. Blocken, B.; Derome, D.; Carmeliet, J. Rainwater runoff from building facades: A review. *Build. Environ.* **2013**, *60*, 339–361. [CrossRef]
67. Júnior, C.M.M.; Carasek, H. Relationship between the deterioration of multi story buildings facades and the driving rain. *Rev. Constr.* **2014**, *13*, 64–73.
68. National Precast Concrete Association. *Precast Concrete Architectural Repair Guide*; National Precast Concrete Association: Carmel, IN, USA, 2013.
69. Gaspar, P.L. End of the service life of ceramic cladding: Lessons from the Girasol Building in Madrid. *J. Perform. Constr. Facil.* **2017**, *31*, 04016088. [CrossRef]

© 2020 by the authors. Licensee MDPI, Basel, Switzerland. This article is an open access article distributed under the terms and conditions of the Creative Commons Attribution (CC BY) license (http://creativecommons.org/licenses/by/4.0/).

Article

Investigation of the Effect of Operational Factors on Conveyor Belt Mechanical Properties

Anna Rudawska [1], Radovan Madleňák [2,*], Lucia Madleňáková [2] and Paweł Droździel [1]

[1] Faculty of Mechanical Engineering, Lublin University of Technology, Nadbystrzycka 36, 20-618 Lublin, Poland; a.rudawska@pollub.pl (A.R.); p.drozdziel@pollub.pl (P.D.)
[2] Faculty of Operation and Economics of Transport and Communications, University of Zilina, Univerzitna 1, 01026 Zilina, Slovakia; lucia.madlenakova@fpedas.uniza.sk
* Correspondence: radovan.madlenak@fpedas.uniza.sk; Tel.: +421-41-513-3124

Received: 2 June 2020; Accepted: 17 June 2020; Published: 19 June 2020

Abstract: This paper aims to present the effect of specific operational factors (temperature and humidity) on the selected mechanical properties of a conveyor belt. The tests were conducted in a climatic chamber, simulating the effect of both minus and plus temperatures −30 °C to 80 °C (243 K to 353 K) at specific humidity, and in a thermal shock chamber where a varying number of ageing cycles was applied for a specific range of thermal shocks. Six different tests in the climatic chamber and four different tests in a thermal shock chamber were conducted. The results of the climatic chamber tests demonstrate that many strength parameters have undesired values at a temperature of 10 °C (283 K) and 80 °C (353 K) at a relative humidity of 80%. Interestingly, the results revealed that tensile strength, tensile modulus and yield strength are higher at below 0 °C temperature than at above 0 °C temperature. For example, comparing the temperature −30 °C (243 K) and +30 °C (303 K) obtained a difference of tensile modulus of nearly 10%, and comparing the temperature −30 °C (243 K) and +10 °C (283 K) the differences were 22%.

Keywords: conveyor belt; climatic factors; thermal shocks; mechanical properties

1. Introduction

The use of primitive versions of conveyor belts by manufacturers dates back to the 19th century. For the purpose of carrying coal, ores and other products, Thomas Robins invented a conveyor belt in 1892. The Swedish company Sandvik invented and started producing steel conveyor belts in 1901. Conveyor belts, invented by Richard Sutcliffe in 1905, were used in coal mines and revolutionized the entire mining industry. Henry Ford introduced assembly lines at his factory in 1913 that used conveyor belts. Mail volumes in the postal system were snowballing, and postal operators had to deal with this situation by installing automatic sorting machines. The first semiautomatic sorting system was installed in the USA in 1957. This Letter Sorting Machine was made of an upper and lower section, a set of five sorting keyboards and a conveyor belt. That was the first usage of a conveyor belt in an automatic postal processing system.

The issue of testing the properties of materials, and thus the conveyor belts (and components of conveyor belts [1–5]) in varying weather conditions and under different loads is not new [6–10]. The conveyor belts are exposed to different factors such as changes in temperature, humidity or air pressure during their service life [9,11–13]. Several authors have examined how these factors affect the properties of the materials forming the conveyor belts. Bocko et al. [14], in their article, presented numerical analyses using the component testing of conveyor belts and information about complex materials. Their article contains a definition of methods used for the calibration of materials defined by the user and the building of numerical models. The material tests included the determination of the mechanical properties of plies of rubber cover, textile reinforcements, and plies of textile with adjacent rubber. Goltsev and Markochev [15] described the analysis of fatigue testing methods. In their article, these authors compared the conventional testing methods for fatigue cracking resistance with tests of constant amplitude of stress intensity factor. For the analysis of belt–bulk material interactions and the conveyor belts' engineering design, the determination of pressure distribution is crucial. This was the objective of the study by Liu, Pang, and Lodewijks [16]. They developed and verified a theoretical model for predicting the distribution of pressure on conveyor belts. An essential process in the transport industry is impact loading, because this causes wear and failure in important components. Conveyor belts are critically important because of their usage in virtually every industry in which quantities of goods must be moved over short or long distances. Molnar et al. [17] investigated the effects of impact loading to the surface of a rubber conveyor belt and the stress levels inside the material. They tested the impact of loading things onto the conveyor belt, but this was only a partial test in an ideal atmospheric condition.

Another problem with conveyor belts is ageing. The rubbers of conveyor belts used during transport age, because during their operation they are exposed to the aggressive characteristics of the external environment. Significant changes caused by the ageing process appear on the surface of the conveyor belts, primarily due to immediate contact with the transported cargo or due to contact with the ambient air. Nedbal et al. [18] used mechanical and dielectric spectroscopy to determine the physical properties of rubbers used for conveyor belts. One way to assess changes is to test the internal structure of the conveyor belt. Barburski [12], in his analysis, which was dedicated to the measurement of the primary parameters that characterize the internal structure of the fabric of conveyor belts, points out that the weave (heat treatment at 160 °C (433.15 K) for 5, 10 and 15 min) affects the degree of the changes in conveyor belts during exposure to high temperatures. Petrikova's paper [19] presented another numerical and experimental study dedicated to the behavior of the conveyor belt, in which the temperature dependence of the dynamic and tribological behaviors of the belt were investigated.

In summary, none of the authors above tested conveyor belts under changing temperature and humidity conditions. This paper aims to present the effect of these specific operational factors (temperature and humidity) on selected strength parameters of the conveyor belts that comprise the main element of the sorting line in a Slovakian central sorting center. The tests were conducted in a climatic chamber, simulating the effect of both temperatures both above and below 0 °C at specific humidity, and in a thermal shock chamber where a varying number of ageing cycles was applied for a specific range of thermal shocks.

Another important reason for the implementing this test is that the authors see practical benefits. Automated sorting systems currently play a significant role in the environment of postal and logistical businesses. The reason for this is an increase in the volume of shipments originating in e-commerce needing to be processed and delivered to the addressee within a set timeframe. The reliability of sorting systems does not depend only on the selection of the correct sorting program, but also on the construction and functionality of individual technical components of the whole system. In this context, this study examines the effect of various environmental effects (temperature, humidity, pressure, etc.) on a critical part of the sorting system; specifically, the conveyor belt. The test results (which enable the search for dependencies between changes in the mechanical attributes of the belt and changes in atmospheric conditions) are significant in terms of the effective usage and operation of a device in two respects. Firstly, with respect to the organization of activities and the definition of processes within the facility itself. This is mainly related to the setting, for example, of the speed of the conveyor belt during loading, offloading and further handling of shipments relative to the number, weight and frequency of shipments on the belt. Together with other tested attributes, this significantly affects the lifespan and maintenance of the belt, as well as the overall sorting system. Secondly, with respect to the management of costs related to ensuring regular operation and maintenance, but also in terms of additional costs that can be incurred by the operator due to failures of the conveyor belt. These include not only the costs of repairing the damaged or a non-functional belt, or the procurement of replacement belts, but also the significant financial burden occurring due to the non-functionality of the device itself, i.e., the inability to perform sorting and routing of shipments in an automated manner. In this case, manual sorting takes place, which is time-consuming and prone to error, as well as various discrepancies related to human error. In addition to the financial burden related to manual labor, there are also costs related to missing the deadline for transportation, especially in time-warranted shipments, or the costs of damage caused by loss of or damage to the shipment, or its content. The functionality of the conveyor belt (maintaining its mechanical properties or anticipating changes under changing atmospheric conditions) and of course the whole sorting device, is an essential part of guaranteeing the reliability of the whole distribution process.

In many companies, the operation of a conveyor belt is significantly affected by the temperature and humidity of the hall in which the conveyor belt is mounted, as well as by the conveyor's exposure to variations in temperature. In the studied case, a specific fragment of a conveyor was operated at various temperatures. At parcel loading and unloading, the conveyor belt was operated within a temperature range from −16 °C (257 K) to +30 °C (303 K), with a mean temperature of 11 °C (284 K), and the relative humidity (RH) ranging between 20% and 80%. These conditions are connected with, e.g., season changes.

2. Research Methodology

2.1. Experimental Material

The sorting line, the main element of which is the conveyor belt, is presented in Figure 1. The intended environmental conditions for the operation of the conveyor belt is a temperature range from 0 to +40 °C (273 K–313 K) and RH ranging from 20% to 80%, with no condensation.

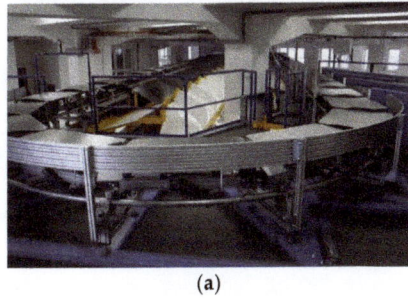 

Figure 1. (a) Sample sorting line; (b) fragment of the conveyor belt in the sorting line.

The parts of the sorting line should not be exposed to corrosive gases, organic solvents or any other chemicals. The weather conditions for the first year of sorting line operations were: average temperature: 11 °C (284 K); minimum temperature: −16 °C (257 K); maximum temperature: 30 °C (303 K); and RH: 20–80%.

The material specifications of the conveyor belt are presented in Tables 1 and 2. The information applies at approximately 20 °C (293 K). The tested conveyor belt consists of two fabric plies made from polyester (Figure 2). The topside is made from Flexam polyvinyl chloride (PVC), and the bottom side is made from fabric. The conveyor belt was produced by Ammeraal Beltech Company (Heerhugowaard, The Netherlands) [20]. The shape and dimensions of the tested specimens are shown in Figure 3.

Table 1. The characteristics of the conveyor belt (based on [20]).

Type of Elements	Material	Characteristic of Materials	Others
Fabric tension layer	polyester	stable	2-ply
Topside	Flexam PVC	smooth	green
Bottom side	fabric	low friction	

Table 2. Technical belt data (based on [20]).

Characteristic	Value
Hardness topside (DIN 53505)	80A Shore
Force at 1% elongation break (ISO 21181)	10.0 N/mm
Belt thickness (internal AB method KV.002)	2.30 mm
Weight (internal AB method KV.004)	2.50 kg/m²
Thickness of top cover	0.70 mm
Temperature range	−15 to 80 °C (258 K–353 K)
Temperature range short	−15 to 100 °C (258 K–373 K)

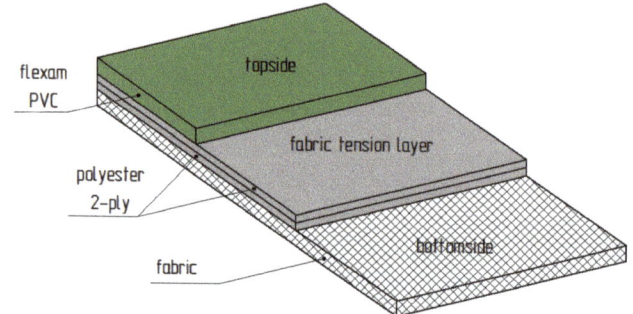

Figure 2. Structure of the tested conveyor belt.

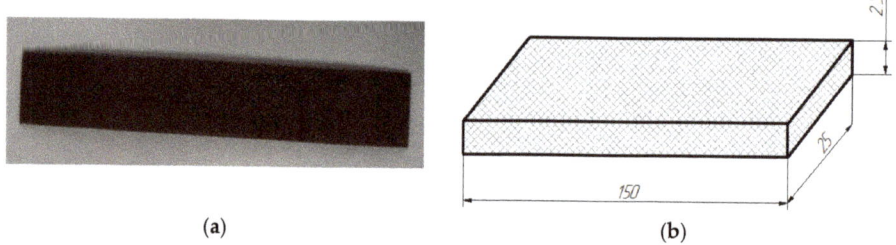

Figure 3. Test samples: (**a**) view of a real sample; (**b**) shape and dimensions of a sample (in mm).

2.2. Testing Specimens

The test samples were exposed to different operational conditions in a climatic chamber SH-661 (product of ESPEC, Klimatest, Poland [21]) and in a thermal shock chamber STE 11 (product of ESPEC, Klimatest, Poland [21]). Detailed information on sample exposure and denotations of test variants in given operational conditions are listed in Tables 3 and 4. A photograph of the screen showing a tested cycle is shown in Figure 4. The residence time at the temperature below 0 °C was 15 min, and the residence time in the warm chamber at the temperature above 0 °C was 15 min.

Six different tests were carried out in the climatic chamber and four different tests were carried out in the thermal shock chamber. One batch of samples was used as a reference batch and was not exposed to ageing (operational) factors. Each batch of samples subjected to tests contained 10–11 pieces.

Table 3. Sample exposure data—climatic chamber.

Test Variant	Temperature	Relative Humidity	Time
KKA-30-20	−30 °C (243 K)	20%	2 weeks
KKA-20-20	−20 °C (293 K)	20%	2 weeks
KKB-10-80	+10 °C (283 K)	80%	2 weeks
KKB-30-80	+30 °C (303 K)	80%	2 weeks
KKB-80-80	+80 °C (353 K)	80%	2 weeks
KKC-30-50	+30 °C (303 K)	50%	2 weeks

Table 4. Sample exposure data—thermal shock chamber.

Test Variant	Temperature	Number of Cycles
KSZ A	−20 °C/+60 °C	600
KSZ B	−20 °C/+60 °C	1200
KSZ C	−20 °C/+60 °C	1800
KSZ D	−20 °C/+60 °C −40 °C/+60 °C	1800 500

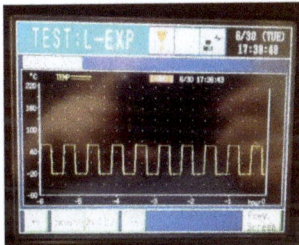

Figure 4. Photograph of the thermal shock chamber screen showing a test cycle.

Following curing in the chambers, the samples were subjected to exposure under ambient conditions (temperature 20 ± 2 °C (293 ± 2 K), humidity 38–40%) for 24 h. After that, they were subjected to strength tests on a ZWICK/ROELL Z150 testing machine in compliance with the standard DIN EN 10002-1 [22]. The tests were conducted using the following parameters: tensile modulus testing rate, 60 MPa/s; yield strength testing rate, 60 MPa/s; and testing rate, 0.008 MPa/s. To perform strength tests, the prepared samples were mounted in helical-wedge chucks of the testing machine [23]. The fixed joints were then subjected to tensile strength testing until their failure. The STATISTICA software (Timberlake Consultants, Ltd., Warsaw, Poland) was used for preparing adequate statistical results.

3. Results

The following strength test results were evaluated: E—tensile modulus in MPa, $R_{p\,0.5}$—yield strength in MPa, R_m—maximum tensile strength in MPa, F_m—maximum tensile force in N, A_{gt}—overall elongation at the maximum tensile force in %. For each test, 10–11 repetitions were carried out. To calculate the arithmetic mean of the obtained results, the extreme results were rejected in some variants.

3.1. Reference Specimens

The results obtained for the reference specimens, i.e., the samples that were not subjected to changing climatic ageing factors, like temperature and humidity, are listed in Table 5 and Figure 5. Values that were significantly different from the other results were rejected.

Table 5. Results for mechanical properties (DIN EN 10002-1) of reference samples.

Results	E (MPa)	$R_{p\,0.5}$ (MPa)	R_m (MPa)	F_m (N)	A_{gt} (%)
Mean	427	16	54.74	2833.27	22.26
Standard deviation	13.6	0	1.17	67.55	0.46

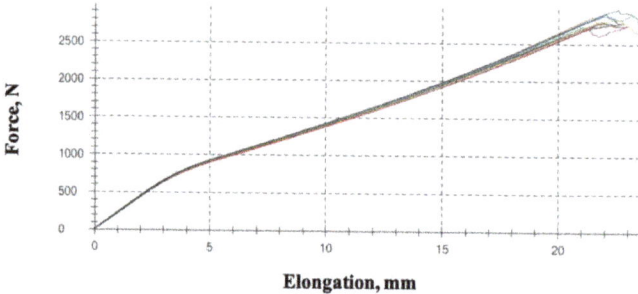

Figure 5. Force versus elongation of conveyor belt samples.

The results of the reference samples will act as a point of reference for the results of the samples subjected to various climatic factors: temperature and humidity.

3.2. Samples Exposed to the Climatic Chamber

The mean results for the studied mechanical properties of the samples subjected to various temperatures and RH values are given in Table 6, while the force versus elongation results for some of these samples are given in Figure 6.

Table 6. Results for mechanical properties (DIN EN 10002-1)—climatic chamber (mean values).

Test Variant	E (MPa)	$R_{p\,0.5}$ (MPa)	R_m (MPa)	F_m (N)	A_{gt} (%)
KKA-30-20	498	15.3	54.0	2710.4	21.42
KKA-20-20	475	15.5	54.6	2761.1	21.55
KKB-10-80	386	13.2	45.4	2340.0	21.28
KKB-30-80	452	15.1	55.0	2738.3	21.22
KKB-80-80	334	16.8	44.7	2238.6	21.30
KKC-30-50	438	14.0	51.0	2557.8	21.21
KKA-30-20	498	15.3	54.0	2710.4	21.42

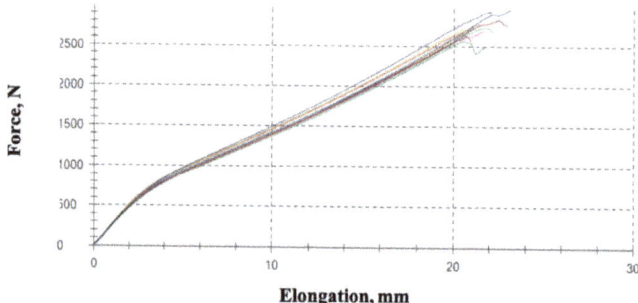

Figure 6. Force versus elongation of conveyor belt samples, KKA-20-20.

The previous results reveal that the mechanical properties are quite similar, independently of the test conditions.

3.3. Specimens Exposed to the Thermal Shock Chamber

The results of strength tests (mean value) for the samples subjected to thermal shock testing are given in Table 7. The results of force versus elongation are given in Figure 7.

Table 7. Results for mechanical properties (DIN EN 10002-1)—thermal shock chamber (mean values).

Test Variant	E (MPa)	$R_{p\,0.5}$ (MPa)	R_m (MPa)	F_m (N)	A_{gt} (korr.)
KSZ A, 600 cycles	394	18.1	54.0	2722.5	22.01
KSZ B, 1200 cycles	376	17.2	52.6	2645.9	21.81
KSZ C, 1800 cycles	394	15.9	51.9	2625.9	21.71
KSZ D, 1800 + 500 cycles	406	14.5	51.7	2638.2	21.49

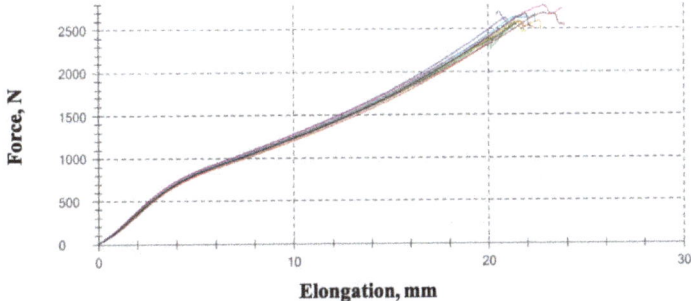

Figure 7. Force versus elongation of conveyor belt samples, KSZ B, 1200 cycles.

The previous results reveal that the mechanical properties are quite similar, independently of the test conditions.

3.4. SEM Analysis Results

The results of SEM analysis for the selected specimens are presented in Figures 8–12. The characteristics of reference samples are presented in the SEM images (Figure 8). The topside is the characteristic view of the Flexam PVC surface (Figure 8a). The bottom side is fabric (Figure 8b), and the cross-section of the conveyor belt sample with two fabric tension layers is presented in Figure 8c.

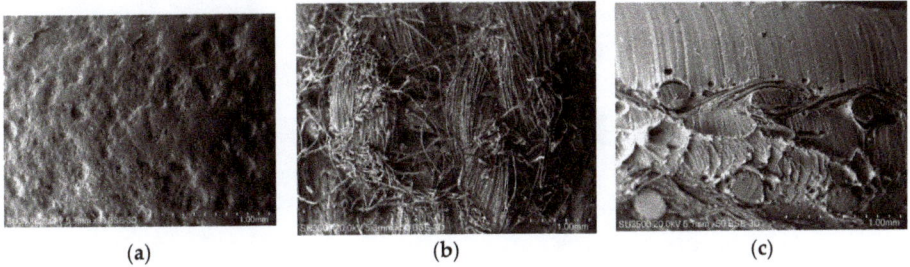

Figure 8. SEM images of: (**a**) topside of Flexam PVC surface; (**b**) bottom side of fabric surface; (**c**) cross-section of a reference sample.

The plain weave of polyester fabric can be noticed in Figure 8. In the plain weave, each warp thread is interwoven and intersects with each weft thread, and thus the fabric is tightly bound and rigid. This is crucial information, because the characteristics of the used fabric (the direction of and the position of the fibers) dramatically affects the strength of the composite materials—in this case, the conveyor belt. Such information regarding the type of the polyester fabric weave was not provided by the producer in the characteristics of the conveyor belt.

The SEM images presented in Figures 9–11 show the cross-section and bottom side of the samples of the materials of the selected variants subjected to exposure to various temperatures and values of humidity. The destruction of some of the external weft of the fabric can be observed in the bottom side materials in the presented variants. The cross-section presents the weave of the polyester fabric and also the Flexam PVC material. The failure of the polyester material (Figure 10a) can be noticed. Much more destruction of the bottom side of samples subjected to the climatic chamber can be noticed compared with the reference samples.

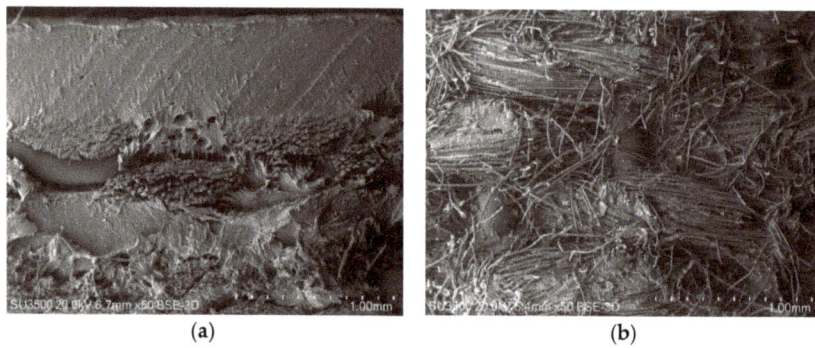

Figure 9. SEM images of: (**a**) cross-section, (**b**) bottom side of material samples for KKA 30-20 test variant.

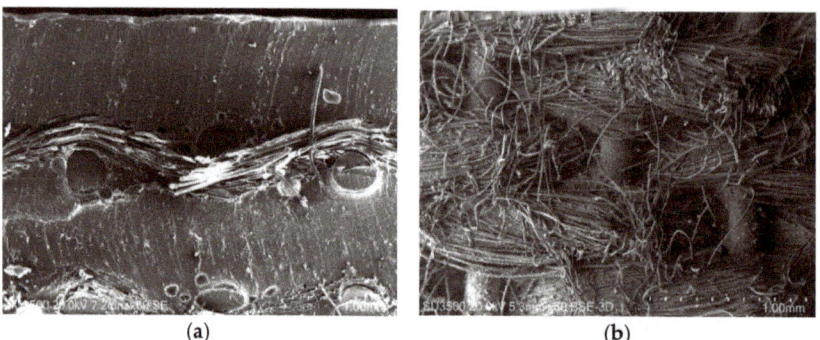

Figure 10. SEM images of: (**a**) cross-section, (**b**) bottom side of material samples for KKA 30-80 test variant.

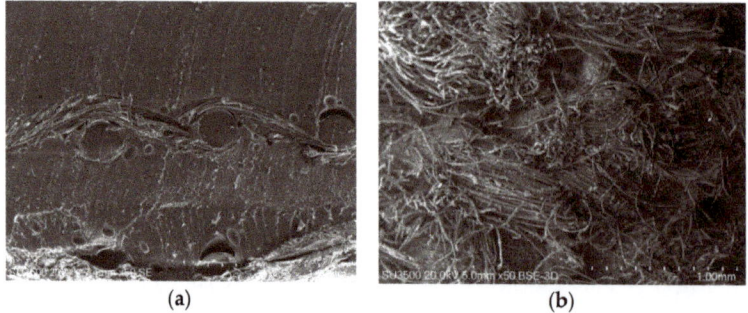

Figure 11. SEM images of: (**a**) cross-section; (**b**) bottom side of material samples for KKA 80-80 test variant.

The SEM images of the cross-section after the failure of the material samples subjected to the thermal shock chamber tests (Figure 12) exhibit much more failure (fiber failure) than the samples subjected to the climatic chamber exposure. The thermal shock affects the failure much more than the climatic conditions. The thermal shock causes more catastrophic failure.

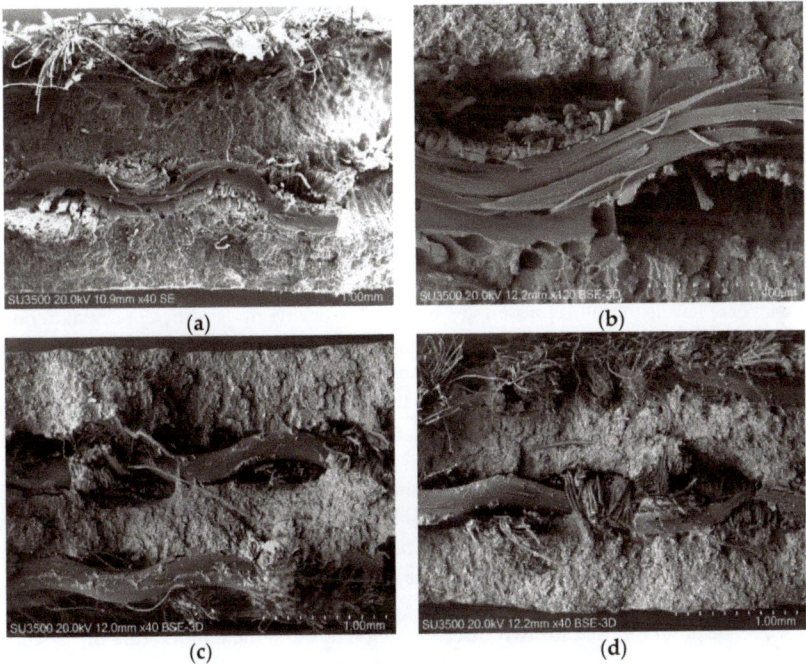

Figure 12. SEM images view of cross-section after the failure of material samples subjected to the thermal shock chamber tests: (**a**) KSZA; (**b**) KSZB; (**c**) KSZC and (**d**) KSZD.

3.5. Views after the Failure

Views of the samples after tensile strength testing are shown in Figures 13 and 14. It can be observed that the failure shown in Figure 13 is almost the same for all the tested samples (except 80-80).

Figure 13. KKA-20-20 sample after tensile strength testing.

Only the samples exposed to the temperature of 80 °C (353 K) and RH of 80% (Figure 14) have a slightly different type of failure (a failure point and a sample shape) compared to that observed in the samples subjected to other ageing variants.

It can be presumed that further studies should be conducted in this direction (lengthwise direction), even though the conveyor belt tape is not subjected to operation at temperatures exceeding 30 °C (303 K).

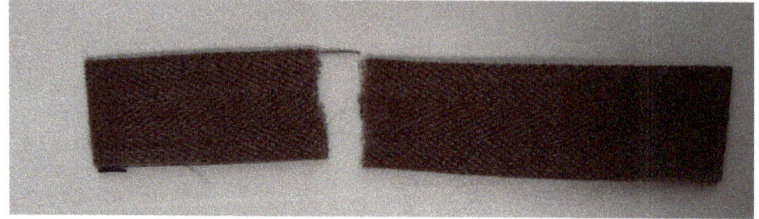

Figure 14. KKB-80-80 sample after tensile strength testing.

4. Comparison of Results and Discussion

4.1. Specimens Exposed in the Climatic Chamber

The test results for the samples exposed to different conditions in the climatic chamber are given in Figure 15.

The results of tensile modulus for samples subjected to ageing under different climatic conditions demonstrate that the highest tensile modulus can be observed for the samples exposed to a temperature set to −30°C (303 K) (KKA-30-20). In the case of samples exposed to a temperature of −20 °C (253 K) (KKA-20-20) for 14 days, E decreased by about 5%. It can be observed that below 0 °C temperatures in the Celsius scale have no negative effects on the tensile modulus compared to the tensile modulus of the reference samples.

The lowest E was recorded for the samples subjected to ageing at a temperature of 80 °C (353 K) and a RH of 80% (KKA-80-80): 334 MPa. A higher value of E was obtained for the samples subjected to exposure to the same RH, but at a lower temperature set to 10 °C (283 K) (KKA-10-80). In the case of KKA-10-80, the value of E was 85% of the value of module E of the samples exposed to a higher temperature (KKA-30-80).

It can also be observed that for two cases of ageing at the same temperature of 30 °C (303 K), but at two different RH values: 80% and 50%, the tensile modulus differed by 3%.

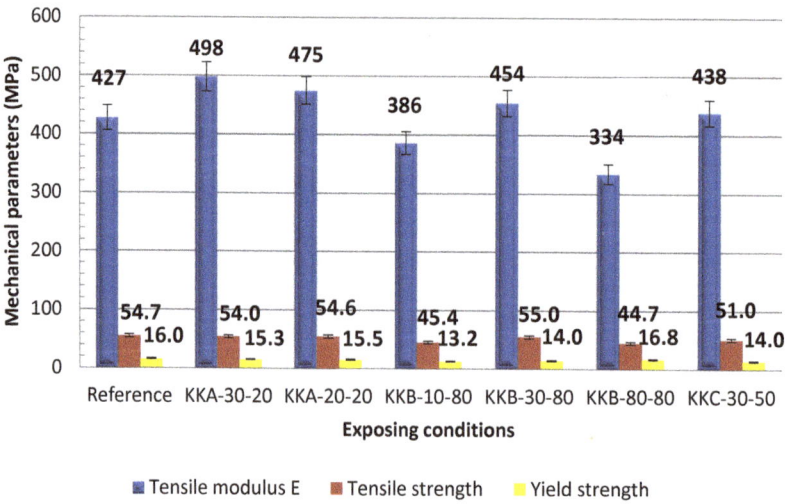

Figure 15. Mechanical parameters of samples tested in the climatic chamber.

Summing up the tensile modulus results, it can be observed that:

- higher values of E were obtained for exposure at below 0 °C temperature in the Celsius scale (decreasing values of the minus temperature does not have a negative effect on tensile modulus),
- among the tested variants, ageing produces the least desired results at a temperature of 80 °C (353 K) and at an RH of 80%;
- an increase in the above 0 °C temperature in the Celsius scale leads to a decrease in tensile modulus,
- a decrease in humidity with the temperature remaining the same leads to a slight decrease in E (KKB-30-80 and KKC 30-50),
- in most cases, the reference samples have a lower E, ranging between 3% and 10%, than samples tested in the climatic chamber, except for two ageing variants—KKB-10-80 and KKB-80-80—where E was lower by 10% and 22% than the reference sample, respectively.

Examining the effect of ageing conditions on tensile strength of the tested conveyor belt samples, it was observed that:

- the highest tensile strength was obtained for three ageing cases, two of them involving exposure at a below 0 °C temperature in the Celsius scale,
- the lowest tensile strength was obtained for samples exposed at a temperature of 80 °C (353 K) and an RH of 80% (KKA-80-80),
- decreasing humidity and maintaining the ageing temperature unchanged (KKC-30-50) leads to a 7% decrease in the tensile strength compared to exposing at a humidity of 80% (KKB-30-80),
- in the majority of cases, the tensile strength is lower compared to the tensile strength of the reference samples; in the three cases KKA-20-30, KKA-20-20 (exposing at minus temperature) and KKB-30-80, the obtained tensile strength had the same value. It can be observed that temperature below 0 °C in the Celsius scale does not have a negative effect on the tensile strength compared to the tensile strength of the reference samples, but it is difficult to determine the effect of temperature or humidity on strength unanimously. The coefficient of correlation (ρ) between the strength and the temperature is $\rho = -0.58$ while that between strength and humidity is $\rho = -0.66$, which can indicate that humidity has a higher effect on the belt's tensile strength than temperature.
- the results for the negative temperature can be explained by the fact that lowering the temperature results in a more exponential behavior of the glassy-brittle state, such as increased stiffness and reduced deformability. A low temperature (above 0 °C) contributes to the fact that polymers behave like vitrified glass bodies with high stiffness, which have already been damaged by a slight deformation and exhibit resilience characteristics, and in particular the presence of reversible deformations, most often increasing proportionally to the increase in stress. The results of this study confirm that the Young's modulus of polymeric plastics is significantly higher at low temperatures compared to room temperature [24–27].

As for yield strength (Figure 15), it can be observed that the reference samples (except one case, KKB-80-80) have a higher yield strength $R_{p0,5}$. The lowest yield strength was obtained for exposure at a temperature of 10 °C (303 K) and a humidity of 80% (KKA-10-80). Therefore, it can be concluded that exposure under these conditions leads to decreased yield strength, thereby decreasing the plastic properties of the belt. In addition, it was observed that the parameter A_{gt}, i.e., total elongation at the highest tensile force, has similar values for the tested variants of ageing.

Most polymers have low humidity, but there are exceptions. Increased humidity causes the consumption of raw material during processing and exposes the producer to losses. Some polymeric materials absorb water or moisture from the air. An increase in the water content of the material increases the impact strength and elasticity but reduces the strength. However, it should be remembered that the absorption of water is also associated with increasing the volume of the material, and thus the dimensions of the elements, which may lead to incompatibility of construction dimensions, adversely affecting the work of the element. It can be seen (Figure 15) that an increase in humidity causes a decrease or increase in some mechanical parameters, but this is a more complex phenomenon, as it is also associated with the seasoning temperature.

Comparing the results of tensile modulus and tensile strength of the samples exposed to the constant humidity of 80% and at different temperatures—10 °C (283 K), 30 °C (303 K), 80 °C (353 K) (Figure 16)—it can be observed that the highest tensile strength was obtained for exposure at a humidity of 80% and at a temperature of 30 °C (303 K). When both lower and higher temperatures are applied, the tensile strength significantly decreases by 15% and by 26%, respectively.

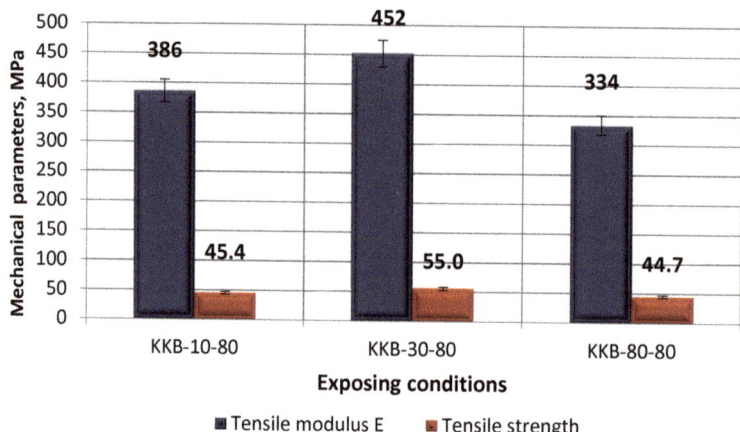

Figure 16. Tensile modulus and tensile strength of samples exposed to selected ageing variants.

Some statistical results (correlation coefficient) of the tensile modulus for the samples subjected to ageing in different climatic conditions and for reference samples are presented in Table 8. In the analysis, the Pearson correlation coefficient was used. Some information regarding the correlation coefficient was presented in [28].

Table 8. The correlation coefficient of the tensile modulus obtained in various ageing variants and in the reference condition (for significance coefficient $p < 0.05$).

Variable	Reference	KKA-30-20	KKA-20-20	KKB-10-80	KKB-30-80	KKB-80-80	KKC-30-50
Reference	1.0000	−0.6423	−0.0329	0.4513	0.5976	−0.5896	−0.7129
KKA-30-20	−0.6423	1.0000	−0.0915	0.1918	−0.6037	−0.0610	0.7848
KKA-20-20	−0.0329	−0.0915	1.0000	−0.0631	−0.1835	−0.4503	0.0010
KKB-10-80	0.4543	0.1918	−0.0631	1.0000	0.5345	−0.7128	0.2988
KKB-30-80	0.5976	−0.6037	−0.1835	0.5345	1.0000	−0.1406	−0.1600
KKB-80-80	−0.5896	−0.0610	−0.4503	−0.7128	−0.1406	1.0000	0.0941
KKC-30-50	−0.7129	0.7848	0.0010	0.2988	−0.1600	0.0941	1.0000

Based on the correlation of the coefficient results presented in Table 8, the following observations can be made:

- comparing the results of the tensile modulus for the samples subjected to ageing under different climatic conditions and for reference samples, in the majority of cases, a correlation between the value of tensile modulus and the ageing conditions (considering both temperature and humidity) could not be observed,
- in some cases, a correlation (both positive and negative) was observed, but the correlation is not strong. For example, the negative correlation of the tensile modulus for the samples KKB-10-80 and KKB-80-80 subjected to ageing at the same relative humidity (80%), but at different temperatures

(10 °C (283 K) and 80 °C (353 K)); the correlation coefficient here was −0.7128, indicating the negative influence of the higher temperature on the value of the tensile modulus. This means that an increase in the temperature leads to a decrease in the tensile modulus in the case of high humidity (80%).

The obtained results indicate that in most comparisons between variants of ageing, a correlation between the tensile modulus results in the various test variants was not observed. In some cases, a correlation was observed, but it was not strong.

The statistical results of the tensile strength for the samples subjected to ageing in different climatic conditions and for the reference samples are presented in Table 9.

Table 9. The correlation coefficient of the tensile strength obtained under various ageing variants and in the reference condition (for significance coefficient $p < 0.05$).

Variable	Reference	KKA-30-20	KKA-20-20	KKB-10-80	KKB-30-80	KKB-80-80	KKC-30-50
Reference	1.0000	0.3303	0.3102	0.7420	0.1880	−0.7522	0.5233
KKA-30-20	0.3303	1.0000	0.6964	0.4763	−0.2325	−0.2274	0.0000
KKA-20-20	0.3102	0.6964	1.0000	0.6423	−0.2495	−0.6713	−0.1515
KKB-10-80	0.7420	0.4763	0.6423	1.0000	−0.4603	−0.6924	−0.1561
KKB-30-80	0.1880	−0.2325	−0.2495	−0.4603	1.0000	−0.2115	0.9190
KKB-80-80	−0.7522	−0.2274	−0.6713	−0.6924	−0.2115	1.0000	−0.3596
KKC-30-50	0.5233	0.0000	−0.1515	−0.1561	0.9190	−0.3596	1.0000

The following observations can be made based on the statistical results presented in Table 9:

- a strong positive correlation (the correlation coefficient is 0.9190) was observed for KKB-30-80 and KKC-30-50, where the samples were exposed to the same temperature (30 °C-303 K), but in different humidities (50% and 80%).
- a correlation between the value of tensile strength and the ageing conditions (considering both temperature and humidity) in the majority of variants of ageing in different climatic conditions cannot be observed.

4.2. Samples Exposed in the Thermal Shock Chamber

A comparison of the results obtained for the samples tested in the thermal shock chamber with the application of a varying number of cycles is shown in Figure 17.

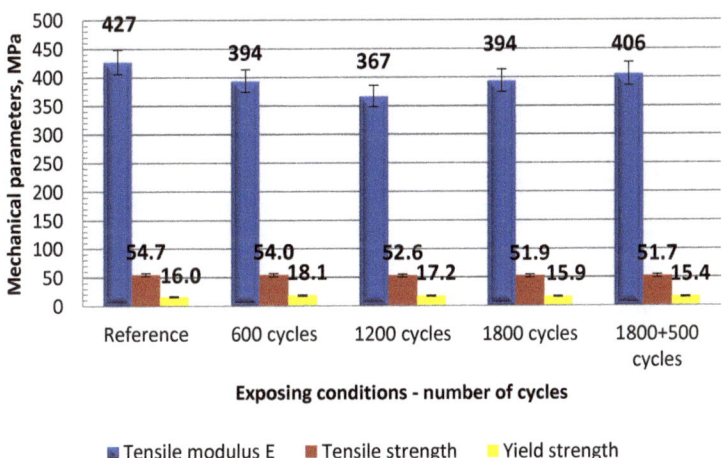

Figure 17. Strength parameters of samples exposed in the thermal shock chamber.

The results (Figure 17) demonstrate that the tensile modulus is similar in the majority of the tested cases. Only when the samples were exposed to 1200 cycles, did this lead to the lowest value of E being achieved, amounting to 367 MPa, which is 90% of the highest tensile modulus obtained for variant IV (1800 cycles and 500 cycles). In other cases, the difference is about 3%. It can, therefore, be observed that the number of curing cycles does not affect the tensile modulus of the tested materials. Nonetheless, compared to the reference samples (i.e., those that were not exposed to varying temperature), the exposure to thermal shock led to a decrease in E by 8%, 14%, 8% and 5%, respectively. In other words, the test samples exposed to thermal shock had lower properties (tensile modulus) compared to the reference samples.

Analyzing the results, it can also be observed that the tensile strength results do not differ much in terms of the number of curing cycles of the samples in the thermal shock chamber. The maximum tensile strength is 54 MPa, and the lowest is 51.7 MPa, which is a difference of over 4%. However, one can see that tensile strength tends to decrease with the increasing number of curing cycles of the samples in the thermal shock chamber, particularly when compared to the tensile strength of the reference samples. It was observed that the tensile strength of the reference samples was higher than that of the samples exposed to thermal shocks. Although the differences were not significant, the above-mentioned trend could be observed.

Concerning the yield strength, a trend can be observed whereby the yield strength decreases with increasing numbers of cycles of the samples exposed in the thermal shock chamber. The difference between the highest and the lowest yield strength is 3.6 MPa, which amounts to almost 20% of the maximum value. The highest yield strength, 18.1 MPa, was obtained for the ageing of the samples in the climatic chamber for 600 cycles (−20 °C/+60 °C) (253–333 K), while the lowest was obtained for KSZ D (1800 + 500 cycles). Comparing these results with those of the reference samples, it can be observed that the yield strength of the samples exposed to a reduced number of cycles was lower, while the yield strength of the test samples is lower than that of the reference samples with increasing numbers of cycles.

The correlation coefficient results of the tensile modulus for the samples subjected to ageing under different thermal shock conditions and for reference samples are presented in Table 10, and in Table 11 the results refer to the tensile strength.

Table 10. The correlation coefficient of tensile modulus obtained in various ageing variants under thermal shock conditions and in the reference condition (for significance coefficient $p < 0.05$).

Variable	Reference	KSZ A	KSZ B	KSZ C	KSZ D
Reference	1.0000	−0.8365	−0.9888	−0.5745	−0.8802
KSZ A	−0.8365	1.0000	0.9089	0.0439	0.5942
KSZ B	−0.9888	0.9089	1.0000	0.4487	0.8329
KSZ C	−0.5745	0.0439	0.4487	1.0000	0.6383
KSZ D	−0.8802	0.5942	0.8329	0.6383	1.0000

Table 11. The correlation coefficient of tensile strength obtained in various ageing variants under thermal shock conditions and in the reference condition (for significance coefficient $p < 0.05$).

Variable	Reference	KSZ A	KSZ B	KSZ C	KSZ D
Reference	1.0000	−0.4861	−0.7497	0.0655	−0.2703
KSZ A	−0.4861	1.0000	0.6259	0.4191	−0.5748
KSZ B	−0.7497	0.6259	1.0000	−0.3298	0.0326
KSZ C	0.0655	0.4191	−0.3298	1.0000	−0.7440
KSZ D	−0.2703	−0.5748	0.0326	−0.7440	1.0000

The following observations can be made based on the statistical results presented in Table 10:

- in comparison to the correlation between the value of tensile modulus in the thermal shock conditions and the reference variant, the strongest correlation can be seen in the case of KSZ B; in this case, a strong negative correlation (the correlation coefficient is 0.9888) was observed for KSZ B and the reference variant. This means that an increase in the number of cycles (1200 cycle variant of thermal shock) leads to a decrease in the tensile modulus,
- in the majority of variants of ageing under the tested thermal shock conditions, the correlation or strong correlation between the value of the tensile modulus and the thermal shock conditions cannot be observed, although in some cases, some correlation (albeit not strong) can be seen; for example, this case is presented in the variant KSZ A and KSZ B (the correlation coefficient is 0.9089).

The correlation or strong correlation between the value of the tensile strength and the thermal shock condition cannot be observed in the majority of ageing variants in the tested thermal shock conditions based on the correlation coefficient results of presented in Table 11.

Temperature, humidity and thermal shock are some of the factors that degrade polymer materials. Degradation, as defined, is a process of structural change that can be the result of physical or chemical changes occurring in polymer materials under the influence of the long-term action of various external factors. It should be noted, however, that in some cases, in the first phase of degradation, the degradation agent improves specific properties of the material, especially mechanical strength. This is done by additional cross-linking of the material structure under the influence of, for example, heat. It is only during the further action of degradation factors that the properties deteriorate due to various processes, e.g., excessive cross-linking or molecular weight reduction.

5. Conclusions

A comparison of the experimental results for the samples subjected to various ageing conditions reveals different relationships regarding the climatic chamber and the thermal shock chamber results, as well as different strength parameters.

The results indicate that:

1. at various values of temperature and humidity, the observed relationships were different from those observed for the reference samples.
2. many strength parameters were undesirable with respect to the required values (in the tested variants) at temperatures of 10 °C (282 K) and 80 °C (353 K) at a humidity of 80%.
3. tensile strength, tensile modulus and yield strength (and other variables) were higher at below 0 °C temperatures than at above 0 °C temperatures in the Celsius scale. It can, therefore, be concluded that below 0 °C temperatures do not have a negative effect on the mechanical properties of the conveyor belt samples. However, it is difficult to unequivocally determine the effects of temperature and humidity with respect to the reference samples, although temperature can have a more significant effect than humidity.
4. thermal shocks affect some of the mechanical properties of the conveyor belt; for example, some mechanical properties of the tested materials decreased with the increasing number of cycles. A trend could be observed in which the mechanical properties decrease with increasing exposure time of the samples to thermal shock. The observed trend of an increase in the tensile modulus can lead to lower elastic deformation of the tested conveyor, which can be reflected in changes in its dimensions, affecting the operation of the conveyor belt throughout the entire line, and also with respect to other devices.
5. the SEM analysis of the cross-section and surface of the bottom side of the fabric indicates that thermal shock affects the failure much more than climatic conditions.

In summary, thermal shock had a more significant effect on the mechanical properties of the samples than temperature and humidity combined. It is worth emphasizing the known fact that a temperature below 0 °C does not lead to a decrease in the mechanical properties of the conveyor belt.

It is recommended that further studies be conducted to determine the effect of higher ranges of factors such as changes in temperature and humidity and thermal shock.

The resulting research could serve as input and impetus for further research and subsequent innovations of the material composition of belts used in similar conditions, such as halls of distribution or logistical centers. The presented research results will also help the users of test belts to regulate atmospheric conditions in a suitable way on their own premises on which conveyor belts are installed, in order to achieve the optimum functionality and lifespan of the conveyor belt.

Author Contributions: Conceptualization, A.R. and R.M.; methodology, A.R.; validation, A.R., L.M. and P.D.; investigation, A.R.; resources, R.M. and L.M.; writing—original draft preparation, A.R.; writing—review and editing, R.M.; visualization, A.R. and P.D.; supervision, A.R.; project administration, L.M.; funding acquisition, R.M. All authors have read and agreed to the published version of the manuscript.

Funding: This research was funded by VEGA, grant number 1/0721/18.

Acknowledgments: The authors thank the Direct Parcel Distribution SK s.r.o. for providing the testing samples and for the technical assistance.

Conflicts of Interest: The authors declare no conflict of interest.

References

1. Andrejiova, M.; Grincova, A.; Marasova, D. Measurement and simulation of impact wear damage to industrial conveyor belts. *Wear* **2016**, *368*, 400. [CrossRef]
2. Anand, S.C.; Brunnschweiler, D.; Swarbrick, G.; Russell, S.J. Mechanical bonding. In *Handbook of Nonwovens*; Russell, S.J., Ed.; Woodhead Publishing Limited: Cambridge, UK, 2007; pp. 255–294.
3. Montgomery, D.C. *Design and Analysis of Experiments*, 6th ed.; WIlley: Hoboken, NJ, USA, 2005.
4. Hicks, C.R. *Fundamental Concepts in the Design of Experiments*, 4th ed.; Oxford University Press: Oxford, UK, 1993.
5. Wootton, D.B. *The Application of Textiles in Rubber*; Rapra Technology Ltd.: Shawbury, UK, 2003.
6. Rudawska, A. Influence of the thickness of joined elements on lap length of aluminium alloy sheet bonded joints. *Adv. Sci. Technol. Res. J.* **2015**, *9*, 35. [CrossRef]
7. Ambrisko, L.; Marasova, D.; Grendel, P. Determination the effect of factors affecting the tensile strength of fabric conveyor belts. *Eksploat. Niezawodn. Maint. Reliab.* **2016**, *18*, 110. [CrossRef]
8. Tausif, M.; Russell, S.J. Characterisation of the z-directional tensile strength of composite hydroentangled nonwovens. *Polym. Test.* **2012**, *31*, 944. [CrossRef]
9. Koo, H.-J.; Kim, Y.-K. Reliability assessment of seat belt webbings through accelerated life testing. *Polym. Test.* **2005**, *24*, 309. [CrossRef]
10. Rudawska, A. Strength of adhesive joint of epoxy composites with dissimilar adherends. *Int. J. Fract. Fat. Wear* **2014**, *2*, 25.
11. Wang, S.; Li, Q.; Zhang, W.; Zhou, H. Crack resistance test of epoxy resins under thermal shock. *Polym. Test.* **2002**, *21*, 195. [CrossRef]
12. Barburski, M.; Goralczyk, M.; Snycerski, M. Analysis of changes in the internal structure of pa6.6/pet fabrics of different weave patterns under heat treatment. *Fibres. Text East. Eur.* **2015**, *23*, 46.
13. Drozdziel, P.; Komsta, H.; Krzywonos, L. An analysis of the relationships among selected operating and maintenance parameters of vehicles used in a transportation company. *Transp. Probl.* **2011**, *4*, 93.
14. Bocko, P.; Marada, O.; Bouda, T. Material and Component Testing of Conveyor Belts and its Numerical Analyses. In Proceedings of the 54th International Conference of Machine-Design-Departments, Hejnice, Czech Republic, 10–12 September 2013; pp. 565–571.
15. Goltsev, V.Y.; Markochev, V.M. Comparative Analysis of Methods for Testing Materials for Fatigue Cracking Resistance. *Atom. Energy* **2015**, *118*, 17. [CrossRef]
16. Liu, X.; Pang, Y.; Lodewijks, G. Theoretical and experimental determination of the pressure distribution on a loaded conveyor belt. *J. Int. Measur. Confed.* **2016**, *77*, 307. [CrossRef]

17. Molnar, W.; Nugent, S.; Lindroos, M.; Apostol, M.; Varga, M. Ballistic and numerical simulation of impacting goods on conveyor belt rubber. *Polym. Test.* **2015**, *42*, 1. [CrossRef]
18. Nedbal, J.; Neubert, M.; Velychko, V.; Valentova, H. Conveyor belt rubber aging measured by dielectric spectroscopy. In Proceedings of the 11th International Multidisciplinary Scientific Geoconference, Albena, Bulgaria, 20–25 June 2011; Volume I, pp. 751–757.
19. Petrikova, I.; Marvalova, B.; Tuan, H.S.; Bocko, P. Experimental evaluation of mechanical properties of belt conveyor with textile reinforcement and numerical simulation of its behavior. In Proceedings of the 8th European Conference on Constitutive Models for Rubbers, San Sebastian, Spain, 25–28 June 2013; pp. 641–644.
20. Ammeraal Beltech's Specific Brochure. Available online: http://www.ammeraalbeltech.com (accessed on 2 September 2018).
21. Klimatest. Available online: http://www.klimatest.eu/katalog/leaflets/espec/ESPEC_seria_S_v2_2.pdf (accessed on 2 September 2018).
22. *DIN EN 10002-1. Tensile Testing of Metallic Materials. Method of Test at Ambient Temperature*; European Committee for Standardization: Brussels, Belgium, 2001.
23. Rudawska, A.; Warda, T. Test System for Lap-Joints Testing. PL Patent W.123189; filled 23 June 2014, 1 March 2016.
24. Fu, S.-Y. Cryogenic Properties of Polymer Materials Chapter 2. In *Polymers at Cryogenic Temperatures*; Kalia, S., Fu, S.-Y., Eds.; Springer: Heidelberg, Germany, 2013; pp. 9–39.
25. Yano, O.; Yamaoka, H. Cryogenic properties of polymers. *Prog. Polym. Sci.* **1995**, *20*, 585. [CrossRef]
26. Gradt, T.; Borner, H.; Hubner, W. Low Temperature Tribology at the Federal Institute for Materials Research and Testing (BAM). Available online: http://esmats.eu/esmatspapers/pastpapers/pdfs/2001/gradt.pdf (accessed on 25 October 2018).
27. Perepechko, I. *Low-Temperature Properties of Polymers*; Pergamon: Moscow, Russia, 1980.
28. Rudawska, A.; Danczak, I.; Müller, M. The effect of sandblasting on surface properties for adhesion. *Int. J. Adhes. Adhes.* **2016**, *70*, 176. [CrossRef]

© 2020 by the authors. Licensee MDPI, Basel, Switzerland. This article is an open access article distributed under the terms and conditions of the Creative Commons Attribution (CC BY) license (http://creativecommons.org/licenses/by/4.0/).

Article

Estimating Thermal Material Properties Using Step-Heating Thermography Methods in a Solar Loading Thermography Setup

Samuel Klein [1,*], Tobias Heib [1] and Hans-Georg Herrmann [1,2]

[1] Chair for Lightweight Systems, Saarland University, Campus E3 1, 66123 Saarbrücken, Germany; tobias.heib@uni-saarland.de (T.H.); hans-georg.herrmann@izfp.fraunhofer.de (H.-G.H.)
[2] Fraunhofer Institute for Nondestructive Testing IZFP, Campus E3 1, 66123 Saarbrücken, Germany
* Correspondence: samuel.klein@uni-saarland.de

Featured Application: Exploring step-heating thermography algorithms applied to solar loading thermography.

Abstract: This work investigates solar loading thermography applications using active thermography algorithms. It is shown that active thermography methods, such as step-heating thermography, present good correlation with a solar loading setup. Solar loading thermography is an approach that has recently gained scientific attention and is advantageous because it is particularly easy to set up and can measure large-scale objects, as the sun is the primary heat source. This work also introduces the concept of using a pyranometer as a reference for the evaluation algorithms by providing a direct solar irradiance measurement. Furthermore, a recently introduced method of estimating thermal effusivity is evaluated on ambient-derived thermograms.

Keywords: solar loading thermography; step-heating thermography; active thermography; thermal effusivity; linear effusivity fit; infrastructure; NDT

1. Introduction

Solar loading thermography has gained recent scientific attention [1–4] and shows the potential to become a proven non-destructive testing (NDT) method for large-scale outdoor structures and infrastructure. As natural heat sources are inherently erratic (affected by cloud coverage, wind gusts and ambient temperature fluctuations), evaluation algorithms must be reviewed on a case-by-case basis. The range of evaluation algorithms discussed in this work is constrained to conventional active thermography approaches, namely, lock-in thermography (LIT) [5], step-heating thermography (SHT) [6] and linear effusivity fit (LEF) [7]. Each of these may be used to determine thermal material properties, in this case, the thermal effusivity of an excited object. Thermal effusivity is a measure of how strongly a material exchanges heat with its surroundings: higher effusivity means a slower response to heat input, whereas lower effusivity means a quicker response. Thermal effusivity is also the main reason why different materials feel cold or hot to the touch, despite being at the same temperature initially.

Determining thermal effusivity using solar loading lock-in thermography was shown in our previous work [1], but empirical calibration was needed to do so. In this work, we propose an alternative approach using a pyranometer to measure the insolation (solar irradiance) directly. This allows the experimenter to directly assess the irradiation shape and amplitude at each point in time, which, in turn, allows the creation of analogies to conventional, laboratory-scale active thermography methods.

2. Materials and Methods

This work used a pyranometer to measure insolation. The model used was an "LP PYRA 03" from Delta Ohm (Padua, Italy), This model is a second-class pyranometer, which means that it satisfies the ISO 9060 Second-class accuracy class definition. It is connected to the measurement card consisting of the ADS1256 ADC, which has sufficient accuracy and resolution to digitize the signal generated by the pyranometer appropriately.

A custom data acquisition platform was developed, powered by a rechargeable battery and consisting of an embedded Linux single board computer (RaspberryPi 3B) and the FLIR Boson 640 in USB Video (UVC) mode. The FLIR long wave infrared (LWIR) thermal imaging camera, which can directly send raw image data over USB, was used in all further processing. This camera is sensitive between 7.5 µm and 13.5 µm.

The camera was calibrated once within the approximated expected temperature range using a large-aperture blackbody source (Mikron M345X4). All thermograms were corrected using a two-point calibration scheme based on a radiometric calibration. All algorithms and figures were created using the free software "GNU Octave" and commercial software "MATLAB".

3. Theoretical Framework

Solar loading lock-in thermography does not use artificial heat sources, such as heat lamps; its only heat processes are naturally occurring. These consist of the following: heat transfer within the structure (conduction); heat transfer with the surrounding air, which is free to move (convection); and heat transfer via radiation (solar irradiation, radiative heat loss). The fourth heat process, evaporative cooling, was not included in this study.

To estimate thermal material properties, the value of the heat flux input is needed. Therefore, a direct measurement of the insolation was conducted using a pyranometer.

As described in our previous work [1], the infrared camera was set up in a stationary position to capture a set of thermal images regularly spaced in time. These images represented a 2D temperature field that was assumed to capture the surface temperature of a structure after radiance calibration with a blackbody source. Effects such as emissivity/reflectivity can be corrected to improve absolute measuring accuracy for a given surface. These kinds of corrections were not performed in the scope of this work but are hypothesized to further increase measurement accuracy in future research [1].

The pyranometer was fixed horizontally, effectively measuring global horizontal irradiation (GHI). As different captured structures faced different angles, a correction from GHI to the respective vertical angle surfaces was conducted. This correction was performed via a two-step process:

1. Calculate the theoretical values for direct horizontal irradiation (DHI) and direct vertical irradiation (DVI) at the angle of interest.
2. Correct the pyranometer GHI data using the factor DVI/DHI.

The correction can be found in Figure 1.

Both DHI and DVI can be determined using the air mass coefficient method [8], where DVI must be calculated for each angle of interest (e.g., 180° = facing south, 225° = facing south west).

For horizontally oriented surfaces, no correction to the pyranometer data was necessary, as they were recorded horizontally.

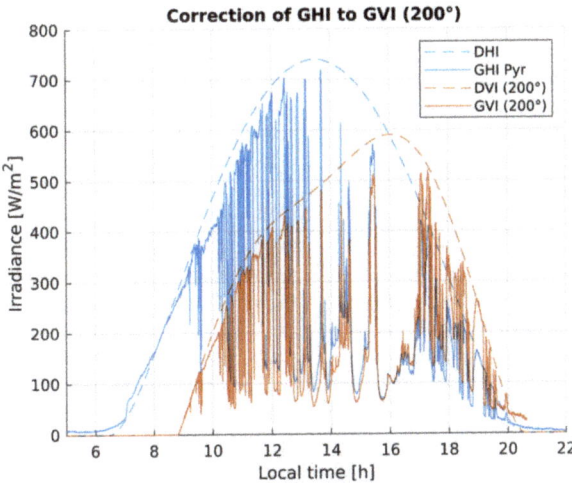

Figure 1. Correction of GHI to GVI on a 200° facing angle. Note that the theoretical values for DHI and DVI were calculated according to [8] and were used to correct the measured GHI data to infer the irradiance on the vertical standing wall structures (GVI).

After correction of the irradiation value, multiple approaches from active thermography to determine thermal material properties were evaluated:

1. Determining thermal effusivity b via step-heating thermography approach [6];
2. Determining thermal effusivity b via linear effusivity fit approach [7];
3. Determining surface heat capacity c_S [9].

The results of these three methods are discussed in the next section.

3.1. SHT Approach

According to [6], the temperature rise of thermally thick sample is given by

$$\Delta T = 2\dot{q} \cdot \sqrt{\frac{t}{\pi \cdot \lambda c_P \rho}} \quad (1)$$

where $\dot{q}$ is the step heating irradiation in $[W/m^2]$, t the irradiation time, λ the heat conductivity in $[W/mK]$, c_P the specific heat in $[J/kg]$ and ρ the density of the material in $[kg/m^3]$.

Considering the thermal effusivity, which is given by:

$$b = \sqrt{\lambda c_P \rho} \quad (2)$$

The formula can be rewritten as:

$$b = \frac{2\dot{q}}{\Delta T} \cdot \sqrt{\frac{t}{\pi}} \quad (3)$$

This formula calculates the thermal effusivity in a step-heating context on thermally thick materials (materials where the thermal wave can properly develop and travel without reflecting). Generally, thermally thick materials have a thickness $d \gg \mu$, where μ is the thermal diffusion length in $[m]$. The thermal diffusion length is given by:

$$\mu = \sqrt{\frac{2\lambda}{\rho c_P \cdot \omega}} \quad (4)$$

where ω is the angular frequency of the incident thermal wave in $[s^{-1}]$.

In this work, the angular frequency of the step-heating pulse was determined using $\omega = \pi/t$. In practice, objects with a thickness of $d > 1.5\mu$ show negligible error when considered thermally thick [6]. This cutoff value of 1.5μ is not universally applicable and was only used in this context because it generated reasonable results. In other setups with different excitation conditions and materials, other values distinguishing thermally thick from thermally thin materials may be necessary.

3.2. Surface Heat Capacity (SHC) Approach

Contrary to thermally thick materials, thermal "thinness" can be assumed when $d < 0.5\mu$. Again, the same restrictions as above apply to this empirically generated value.

When measuring a thermally thin object, thermal effusivity is not a valid material property to determine, because the formulas calculating it consider the material to be fully extended as a half space (infinitely thermally thick). Instead, surface heat capacity c_S in $[J/(m^2 K)]$ can be calculated and used to determine one of three values (ρ, c_P or d) when the other two are either known or assumed.

The surface heat capacity is defined as the ratio of the radiant exposure q in $[J/m^2]$ and the resulting temperature difference ΔT:

$$c_S = \frac{q}{\Delta T} \tag{5}$$

The radiant exposure of a step is the product of the heat flux density $\dot{q}$ and the corresponding effective time t and leads to:

$$c_S = t \cdot \frac{\dot{q}}{\Delta T} \tag{6}$$

In thermally thin materials, the surface heat capacity can be derived from the volumetric heat capacity s, as thermally thin materials act as if the heat spreads instantly in the direction of their thickness. In this case, c_S can also be determined via:

$$c_S = \overbrace{\rho c_P}^{s} \cdot d \tag{7}$$

This allows the determination of ρ, c_P or d if both other values are either known or assumed.

3.3. LEF Approach

The linear effusivity fit is a method introduced by Suchan and Hendorfer [7], using the Laplace transform to transform both the irradiation pulse and the temperature response from the surface into the frequency domain to determine the thermal effusivity by fitting a linear function to the quotient of the transformed data. It can be briefly described as a four-step process:

1. Transform both irradiation pulse $q(t)$ and surface temperature $T(t)$ using the discrete Laplace transform into the complex frequency domain to obtain $\bar{q}(s)$ and $\bar{T}(s)$, respectively. Note that s denotes the complex Laplace operator value.
2. Calculate the quotient of the aforementioned signals:

$$\bar{Z}(s) = \frac{\bar{T}(s)}{\bar{q}(s)} \tag{8}$$

3. Thermal effusivity is given by:

$$\bar{Z}(s) = \frac{1}{b \cdot \sqrt{s}} \tag{9}$$

Plotting $\overline{Z}(s)$ over $\left(\sqrt{s}\right)^{-1}$ results in a linear relationship: $\overline{Z}(s)/\sqrt{s} = b^{-1}$, the slope m of this function being b^{-1}.

4. A linear fit using $\overline{Z}(s)$ over $\left(\sqrt{s}\right)^{-1}$ is performed and the slope of this fit is inverted to obtain

$$b = 1/m_{Fit} \qquad (10)$$

The linear fit is necessary to suppress noise and high-frequency content in the resulting transformed values, but in theory, two points from $\overline{Z}(s)$ could be used to determine b.

This approach was employed in a region of interest-based context, where a single region of interest (ROI) was given as a mean value time vector to the algorithm to calculate the material properties of the whole ROI. Further, it was conducted on a pixel basis, where the LEF method was performed on each pixel to obtain an "effusivity image".

3.4. Albedo Factor and Surface Emissivity

Figure 2 shows that the heating step had an amplitude of ~320 W/m² for horizontal surfaces and ~300 W/m² for vertical surfaces at 200°. These values represent the total amount of incident radiation. However, as every surface absorbed only a part of the incoming energy (called the emissivity factor), this irradiation was only absorbed by blackbodies. As incident radiation energy (sunlight) is primarily visual and near infrared (NIR) energy [10], the factor used in this work to refer to the amount of reflected sunlight was albedo α, and the term describing the energy factor absorbed (and emitted) in the LWIR range was determined as emissivity ε.

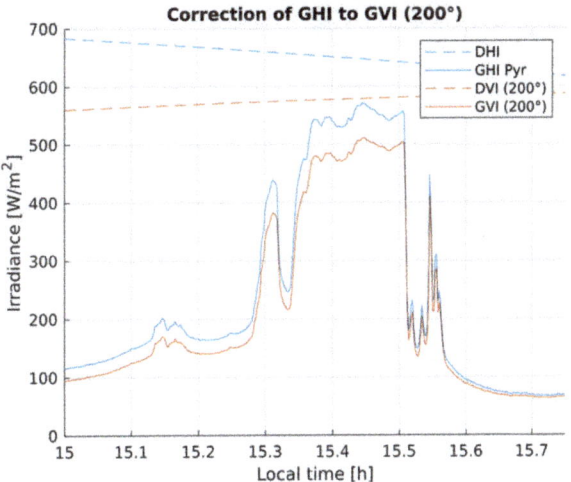

Figure 2. Correction of GHI to GVI as before, zoomed in to show the relevant part of the dataset containing step heating.

As the albedo of white plaster is approximately ~0.65 [11], the actual absorbed energy is 35% of the total incident irradiation. On grey surfaces, the albedo is ~0.55 [11], which results in an energy absorption of 45%. The albedo of grey pavement is ~0.60 [11].

As common building materials such as plaster and concrete are considered to have high emissivity (≥ 0.95) in the LWIR range [12] p. 316, an emissivity correction for the LWIR camera was not performed. Instead, the factors researched imposed an albedo correction first.

For Table 1 and all corrected figures, the albedo correction was performed using the aforementioned values.

The scene in Figure 3 shows the composition of a common street with houses, garages, trees, a car and a paved road. Due to this composition, not all of the four selected regions of interest are perpendicular to the optical axis of the thermography camera. For further inspections, it was necessary to perform a projective transformation from the observed angle to a plane that is perpendicular to the optical axis. For the creation of the transformation matrix, two assumptions were made. First, it was assumed that the observed ROIs were flat surfaces, and second, the two building walls and the garage sidewall in particular had a rectangular shape. For each ROI, the four edge points of the studied object were selected and registered to a rectangular geometry with the aspect ratio of the observed object [13]. The projective mapping was carried out with MATLAB, using the "image processing toolbox". Figure 4 shows a typical thermogram of the scene, with an overview over the ROIs chosen.

Figure 3. Optical overview over the scene.

Table 1. Overview of the different estimations. Note this table corrects for the albedo factor of the different image parts. [†] The reference values are taken from [14–16] [‡] as the concrete road lay horizontally. GHI is the excitation intensity (see Figure 2).

	Temperature		Excitation Intensity	Surface Heat Capacity	Effusivity		
	Difference	Gradient			SHT	LEF	Reference [†]
Region of interest	K	K/min	W/m^2	kJ/(Km^2)		J/(Km$^2\sqrt{s}$)	
(1) Wall left	7	0.6	300	6	324	1024	490–780
(2) Wall right	5	0.4	300	14	454	1061	490–780
(3) Concrete road	4	0.3	320 [‡]	27	2132	1971	1730–2120
(4) Garage	3	0.25	300	19	1362	1581	1100–1400

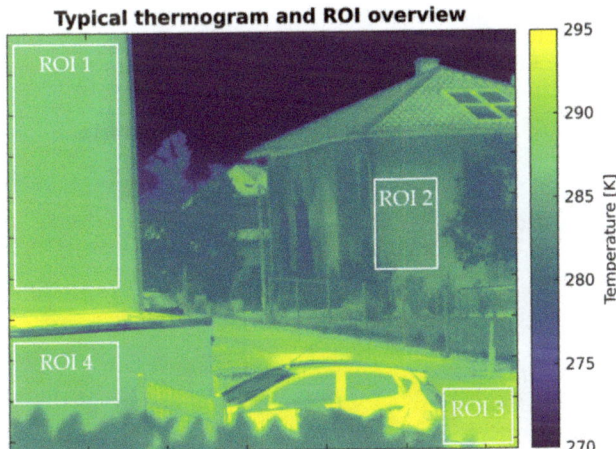

Figure 4. Typical thermogram of the scene with the selected ROIs of the experiment. ROI 1 is the left building wall, ROI 2 the right building wall, ROI 3 is the paved road, and ROI 4 the garage sidewall.

4. Results and Discussion

The experiment site was set up by measuring two building walls, a concrete-paved road and a garage sidewall. The general setup of the scene can be observed in Figure 3. The pyranometer data are shown in Figure 1 and constitute an excerpt from a much larger dataset, which was acquired over multiple days. However, for this work, the aforementioned algorithms required only a small portion of the whole dataset. Figure 2 shows the data section that is used in this work.

Figure 2 shows the relevant section from 15:00 h to 15:45 h with approximate step heating, generated by variation in cloud coverage. The blue curve (GHI) shows the irradiance incident to all horizontal surfaces in the scene, in this case, predominantly the paved road (ROI 3). The orange curve (GVI) was calculated from the GHI data to represent the incident irradiance on all 200° facing vertical structures, such as the two building sidewalls (ROI 1 and ROI 2) and the garage wall (ROI 4) shown in Figure 3. This explains the difference between the excitation intensities in Table 1.

Figure 5 was generated using a pixel-wise subtraction of the last image in the heating phase with the first image thereof. Note the uneven heating on the building on the right due to shadows being cast by the roof onto the wall below. This difference image was the basis for both the SHT effusivity estimation and SHC images below.

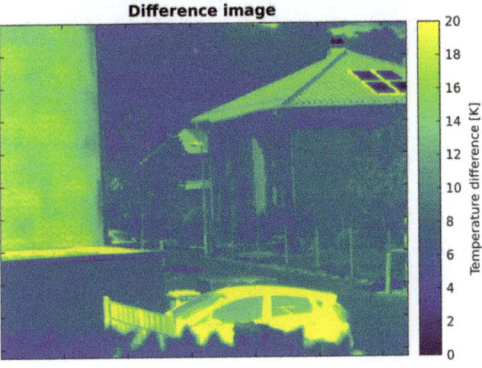

Figure 5. Difference image: last image in the heating phase subtracted by the first image thereof.

Figure 6 shows the uncorrected thermal effusivity. The shown image assumes a heat flux of 320 W/m² for all surfaces, but the real heat flux (albedo correction) must be estimated on a case-by-case basis to obtain correct estimates of thermal effusivity.

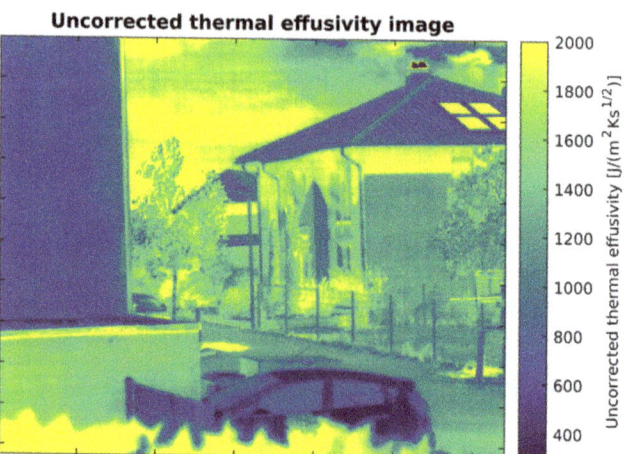

Figure 6. Effusivity image resulting from the SHT approach.

Figure 7 shows the uncorrected heat capacity, neglecting the albedo factor of the objects in the scene. The real surface heat capacity was reduced by a factor of $(1 - \alpha)$.

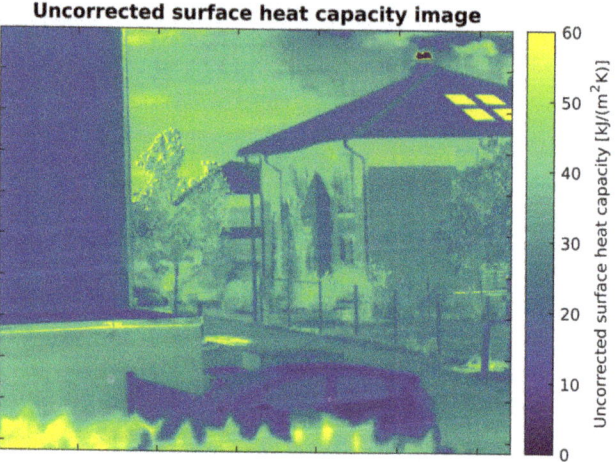

Figure 7. Surface heat capacity image.

The LEF algorithm was performed on each pixel of Figure 8 separately. Note that as before, no albedo correction was performed on this image, and each pixel was assumed to have absorbed the full incident irradiation.

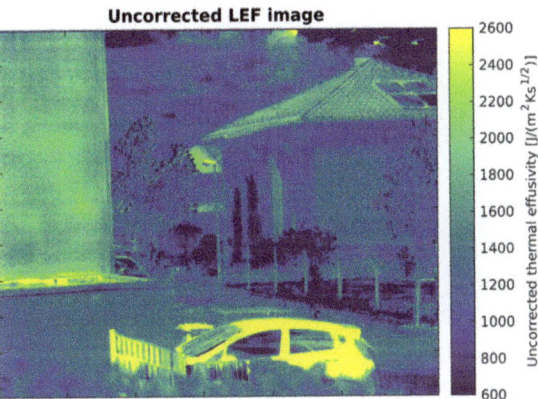

Figure 8. Linear effusivity fit image.

Note that this effusivity image (see Figure 8) differs from the image generated by the SHT effusivity estimation method (see Figure 6). This is, in part, because the underlying algorithm to determine surface effusivity by LEF was noise sensitive [7], and reducing noise by averaging a whole ROI delivered more reasonable values.

Table 1 summarizes the results of the different methods for the observed ROIs.

The garage wall consisted of hollow concrete blocks covered with a layer of gypsum. The two building walls consisted of an unknown type of insulating gypsum covering a conventional building insulation material, such as mineral wool or polystyrene.

Separating ROIs and correcting the trapezoidal distortion generated by the camera viewing each ROI surface from an angle helped to improve the visual clarity of the image. The results with separation and perspective correction are shown in Figures 9–12. Especially noteworthy are the following distinctive features:

- Wave-like structure in ROI 1—possible remnant of the insulation installation;
- Hot spot in the top middle of the garage (ROI 4);
- Apparent recognition of the floor in ROI 2.

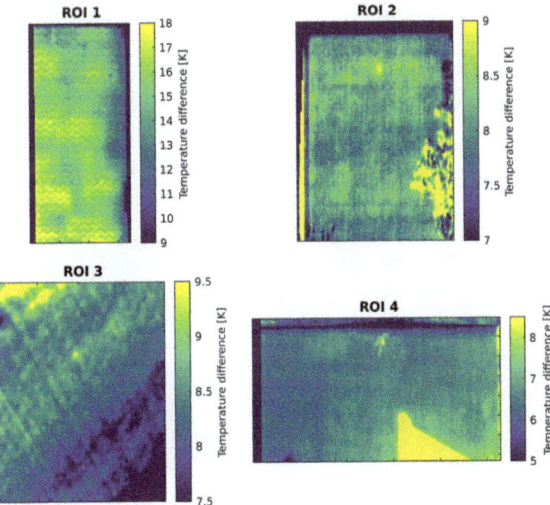

Figure 9. Perspective-corrected difference images.

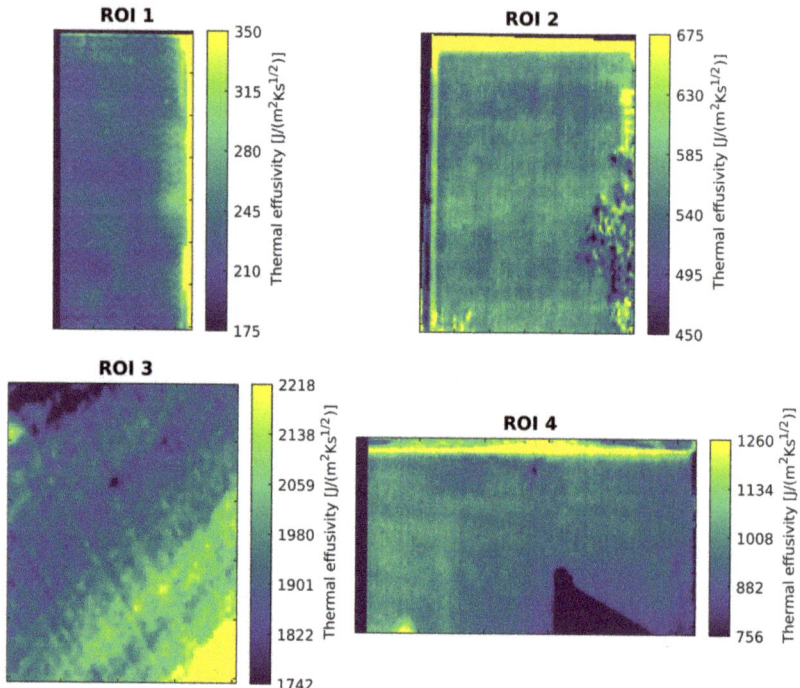

Figure 10. Perspective- and albedo-corrected SHT effusivity images.

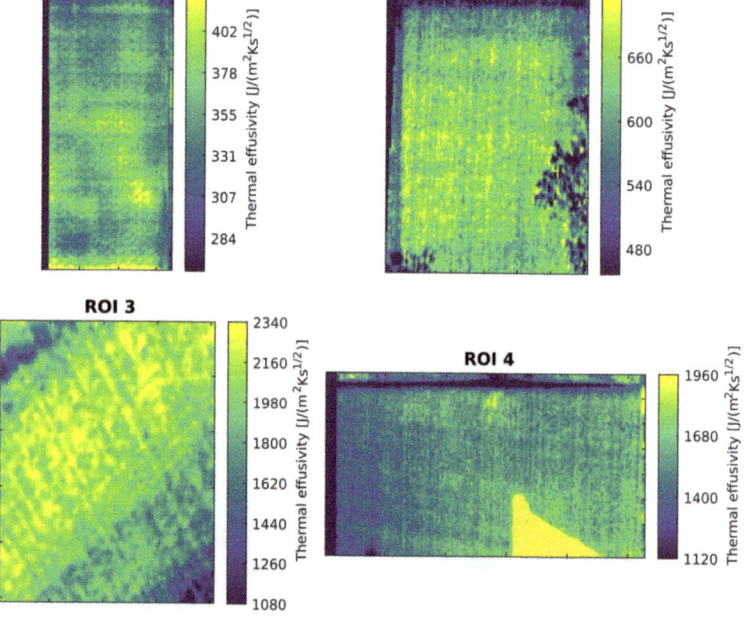

Figure 11. Perspective- and albedo-corrected LEF effusivity images.

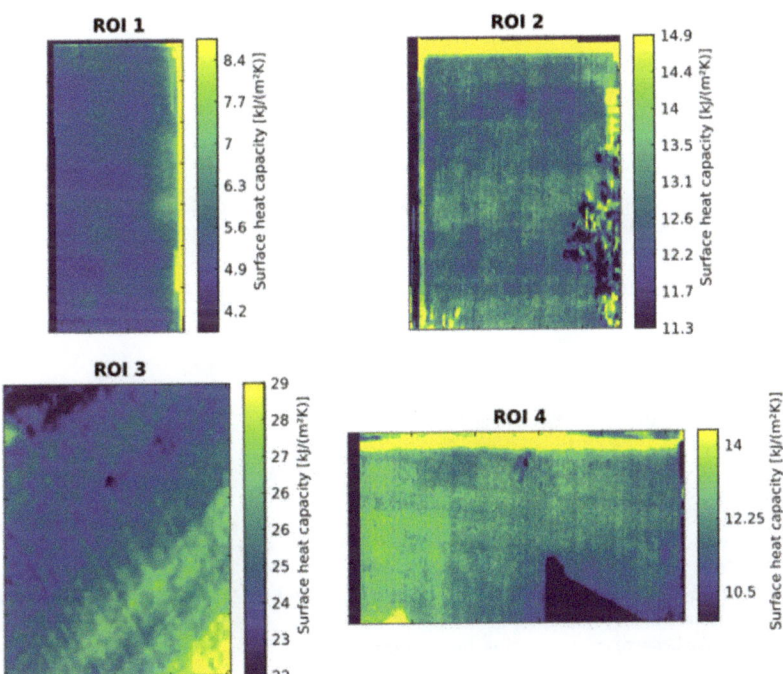

Figure 12. Perspective- and albedo-corrected surface heat capacity images.

5. Conclusions and Further Work

This study demonstrated that the methods used by laboratory-scale active thermography can be evaluated on solar loading thermograms. Depending on location conditions, such as weather and cloud coverage, both strong and weak resemblances to different testing methodologies, such as lock-in or step-heating thermography, can be observed. In general, signal-to-noise ratio is satisfactory in a solar loaded context, but the erratic excitation conditions, such as fluctuating irradiation and changing wind speeds, hamper the accuracy of the previously established assumptions made by the evaluation algorithms.

It was found that ideally, the experimental setup should record the region of interest perpendicularly in the first place. As this is not feasible in every experiment, separating ROIs from the whole image and correcting the perspective error can improve the apparent contrast and enhance visual clarity on the images.

Further work should evaluate how to take into account the given environmental conditions to better fit a specific test cast. For example, manual shading of a particular ROI may generate a step-heating scenario in strong sunlit conditions without the need for specific cloud coverage. Wind speed and ambient temperature could also be measured and evaluated to calculate the heat flow of the specific structures at all points in the experiment. However, specific algorithms and methods must be devised to correct the influences of these unpredictable factors.

Author Contributions: Conceptualization, S.K. and H.-G.H.; methodology, S.K.; software, S.K. and T.H.; validation, S.K. and H.-G.H.; writing—original draft preparation, S.K.; writing—review and editing, S.K., T.H. and H.-G.H.; visualization, S.K.; supervision, H.-G.H.; project administration, H.-G.H.; funding acquisition, H.-G.H. All authors have read and agreed to the published version of the manuscript.

Funding: The European Regional Development Fund (ERDF), especially grant number 14.2.1.4-2016-1, funded this research. Furthermore, we acknowledge support of the Deutsche Forschungsgemeinschaft (DFG, German Research Foundation) and Saarland University within the funding program Open Access Publishing.

Institutional Review Board Statement: Not applicable.

Informed Consent Statement: Not applicable.

Acknowledgments: Gratitude goes to Josef Suchan and Günther Hendorfer for excellent practice in scientific exchange. Furthermore, the authors thank Jessica Jacob for her help with experimental setup and conduction.

Conflicts of Interest: The authors declare no conflict of interest. The funders had no role in the design of the study; in the collection, analyses, or interpretation of data; in the writing of the manuscript; or in the decision to publish the results.

References

1. Klein, S.; Fernandes, H.; Herrmann, H.-G. Estimating thermal material properties using solar loading lock-in thermography. *Appl. Sci.* **2021**, *11*, 3097. [CrossRef]
2. Tu, K.; Ibarra-Castanedo, C.; Sfarra, S.; Yao, Y.; Maldague, X.P.V. Multiscale analysis of solar loading thermographic signals for wall structure inspection. *Sensors* **2021**, *21*, 2806. [CrossRef] [PubMed]
3. Ibarra-Castanedo, C.; Sfarra, S.; Klein, M.; Maldague, X. Solar loading thermography: Time-lapsed thermographic survey and advanced thermographic signal processing for the inspection of civil engineering and cultural heritage structures. *Infrared Phys. Technol.* **2017**, *82*, 56–74. [CrossRef]
4. Bagavathiappan, S.; Lahiri, B.B.; Saravanan, T.; Philip, J.; Jayakumar, T. Infrared thermography for condition monitoring—A review. *Infrared Phys. Technol.* **2013**, *60*, 35–55. [CrossRef]
5. Breitenstein, O.; Warta, W.; Langenkamp, M. *Lock-in Thermography*; Springer: Berlin/Heidelberg, Germany, 2010; ISBN 970-3-642-02416-0.
6. Boué, C.; Fournier, D. Infrared thermography measurement of the thermal parameters (effusivity, diffusivity and conductivity) of materials. *Quant. InfraRed Thermogr. J.* **2009**, *6*, 175–188. [CrossRef]
7. Suchan, J.; Hendorfer, G. Thermal effusivity determination of carbon fibre-reinforced polymers by means of active thermography. *Quant. InfraRed Thermogr. J.* **2020**, *17*, 210–222. [CrossRef]
8. Meinel, A.B.; Meinel, M.P. *Applied Solar Energy: An Introduction*; Addison-Wesley Publishing Company: Reading, MA, USA; New York, NY, USA, 1977; ISBN 9780201047196.
9. Logvinov, G.N.; Gurevich, Y.G.; Lashkevich, I.M. Surface heat capacity and surface heat impedance: An application to theory of thermal waves. *Jpn. J. Appl. Phys.* **2003**, *42*, 4448. [CrossRef]
10. Department of Agronomy, Iowa State University. Thermal Energy-Radiation Spectrum. Available online: http://agron-www.agron.iastate.edu/courses/Agron541/classes/541/lesson09a/9a.3.html (accessed on 1 July 2021).
11. Reagan, J.A.; Acklam, D.M. Solar reflectivity of common building materials and its influence on the roof heat gain of typical southwestern U.S.A. residences. *Energy Build.* **1979**, *2*, 237–248. [CrossRef]
12. Vavilov, V.; Burleigh, D. *Infrared Thermography and Thermal Nondestructive Testing*; Springer International Publishing: Cham, Germany, 2020; ISBN 978-3-030-48001-1.
13. Heckbert, P. Fundamentals of Texture Mapping and Image Warping. Master's Thesis, Berkeley University, Berkeley, CA, USA, 1989.
14. John, H.; Lienhard, I.V. *A Heat Transfer Textbook*, 5th ed.; Phlogiston Press: Cambridge, MA, USA, 2020.
15. Ouakarrouch, M.; Azhary, K.E.; Laaroussi, N.; Garoum, M. Thermal Performances of Hollow Concrete Blocks based on ISO Norm Calculations. In *Thermal Performances of Hollow Concrete Blocks based on ISO Norm Calculations*; IEEE: Piscataway, NJ, USA, 2019; ISBN 978-1-7281-5152-6.
16. American Society of Heating, Refrigerating and Air-Conditioning Engineers. In *2017 ASHRAE Handbook: Fundamentals*; ASHRAE: Atlanta, GA, USA, 2017; ISBN 9781939200570.

Article

Quantitative Evaluation of Unfilled Grout in Tendons of Prestressed Concrete Girder Bridges by Portable 950 keV/3.95 MeV X-ray Sources

Mitsuru Uesaka [1,*], Jian Yang [2], Katsuhiro Dobashi [1], Joichi Kusano [3], Yuki Mitsuya [1] and Yoshiyuki Iizuka [4]

1. Nuclear Professional School, School of Engineering, The University of Tokyo, Ibaraki 319-1188, Japan; kdobashi@nuclear.jp (K.D.); y.mitsuya@sogo.t.u-tokyo.ac.jp (Y.M.)
2. Department of Nuclear Engineering and Management, School of Engineering, The University of Tokyo, Tokyo 113-8654, Japan; yang.jian@nuclear.jp
3. Accuthera Inc., Kanagawa 215-0033, Japan; kusano@accuthera.com
4. Atox Co., LTD., Tokyo 108-0014, Japan; yoshiyuki_iizuka@atox.co.jp
* Correspondence: uesaka.mits@gmail.com

Abstract: We have developed porTable 950 keV/3.95 MeV X-band (9.3 GHz) electron linear accelerator (LINAC)-based X-ray sources and conducted onsite prestressed concrete (PC) bridge inspection in the last 10 years. A T-shaped PC girder bridge with a thickness of 200–400 mm and a box-shaped PC girder bridge with a thickness of 200–800 mm were tested. X-ray transmission images of flaws such as thinning, fray, and disconnection caused by corrosion of PC wires and unfilled grout were observed. A three-dimensional structural analysis was performed to estimate the reduction in the yield stress of the bridge. In this study, we attempted to evaluate the unfilled grout quantitatively because it is the main flaw that results in water filling and corrosion. In the measured X-ray images, we obtained gray values, which correspond to the X-ray attenuation coefficients of filled/unfilled grouts, PC wires (steel) in a sheath, and concrete. Then, we compared the ratio of the gray values of the filled/unfilled grouts and PC wires to determine the stage of the unfilled grout. We examined this quantitative evaluation using the data obtained from a real T-shaped PC girder bridge and model samples to simulate thick box-shaped PC girder bridges. We obtained a clear quantitative difference in the ratios for unfilled and filled grouts, which coincided with our visual perception. We synthesized the experience and data and proposed a quantitative analysis for evaluating the unfilled grout for subsequent steps such as structural analysis and destructive evaluation by boring surveys.

Keywords: onsite X-ray bridge inspection; 950 keV/3.95 MeV X-ray sources; PC bridge; unfilled grout; quantitative evaluation of stage of unfilled grout

1. Introduction

Many concrete structures are facing aging problems. Most social infrastructure components, including bridges and tunnels, were built during the rapid economic boom of the 1960s. Since then, the aging and degradation effects have gradually proceeded. Thus, the need for maintenance of these structures is growing exponentially. Due to the legendary growth boom that started after WWII, 16% of the overall concrete social infrastructure was over 50 years old in 2011. The lifespan of these structures is generally 50 years [1]. Only 10 years later, in 2021, this ratio has reached 42% and will increase to 63% by 2031, as shown in Figure 1. The number and rate of growth of aging structures are astounding, and this constitutes a critical problem for our society.

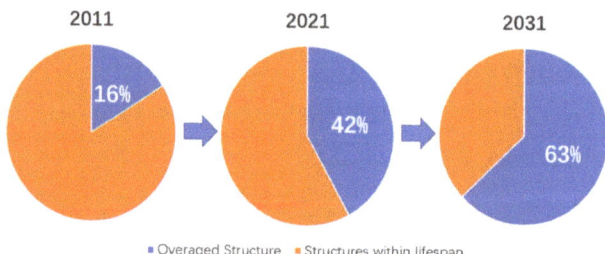

Figure 1. Changes in the ratio of overaged concrete structures in Japan by decade.

Any major malfunction in these infrastructures may have a profound influence on our industrial system and social life. It is very costly to rebuild anything that seems to have been damaged. Only if proper management is carried out can the lifespans of these structures be extended. Even though deterioration has already proceeded, appropriate assessment and management are necessary to ensure their safety. After the accident caused by the fall of a concrete ceiling at a Japanese highway tunnel in 2012, the Japanese government approved a law to maintain and manage such infrastructure every five years in order to secure their continuous and long-term use. However, it is difficult to determine the state of deterioration to prevent a collapse during maintenance. According to the governmental law, the current inspections are mainly carried out by visual inspection, hammer-sounding inspection, and palpation. The surface condition is observed visually, and the wall surface is evaluated by knocking on it with a special hammer to examine the concrete lifting and peeling. Palpation is mainly used to detect loose bolts and nuts. Only superficial conditions can be confirmed by these methods. Damage, degradation, and deterioration proceed continuously, not only outside but also inside the structure. How the inner structure is damaged, especially the metal rods on which the strength and load of the whole structure rely to a large extent, remains unknown from outside. An appropriate evaluation technology for structural integrity is necessary for the assessment of deep inside degradation. Figure 2 shows the number of bridges built per year and the percentage by structure in Japan. From 1950 to 1960, the number of newly constructed bridges increased significantly each year. During the economic boom, most bridges were built with prestressed concrete (PC) [2].

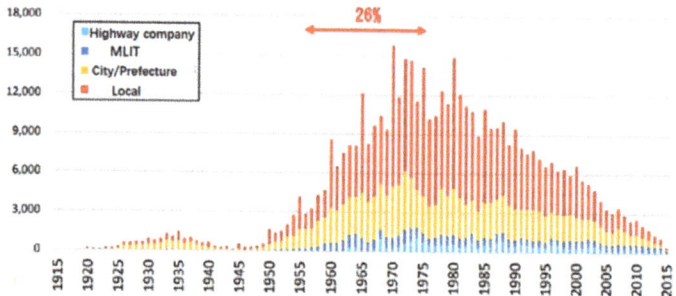

Figure 2. Bridges built in Japan by year (MLIT is the Ministry of Lands, Infrastructure, Transportation, and Tourism).

The PC bridge technology was first applied in construction in 1951 in Japan and subsequently became widespread owing to its ability to meet the high strength required by the design standard at a low construction cost. However, as the scale of bridges expanded, these structures became more complicated [3]. The technology was relatively new, and knowledge had not yet accumulated. As these bridges aged, degradation damage caused by the environment, corrosion, and rupture appeared in the late 1980s and 1990s. These problems are occurring now and will continue in the future, leading to critical situations, as shown in Figure 3. To demonstrate the inspection method, it is necessary to explain the types of concrete bridges while underlining the primary inspection target, namely, PC bridges.

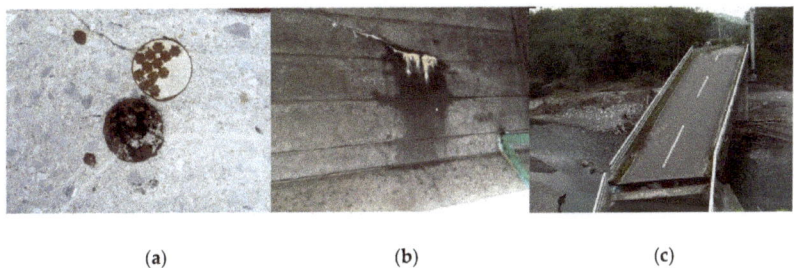

(a)   (b)   (c)

Figure 3. Degraded and collapsed bridges. (**a**) Cross section of PC sheaths (**b**) Concrete surface crack (**c**) collapse [4].

As shown in Figure 4, there are several types of PC bridges. T-shaped PC girder bridges have cast-in-place concrete beams with designed sections on both sides of the beams. The beams are more profound than the deck sections of their cross sections. Thus, it is called a T-shaped PC girder bridge. Box-shaped PC girder bridges are bridges in which the main beams comprise girders in the shape of a hollow box, which is typically rectangular in a cross section. This type is typical of highways and other elevated structures. As it is cost-effective and superior in strength, it can be a precast offsite to assure quality. Slab bridges are monolithic and consist of simple flat concrete slabs with twisted or roughened reinforcing rods concentrated in the lower portion at both ends of the slab. Regardless of its type, a bridge relies on the built-in prestressed steel reinforcement.

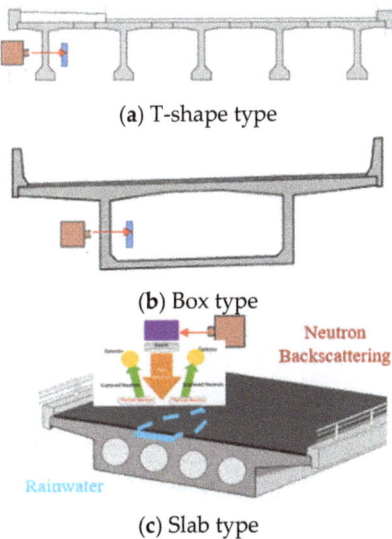

Figure 4. Types of PC bridges. (**a**) T-shape girder type; (**b**) box-shape girder type; (**c**) Slab type.

Reinforcements are generally distributed and fixed at the two ends of the concrete bridge, as shown in Figure 5. Steel reinforcements are generally placed inside a hollow sleeve, called a sheath. The reinforcements are fixed inside the sheath using a material called grout, as shown in Figure 6. The grout inside a sheath may suffer from damage because it is a material different from concrete and is filled into the sheath afterward. Several factors may cause grout failure, such as construction negligence and neutralization with CO_2, as the grout consists of alkali. In addition, water, moisture, and salt invasion are the other main reasons for failure. Rainwater invades the PC sheath from the edge. The situation of filled and unfilled grout in two types of PC is schematically depicted in Figure 7. If there is unfilled grout, a large amount of water remains. Then, it gradually induces corrosion, thinning, and cracking of the PC sheath and wires. The water exudes from the corroded sheath to the surface of the concrete. Thinning, fray, and disconnection of PC wires occur, which leads to a significant reduction in mechanical strength. The unfilled grout itself is a longitudinal discontinuity of the mechanical stiffness. Therefore, it becomes a reason for cracks in the nearby concrete, even at an early stage of use. Finally, rainwater and humidity invade the concrete from the surface.

Unfilled grout of the tendons of PC bridges can cause corrosion and, consequently, provoke significant prestress losses, which, in turn, can induce cracking and excessive deflections during service of the PC bridge girder [5–9].

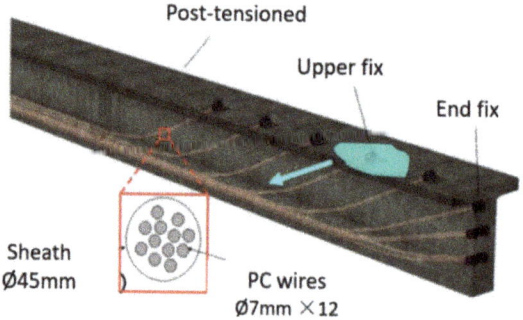

Figure 5. Reinforcement distribution within a structure.

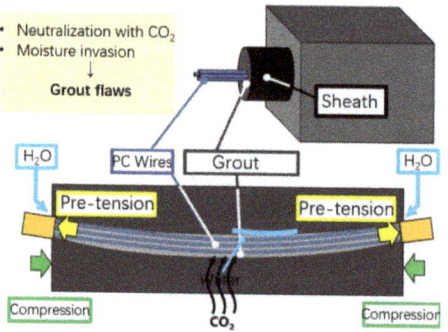

Figure 6. Side cross-section of a sheath inside a PC structure.

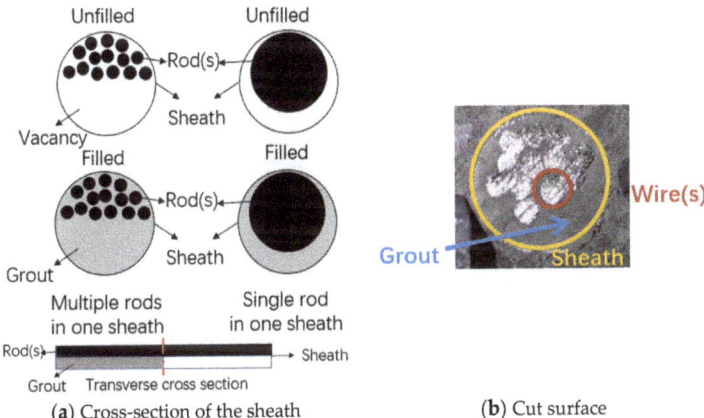

Figure 7. (a) Schematic of the cross-section of a sheath; (b) cut surface of a sheath.

2. Review of Existing Methods

There are several nondestructive evaluation (NDE) methods for detecting poor construction, such as unfilled grout in the PC sheath, and degradation, such as thinning and disconnection of PC wires. RADAR [10], ultrasonic testing [11], magnetic testing [12], and 200–400 kV X-ray tubes [13] are shown in Figure 8. RADAR can detect and visualize 3D iron structures up to 300 mm thick, but the reconstructed image is distorted by a few milliseconds. Ultrasonic testing is available up to 200 mm in thickness, but it is difficult to use to reconstruct the shape of an iron structure. Magnetic testing can detect the disconnection of iron rods or wires up to 300 mm in thickness. With a 400 kV X-ray tube, transmitted images of the PC sheath, wires, rods, and grout can be obtained up to 400 mm in thickness, but it takes approximately 1 h.

Compared with the above existing methods, we explain and emphasize the advantage of our method using 950 keV/3.95 MeV X-ray sources [14–16]. With them, clear detection and visualization with 1 mm resolution of PC iron structures and grout filling at a depth of 400–1000 mm are possible within minutes.

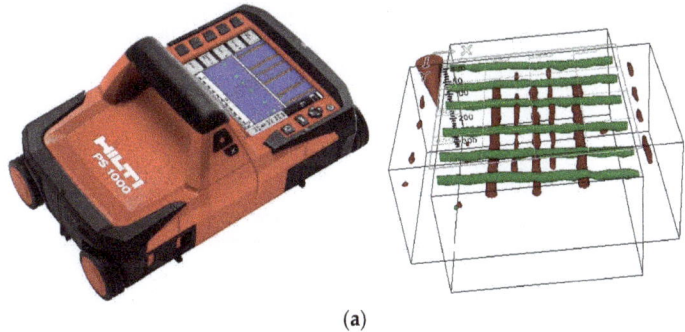

(a)

Figure 8. Cont.

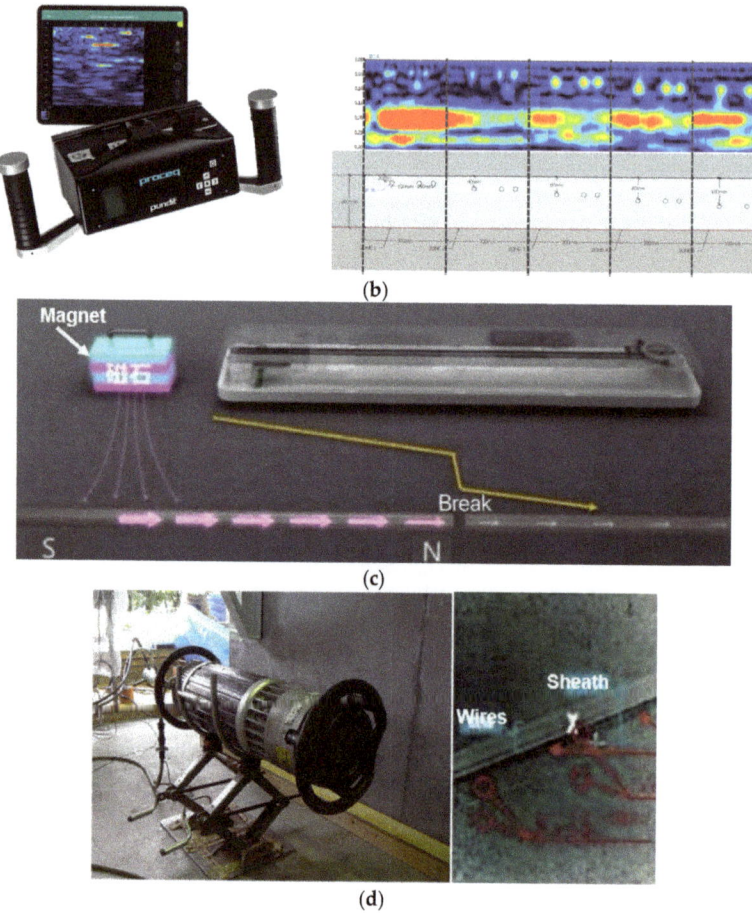

Figure 8. Other inspection methods for the inner structure of bridges. (**a**) RADAR (1.6–3 GHz) [10]; (**b**) ultrasonic testing (50 kHz) [11]; (**c**) magnetic testing [12]; (**d**) X-ray tubes (200–400 keV) [13].

3. Proposed Methodology

3.1. 950 keV/3.95 MeV X-ray Sources

We used an X-band (9.3 GHz) LINAC-based 950 keV/3.95 MeV X-ray source to inspect the actual bridge [14–16]. These systems are shown in Figure 9a,b, respectively.

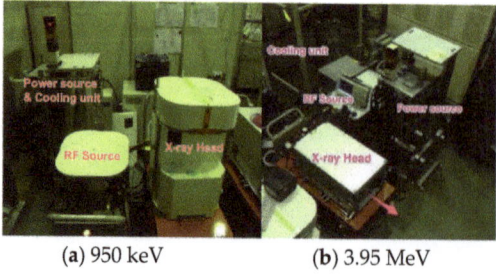

(**a**) 950 keV　　　　(**b**) 3.95 MeV

Figure 9. Portable X-band LINAC-based X-ray sources. The systems are composed of three units: X-ray head, magnetron, and power units. (**a**) 950 keV; (**b**) 3.95 MeV.

The electrons are accelerated up to 950 keV by radio frequency (RF) fields in the first source. We also adopted a side-coupled standing-wave-type accelerating structure. The electrons are injected into the tungsten target, which generates bremsstrahlung X-rays. A tungsten collimator makes the generated X-rays to the shape of a cone with an opening angle of 17°. The X-ray intensity is 50 mSv/min at 1 m for a full magnetron RF power of 250 kW. The system consists of a 50 kg X-ray head, a 50 kg magnetron box, and a stationary unit of an electric power source and water chiller. The X-ray head and the magnetron box are portable. Because a flexible waveguide connects them, only the position and angle of the X-ray head can be finely tuned. We optimized the design based on the X-ray intensity, compactness, and weight. The parameters of the 950 keV X-ray source are listed in Table 1. We placed an X-ray detector on the opposite side of the object to detect the transmitted X-rays. We used a flat panel detector (FPD) of a 0.4 thick GOS scintillator, with 16″ × 16″ detector size and 200 μm pixels.

Table 1. Specifications of X-ray sources.

	950 keV	3.95 MeV
Operating Frequency	9.3 GHz	9.3 GHz
Beam Energy	950 keV	3.95 MeV
Beam Current	130 mA	100 mA
Electron Gun Voltage	20 kV	20 kV
Electron Gun Current	300 mA	300 mA
Pulse Width	2.5 μs	4 μs
Pulse Frequency	330 pps	200 pps
RF Power	250 kW	1.5 MW

The 3.95 MeV system appears in Figure 9b. This system consists of a 62 kg X-ray head with a target collimator of 80 kg, a magnetron box weighing 62 kg, electric power sources at 116 kg, and a water-cooling system weighing 30 kg. The X-ray head and magnetron box are portable, and the position and angle of the former are finely tuned. The X-ray intensity of the system was 2 Gy/min at 1 m.

The calculated attenuation in concrete for the X-rays from the 950 keV/3.95 MeV sources are shown in Figure 10. The results indicate that concrete with thicknesses up to 400 mm and 800 mm can be penetrated by the 950 keV/3.95 MeV sources, respectively.

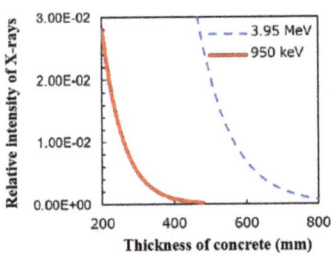

Figure 10. Calculated results of attenuation of the X-rays in concrete from the 950 keV/3.95 MeV X-ray sources.

Figure 11 depicts the procedure of X-ray inspection, flaw evaluation, and structural analysis. Poor construction of unfilled grout in PC sheaths, early degradation such as rainwater intrusion, and finally serious degradation of thinning and disconnection of PC wires are detected with a spatial resolution of 1 mm within minutes by X-ray transmission imaging inspection. The iron components of PC, such as wires, can be clearly seen with a good contrast to concrete. Thinning and disconnection were observed with a spatial resolution of 1 mm. Measured flaws such as unfilled grout and thinning/disconnection are input to the structural analysis described in References [16,17]. Thus, the initial poor construction of unfilled grout and serious thinning and disconnection of PC wires are diagnosed, and their effect on the lack and degradation of strength is quantitatively evaluated by structural analysis. Finally, maintenance, repair, and reconstruction were planned.

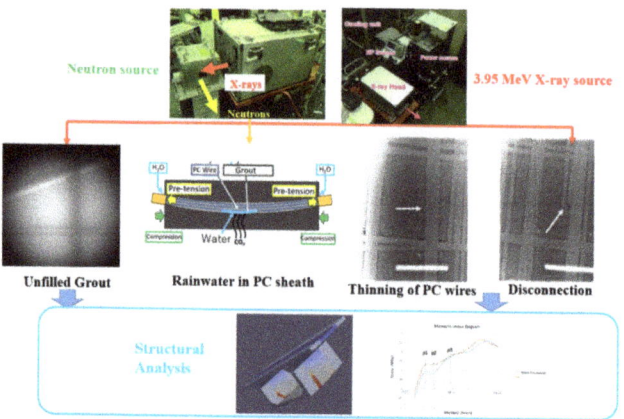

Figure 11. Procedure of X-ray inspection, flaw evaluation, and structural analysis.

3.2. Radiation Safety Control

We complied with the law (including the Law Concerning Prevention from Radiation Hazards due to Radioisotopes) and regulations (Regulations on Prevention of Ionizing Radiation Hazards) [18] when we used the 950 keV/3.95 MeV X-ray sources outside for onsite bridge inspection in Japan. According to the law, an electron beam source below 1 MeV is not defined as an accelerator. Thus, we complied with these regulations. The 950 keV X-ray source was registered at the local agency of labor supervision. Following the regulations, the source was usually operated in a radiation-controlled area with a radiation safety system. The use of sources outside the controlled area was also allowed. In this case, we temporally set a restricted area at the measurement site and put sufficient shielding around the source and object to suppress the air dose rate below 1.3 mSv/3 months. Moreover, we set the facility boundary at 250 µSv/month temporally. An amendment of the law allowing the use of less than 4 MeV accelerators for onsite bridge inspection was approved in Japan in 2005. After we received the governmental registration as a radiation source, we submitted permission for its use outside the radiation-controlled area. Finally, we performed onsite inspections under the same regulations for the 950 keV case.

4. Experimental Results

4.1. WEB Part of T-Shape PC Girder Bridge

X-ray transmission images for a typical T-shaped PC girder bridge were acquired onsite, as shown in Figure 12a. The portable X-ray source box and magnetron box were lifted and finely positioned to the WEB part of the T-shaped PC girder bridge. The electric power source and water chiller are installed in a special vehicle with a 20 kVA electric diesel generator. The X-ray FPD was attached to the opposite side of the WEB part with respect to the X-ray source. Based on visual inspection, the WEB part indicated by the lines became the target ((b)). We found large surface concrete cracks and an exuded white Ca water stain. The inspection region of the PC wires, shown in (b), was examined using an X-ray source. Figure 12c shows the surface cracks and the stain and X-ray transmitted images at the two flaws. In the two X-ray images, we observed declined PC sheaths, where PC wires and grout were inserted, and vertical/horizontal reinforcing rods. In the upper and lower cases, the lower halves beneath the PC wires appear dark/black and bright/white, respectively. In particular, it was speculated that the grout was unfilled and as a result, rainwater was retained in (c). Then, the PC sheath was corroded, and water exuded from the corroded cracks of the sheath to the surface via cracks in the concrete. The obtained X-ray images were attached to the design drawing, as shown in Figure 12d. The inner situation of the PC sheaths in a rather wide region was then visualized.

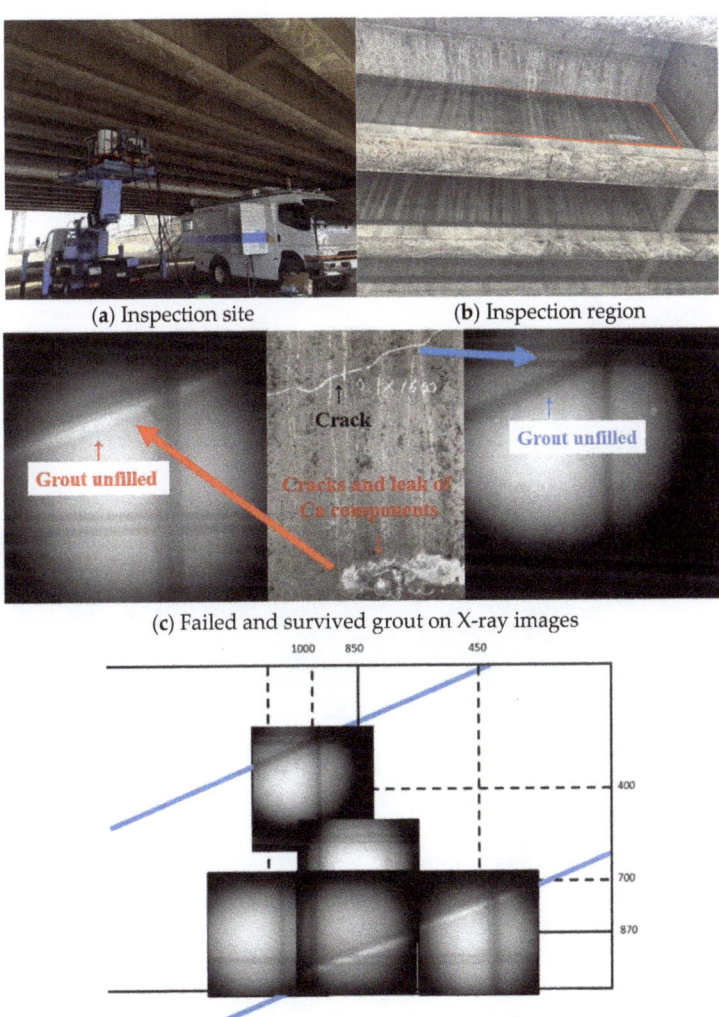

Figure 12. Photographs of onsite inspection and measured X-ray images by 950 keV X-ray source for a T-shape PC girder bridge. (**a**) Way of inspection; (**b**) inspection region; (**c**) filled and unfilled grouts in the X-ray images; (**d**) overall situation composed by several images.

We then evaluated the stage of unfilled grout in the PC sheath by gray value plotting, using the ratio of the gray values of unfilled grout and PC wires from the measured X-ray images. Here, we needed to consider and calibrate the uniformity of the background X-ray intensity distribution.

The X-rays used were emitted from a point with diameter of 2 mm at the W target and collimated by 17° by the W collimator. Therefore, the X-ray intensity at the FPD had an axisymmetric distribution. As the radius increased, the intensity decreased. As shown in Figure 13, if the source releases the X-ray beam in parallel, the dose is equally received by the detector at every location. The beam distribution is uniform. However, the X-rays from our point-source have a cone shape and radial intensity distribution at the FPD, as shown on the right-hand side of the figure.

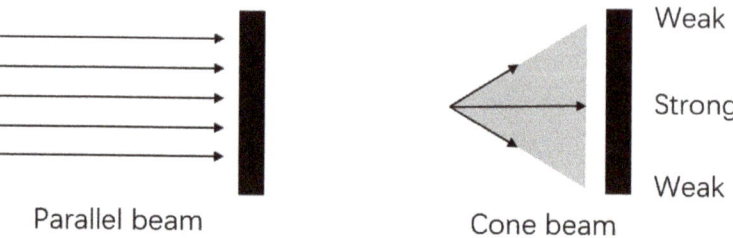

Figure 13. Parallel beam vs. cone beam in the dose regarding location.

A typical X-ray image is shown in Figure 14. This was acquired from a 200 mm-thick WEB part of the T-shaped PC girder bridge using the 950 keV X-ray source. The PC sheath, wires, and other rods are observed. The definition of the gray value is shown in Figure 15. In this measurement, the black and white parts correspond to highly and slightly X-ray-attenuated parts, respectively. According to the X-ray attenuation coefficient, PC wires appear very dark while grout in the PC sheath and concrete appears bright. The locations of typical materials such as PC wires, rods, concrete, and filled/unfilled grouts are shown in Figure 14.

Figure 14. Typical X-ray image of circular shape at the FPD screen and intensity profile evaluating line.

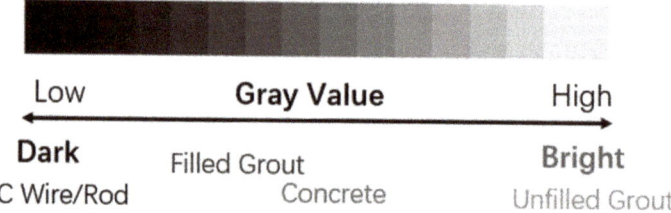

Figure 15. Gray value from low to high and typical materials.

Then, we checked the background X-ray intensity distribution on the line segment, as indicated in Figure 14. Here, we used an 8-bit system so that the full scale corresponds to 256. Darkness, or a lower gray value near the corners, is a result of an ununiformly distributed X-ray dose. The effect must be considered when the evaluation is based on gray values. The gray value of the line segment, where only concrete exists (see Figure 14), is plotted in Figure 16. An 8th-degree polynomial fitting curve was added to the original profile. At that stage, the relative background X-ray intensity distribution was obtained. These data were utilized to compensate for the gray values of unfilled grout and PC wires to obtain their ratio, which is the proposed index for the stage of unfilled grout.

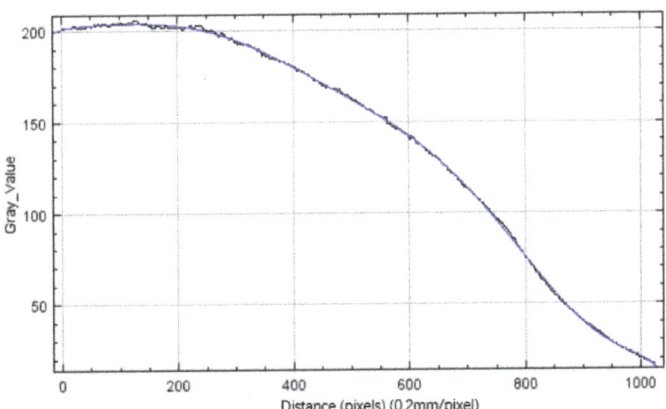

Figure 16. Gray value profile and the polynomial fitting along the line segment in Figure 14 as the background X-ray intensity distribution.

Schematic gray value profiles of the filled and unfilled grout cases are shown in Figure 17a. In fact, these lie on the background of the X-ray point source, as shown in (b). In this measurement by the X-ray FPD, heavy and light materials appear dark and gray, respectively. The gray values of the dark and gray parts correspond to the low and high gray values, respectively. We needed to evaluate the background X-ray intensity distribution first and subtract it from the measured data. We tried to represent the gray value at the plateau, G_w, for PC wires, as shown in (a). The boundary to the grout area appeared to be less dark and its gray value, G_{w_few}, became lower because only a few wires were located. There was gradation in the region of the filled grout as the effective thickness changes (see (a)). Thus, the medium gray value was selected for the filled grout as G_{grout}. The gray value at the high/bright peak was regarded as G_{no_grout}. The ratio, K, of the gray values at filled/unfilled grout and PC wires was calculated for several situations as

$$K_{fw} = \frac{G_{grout}}{G_w} \text{ for filled grout and many wires.}$$

Then,

$$K_{uw} = \frac{G_{no\_grout}}{G_w} \text{ for unfilled grout and many wires,}$$

$$K_{fwf} = \frac{G_{grout}}{G_{w\_few}} \text{ for filled grout and few wires,}$$

$$K_{uwf} = \frac{G_{no\_grout}}{G_{w\_few}} \text{ for unfilled grout and few wires,} \quad (1)$$

$$K_{uf} = \frac{G_{no\_grout}}{G_{grout}} \text{ for unfilled and filled grouts.} \quad (2)$$

Because G_w is lower than G_{grout} and G_{no_grout}, both K_{uw} and K_{fw} must be higher than 1. K_{uw} at the unfilled grout must also be higher than K_{fw} at the filled grout, as explained above.

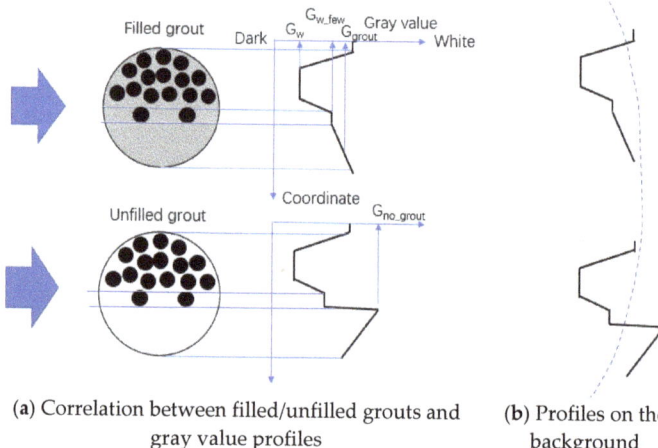

(a) Correlation between filled/unfilled grouts and gray value profiles

(b) Profiles on the background

Figure 17. Schematic drawing of the correlation between filled/unfilled grouts and gray value profiles. (a) Typical gray value profiles for filled and unfilled grout cases; (b) The profiles lie on the intensity background due to point source X-rays.

The inner contents of the image in Figure 14 are explained in Figure 18. The grout appears wholly unfilled in the broken line frame. The bright blank indicates the vacancy at this location, leading to a high dose at the FPD, which reveals a grout flaw. In comparison, the dark straight region above the broken line frame indicates the existence of PC wires, which remarkably attenuate X-rays. To make a quantitative evaluation, it was necessary to establish a method based on sufficient numerical tools. Gray value evaluation of the raw images was expected to be effective on this occasion.

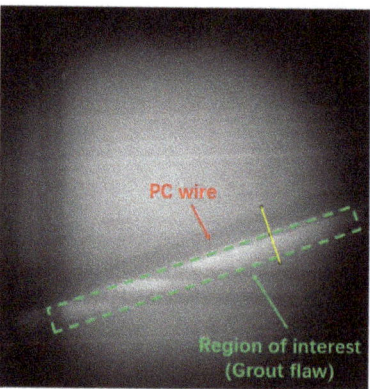

Figure 18. X-ray Image acquired from 200 mm thick WEB of the T-shape PC girder bridge and explanation of the inner structure.

Figure 19 shows the gray value profile plotted along the line segment that crosses the PC sheath transversely. The diameter and thickness of the PC sheath tube were ~38 mm and ~1 mm, respectively. Approximately 15 PC wires of ~7 mm in diameter were installed in a sheath. Each wire can be recognized almost entirely. The magnified profile of the gray value across the PC sheath is shown in Figure 19. We can clearly observe the locations of the PC wires and grout in the concrete. The dark and bright/white parts of the PC wires and grout correspond to the lower and upper gray value peaks, respectively. This quantitative result is consistent with our visual recognition of the original X-ray image. That is, the bright/white part beneath the PC wires appears as a fully unfilled grout. This is supported by gray value analysis. Thus, this case can be evaluated as a fully unfilled grout.

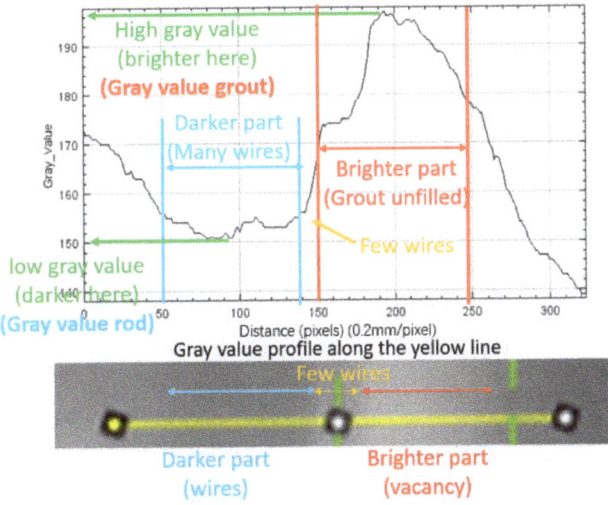

Figure 19. Gray value profile of the line segment in Figure 18.

Next, we proposed a quantitative index of the stage of the unfilled grout. The gray value profile along the line segment across the PC sheath and nearby reinforcing rods (see Figure 20) was obtained, as shown in Figure 21. Regions of "A" and "B" correspond to the PC wires and unfilled grout, respectively. A few wires are seen less to be dark in the region between them, as depicted in Figure 17a. Additionally, two lower peaks attributed to the two rods can be clearly observed as "C" and "D". Because this profile contains an ununiform background distribution of irradiating X-rays, it has to be compensated by using the data shown in Figure 16. The compensated profile is shown in Figure 22. We used these compensated profiles for the gray value analysis to evaluate the filled and unfilled grouts quantitatively.

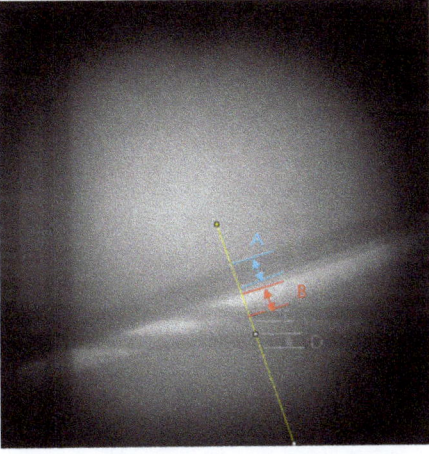

Figure 20. Gray value profile analyzing line.

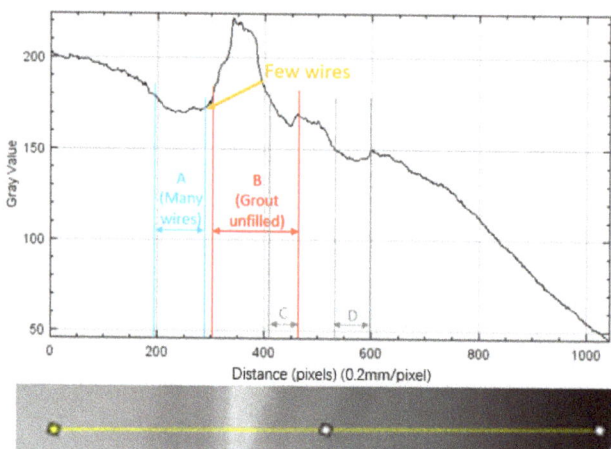

Figure 21. Original gray value profile of the yellow line in Figure 17.

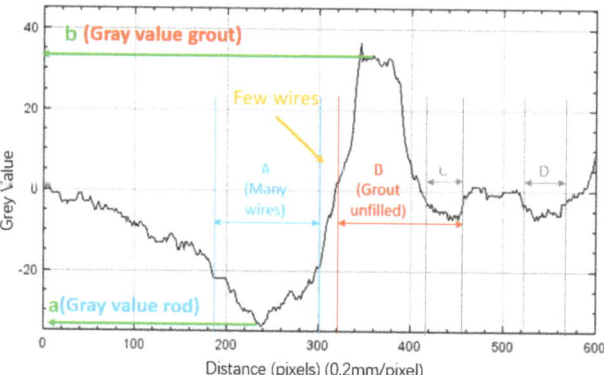

Figure 22. Compensated profile of Figure 20 by subtracting the ununiform background due to the X-ray point source.

Line segments for gray value analysis are indicated in Figure 23 in this X-ray image of the 200 mm thick WEB of the T-shaped PC girder bridge by the 950 keV X-ray source. The gray-value profiles are shown in Figure 24.

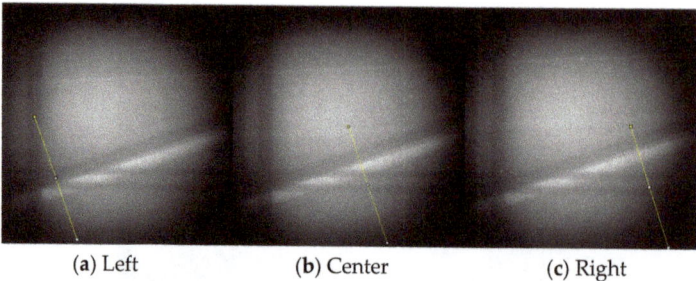

(a) Left   (b) Center   (c) Right

Figure 23. Line segments for gray value analysis for unfilled grout in the X-ray image of 200 mm thick WEB of the T-shape PC girder bridge by the 950 keV X-ray source.

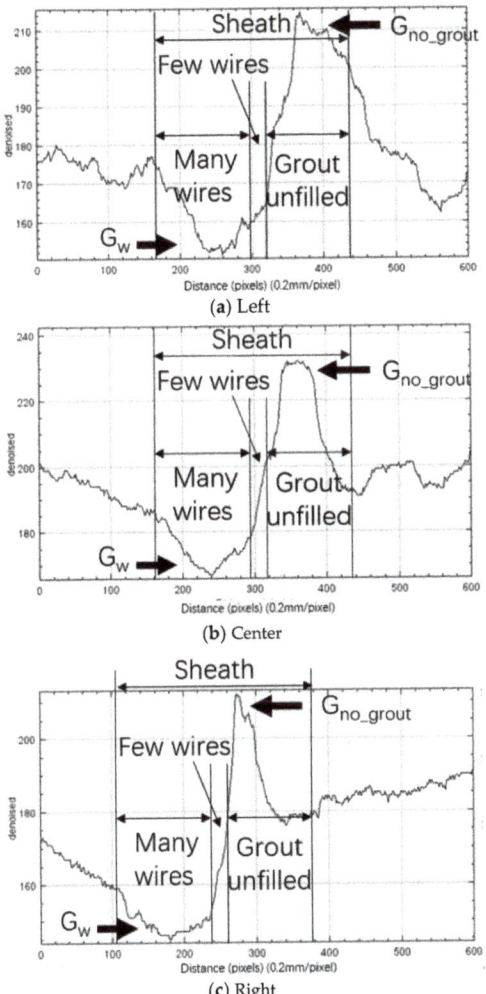

Figure 24. Gray value profiles of (**a**) Left; (**b**) Center; (**c**) Right in Figure 22.

By identifying the sheath range within the gray value profile, gray values could be obtained. The evaluated values are summarized in Table 2. As predicted above, the average ratio of K is clearly higher than 1 and is approximately 1.4.

Table 2. Gray values and K ratios for 200 mm PC, 950 keV and unfilled grout.

Location	G_w	G_{no_grout}	K_{fw} (G_{no_grout}/G_w)
Left	148	210	1.42
Center	167	232	1.39
Right	152	214	1.41
Average	156	228	1.41

Moreover, we plotted another set of gray values for the unfilled grout. Figure 25 shows the original image of the same WEB part (200 mm-thick T-shaped PC girder bridge) taken by the 950 keV source. The inspection location is different from that shown in Figure 23, as exhibited in Figure 12d. Line segments for gray value analysis are also indicated in the figure.

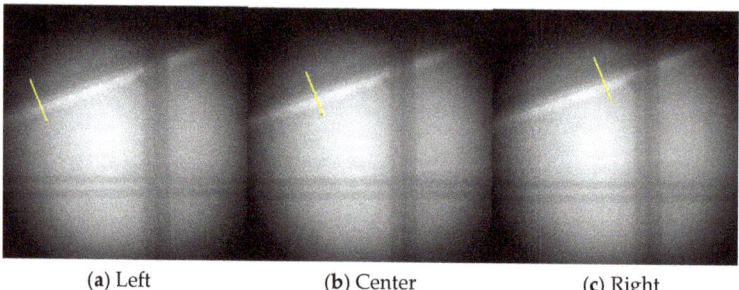

(a) Left   (b) Center   (c) Right

Figure 25. X-ray image of unfilled grout in a 200 mm thick T-shape PC girder bridge taken by the 950 keV source. (**a**) Left; (**b**) center; (**c**) right. Line segments for gray value analysis are also indicated.

The gray value distributions on the line segments in Figure 25 are shown in Figure 26. Table 3 lists the gray values and K ratios. The average ratio of K_{uw} is approximately 1.5. Meanwhile, K_{uw_few} is lower at ~1.1 This is because the parts of many and a few wires look dark and less dark, that is, $G_{w\_few} > G_w$.

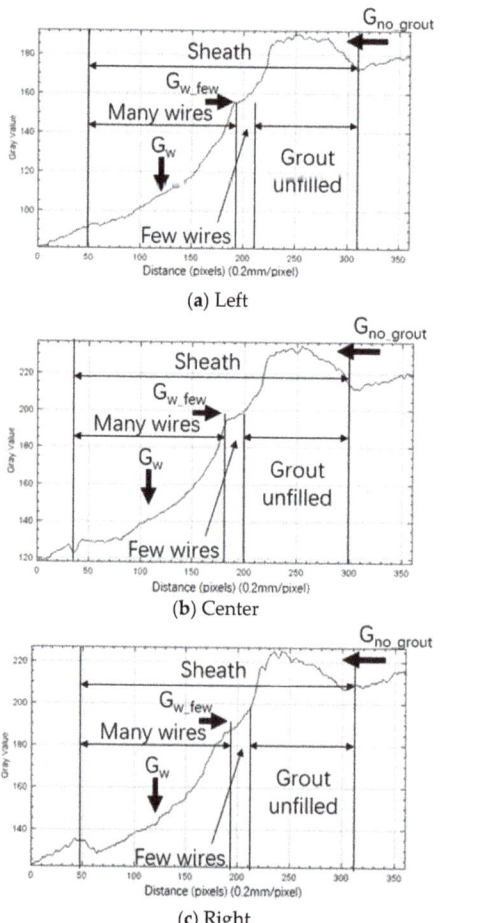

(a) Left

(b) Center

(c) Right

Figure 26. Gray value profiles of Figure 25. (**a**) Left; (**b**) center; (**c**) right.

Table 3. Gray values and K ratios for Figure 26. 200 mm by 950 keV, grout is not filled.

Location	G_w	G_{w_few}	G_{no_grout}	K_{uw} (G_{no_grout}/G_w)	K_{uwf} (G_{no_grout}/G_{w_few})
Left	103	157	170	1.65	1.08
Center	140	200	211	1.51	1.06
Right	139	190	201	1.44	1.06
Average	127	182	194	1.53	1.07

After looking into the unfilled sample, we proceeded to the filled grout case. Figure 27 shows an image taken from the same WEB part (200 mm-thick T-shaped PC girder bridge) by the 950 keV source. The lower half of the PC sheath appears to be dark compared to that in Figures 23 and 25. Line segments for plotting the gray values are also observed.

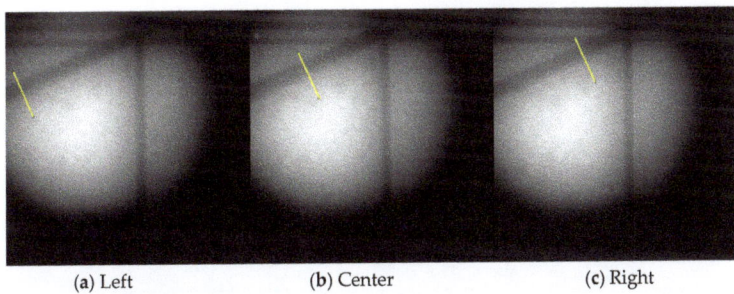

(a) Left   (b) Center   (c) Right

Figure 27. X-ray images of filled grout for 200 mm thick PC bridge taken by the 950 keV source. (a) Left; (b) center; (c) right. Line segments for gray value analysis are also indicated.

In the same way, the profiles were plotted as shown in Figure 28, and the K ratios are calculated in Table 4. The average ratio of K_{fw} is ~1.3, which is lower than the K_{nw} of ~1.4 and ~1.5 indicated in Tables 2 and 3. If we used the gray value at the wire boundary, G_{fw_few}, the ratio of K_{fw_few} became rather low at ~1.1. Again, the gray value at the wire boundary does not appear appropriate for the evaluation of unfilled and filled grouts.

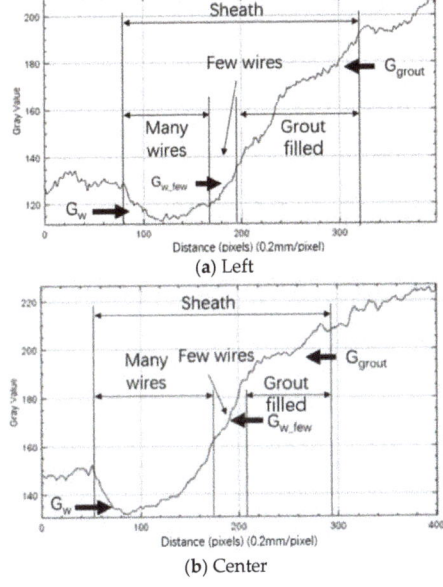

Figure 28. Cont.

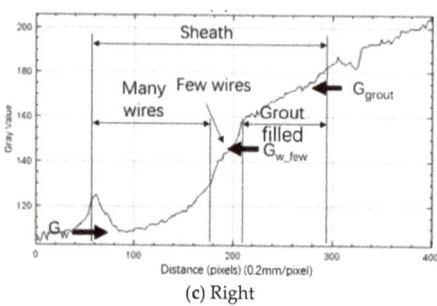

(c) Right

Figure 28. Gray value profiles of (**a**) left; (**b**) center; (**c**) right in Figure 27 for 200 mm thick T-shape PC girder bridge by 950 keV.

Table 4. Gray values and K ratios of Figure 28 for 200 mm thick T-shape PC girder bridge by 950 keV. The grout is filled.

Location	G_w	G_{w_few}	G_{grout}	K_{fw} (G_{no_grout}/G_w)	K_{fwf} (G_{no_grout}/G_{w_few})
Left	115	141	150	1.30	1.06
Center	135	152	175	1.30	1.15
Right	110	134	147	1.34	1.10
Average	120	142	157	1.31	1.10

4.2. Model Samples of 750 mm Thickness for Box-Shape PC Girder Bridge

Now, we explain the similar filled/unfilled grout analysis for the side WEB of a box-shaped PC girder bridge using the 3.95 MeV X-ray source. Because the PC concrete thickness is far greater than 400 mm, we needed to use the 3.95 MeV X-ray source here. Real bridge inspection continued, and the obtained results are currently under evaluation. Hence, the results of the preparatory experiments modeling the real situation are introduced in this paper.

The target was a 750 mm thick PC WEB wall. We constructed model samples using pieces cut from real old bridges. The 750 mm thick assemblies are located between the 3.95 MeV X-ray source and the X-ray FPD detector, as shown in Figure 29a. The photograph is shown in (b). The PC wire used for insertion is shown in (c).

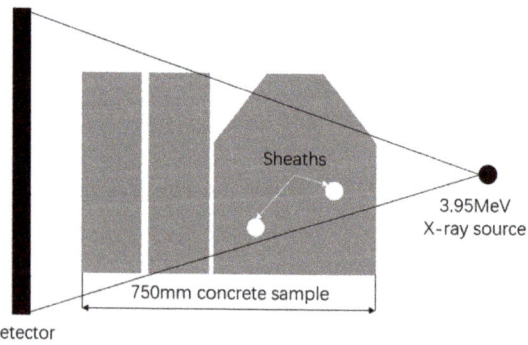

(**a**) Experimental configuration

Figure 29. Cont.

(b) Photograph

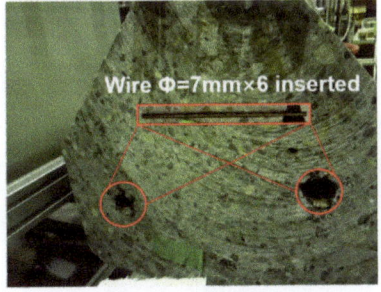

(c) PC wire for insertion

Figure 29. Experimental configuration (a) and photograph (b) for modeling the 750 mm thick side WEB wall of a box-shape PC girder bridge with the 3.95 MeV X-ray source and FPD. (c) PC wire for insertion.

For better image quality, multiple X-ray shots should be averaged into a single image. As shown in Figure 29, the 750 mm-thick assembly is irradiated by the 3.95 MeV X-ray source. Figure 30 shows the X-ray images for 1, 25, 50, and 100 shots, where one shot means 10 s exposure, and their averages are taken for each image to smooth the image [19]. When the exposure time is short and the transmitted X-ray intensity is weak, the image appears spotty and the gray value profile becomes noisy. If the exposure time is increased, the profile is expected to be smoother. Such a denoising effect by stacking 1, 25, 50, and 100 X-ray shots is clear, as shown in Figure 31. The noise attributed to the lack of X-ray intensity is remarkably reduced for many stacking shots. Because the concrete is significantly thick, the intensity of the transmitted X-ray becomes attenuated. Therefore, stacking using appropriate shots is necessary for an accurate quantitative analysis of filled and unfilled grouts based on their gray value difference.

Now, X-ray images taken by the 3.95 MeV source for the model samples of a box-shaped PC girder bridge were analyzed. The inner contents of the samples are shown in Figure 32. Two declined sheaths can be observed. In this case, only a few wires are inserted into the remaining grout in the upper sheath 1, and many wires are located in the lower sheath 2. The gray value profiles were evaluated on the indicated line segments. Gray value profiles and analysis were performed, as shown in Figure 33. A few wires are clearly recognized by the small negative peak in the figure. The calculated values are listed in Table 5. Because there were only a few wires in sheath 2, only K_{uw_few} and K_{fw_few} were evaluated. Additionally, K_{uwf} was calculated.

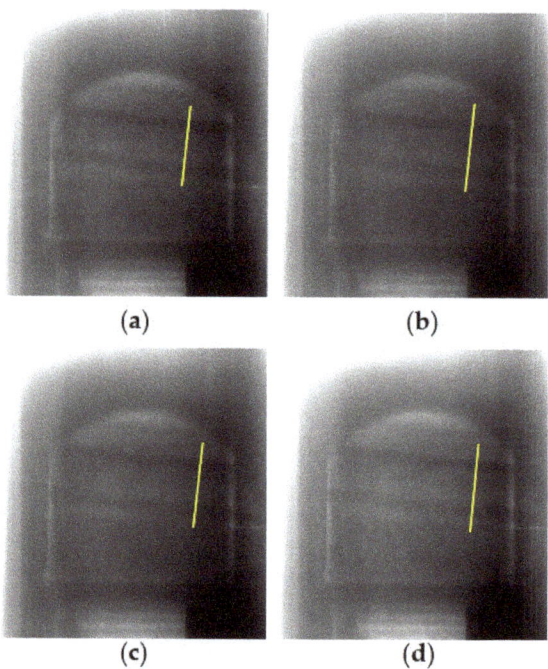

Figure 30. Measured X ray images for 1 (a), 25 (b), 50 (c) and 100 shots (d). Gray value analysis lines are also indicated.

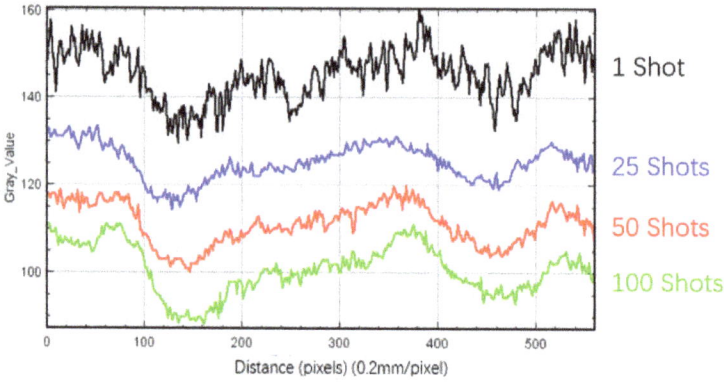

Figure 31. Denoising effects by stacking 1, 25, 50, and 100 X-ray shots.

Table 5. Gray value numbers and K ratios of Figure 33. 750 mm by 3.95 MeV, grout is not filled.

Location	G_{w_few}	G_{no_grout}	G_{grout}	K_{uwf}	K_{fwf}	K_{uf}
Left	66	98	76	1.48	1.15	1.29
Right	92	108	92	1.17	1.00	1.17
Average	79	103	84	1.33	1.08	1.23

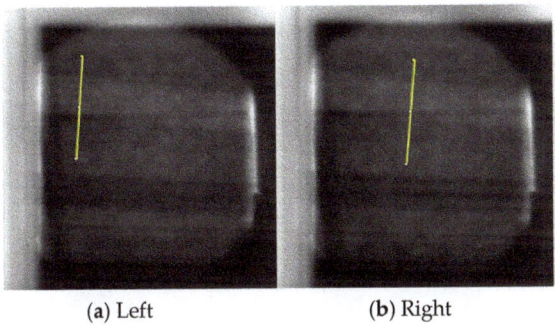

(a) Left (b) Right

Figure 32. X-ray images of the model sample for a 750 mm-thick PC bridge taken by the 3.95 MeV source. The grout is partially filled. Wires are inserted. (**a**) Left; (**b**) Right.

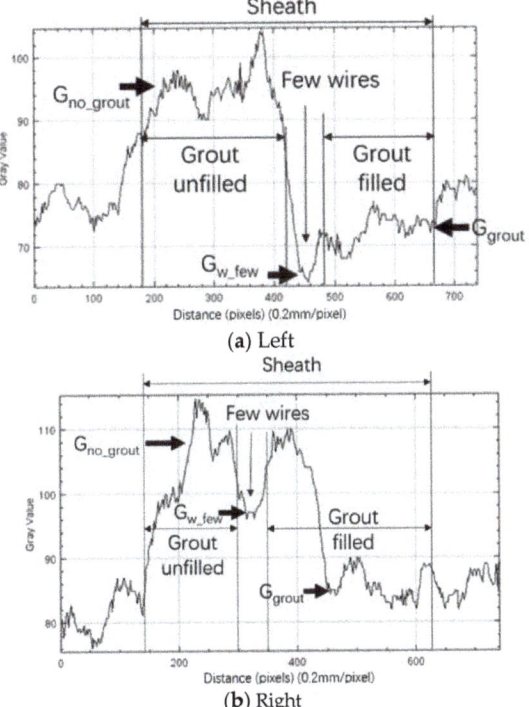

(a) Left

(b) Right

Figure 33. Gray value profiles of Figure 32. (**a**) Left; (**b**) Right.

We then analyzed an additional case for the 750 mm thick concrete of a box-shaped PC girder bridge. We prepared three sheaths, 1, 2, and 3, at the same horizontal level, as shown in Figure 34. In the lower half of sheath 1, several wires were inserted, whereas the upper half was vacant. The grout is filled only in the lower half of sheath 2, and a few wires are placed on it. Sheath 3 is completely empty. First, we carried out a horizontal shot in (a). In this case, only sheath 1, which is the nearest to the X-ray source, could be clearly recognized. We could also observe a few wires in sheath 2 at the wire grout boundary in the X-ray image. Next, the downward shot was reduced to shift the images vertically at the FPD, as shown in (b). The three sheaths were still partially overlapped at their edges.

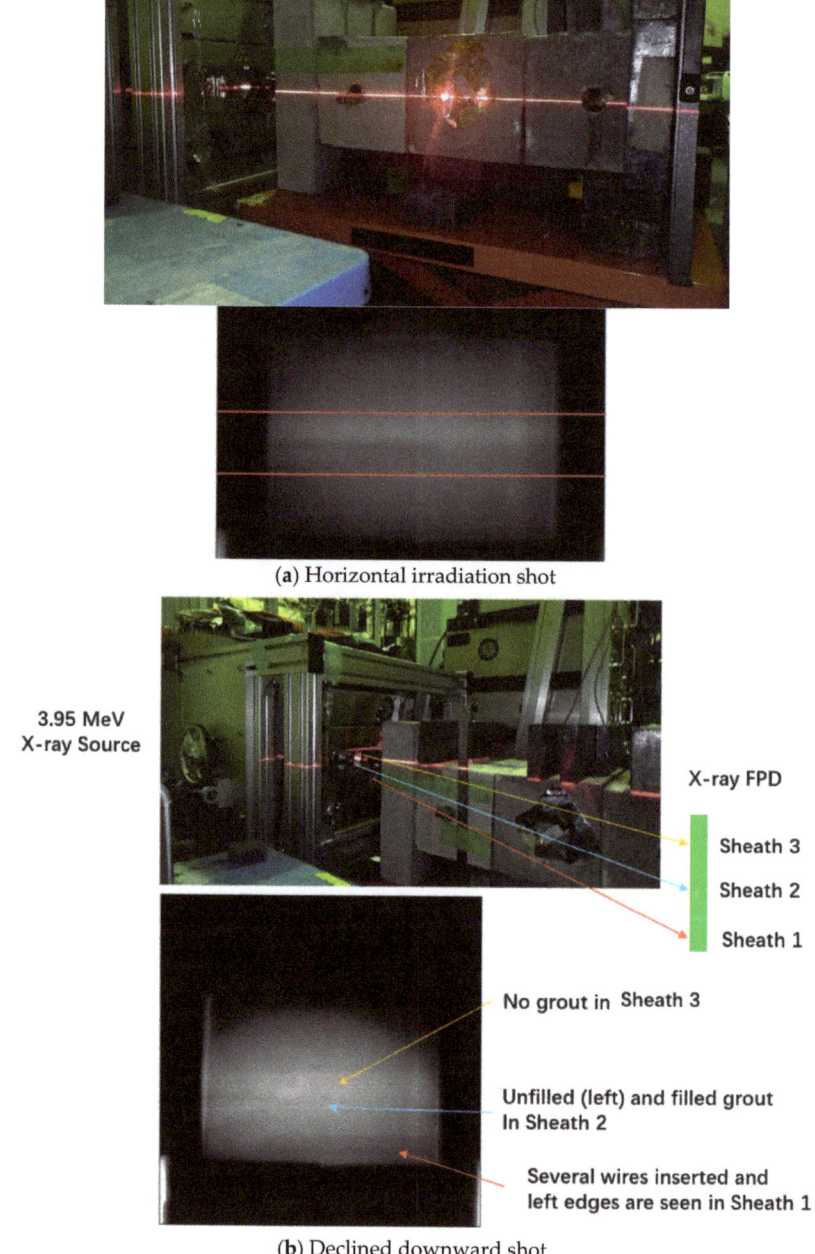

Figure 34. Photograph and X-ray image by 3.95 MeV X-ray source for 750 mm thick model sample for box-shape PC girder bridge. There are three sheaths. (**a**) Horizontal shot. The three sheaths are overlapped and only Sheath 1 can be seen. (**b**) Declined downward X-ray shot to shift the three sheaths at the FPD.

Because the three sheaths partially overlap at each edge, treatment of the regions and gray values was complicated for the gray value analysis in the declined shot case. Thus far, the inner structures were also rather complicated. Therefore, we also adopted a 16-bit system with a full scale of 65,536 to upgrade the resolution of the analysis. Thus, the gray value profile analysis was performed only for sheath 1 in the horizontal shot case, as shown in Figures 35 and 36. We could also observe a few wires in sheath 2 near the wire grout boundary of the sheath 1 image in Figure 35. Therefore, we did not use the gray value at the boundary as G_{w_few} and only calculated K_{nw}. K_{nw} became ~1.2 as given in Table 6.

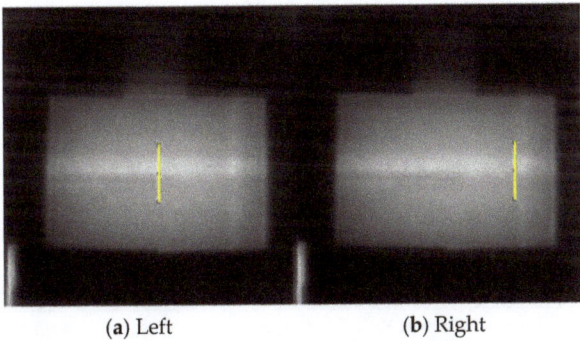

(a) Left   (b) Right

Figure 35. X-ray images of a 750 mm-thick PC bridge taken by the 3.95 MeV source. The grout is not filled. (a) Left; (b) right.

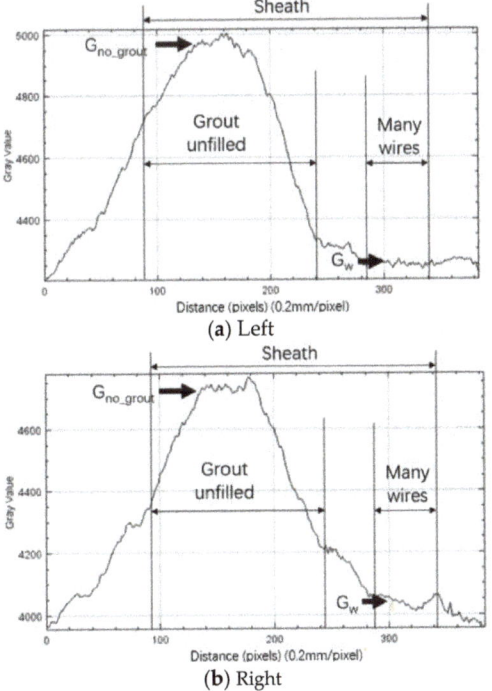

Figure 36. Gray value profiles of Figure 35. (a) Left; (b) right.

Table 6. Gray value numbers and K ratios of Figure 36. 750 mm by 3.95 MeV, grout is not filled.

Location	G_w	G_{no_grout}	K_{uw}
Left	4250	4950	1.16
Right	4050	4750	1.17
Average	4150	4850	1.17

5. Discussion

Table 7 summarizes the *K* ratios obtained thus far. In the case of the T-shape PC girder bridge of 200 mm thickness and 950 keV X-ray source, the *K* ratio, K_{nw}, for unfilled grout is in the range 1.4–1.5, while K_{fw} is approximately 1.3 for filled grout. K_{nw} for the case of the 750 mm thick model samples for a box-shape PC girder bridge and the 3.95 MeV system becomes ~1.2. The ratio between the unfilled and filled grouts was calculated using the averaged K_{uw} of cases 1 and 2 and K_{fw} of cases 3, as presented in the table. In all cases, the difference of the values is at least 10%. It is still early to declare standard values to indicate the status of grout. Further accumulation and verification of tests and data are necessary. However, this analysis is expected to be important for evaluating the grout situation quantitatively during the next steps of structural analysis or destructive inspection.

Table 7. Overall *K* ratio summary in different conditions.

Case	Thickness	X-ray Source	Avg. K_{uw}	Avg. K_{fw}	Avg. K_{uwf}	Avg. K_{fwf}	Avg. K_{uf}	PC Type
1	200	950 keV	1.41					T-shape
2	200	950 keV	1.53		1.07		1.12	T-shape
3	200	950 keV		1.31		1.10		T-shape
4	750	3.95 MeV			1.33	1.08	1.23	Box
5	750	3.95 MeV	1.17					Box

6. Conclusions

We attempted a quantitative evaluation of the stage of unfilled grout in PC bridges using porTable 950 keV/3.95 MeV X-ray sources. We obtained the gray value profiles from the measured X-ray transmitted images and calculated the ratios of the gray values of the PC wires and grout. In this measurement, iron PC wires, filled grout, concrete, and unfilled grout appeared to be black, very dark, dark, and bright, respectively. As the image was darker, the gray value decreased in this analysis. It is possible that the stage of unfilled grout could be quantitatively evaluated as the gray value ratio between the PC wires and grout part by stacking more experiences and data. If the stage looks rather unfilled, further detailed inspection, such as destructive evaluation by boring surveys, for instance, should be performed. This method is applicable to a wide range of scenarios and supports the overall strength evaluation of bridges for safety maintenance [17]. Actual on-site inspection of PC highway bridges and analysis of the results are underway. We proposed a guideline for X-ray inspection using 950 keV/3.95 MeV X-ray sources accompanied by visual and hammering-sound screenings, structural analysis, final repair, and/or reinforcement, as described in Figure 37. The purpose is to extend the lifespan of PC bridges worldwide. An academic and industrial consortium, which included the authors, successfully performed the 3.95 MeV X-ray inspection and evaluation of grout filling for a box-shaped PC girder highway-bridge in Japan in 2020. Detailed results will be presented in the near future.

Figure 37. Proposed guideline for X-ray inspection using 950 keV/3.95 MeV X-ray sources accompanied with visual and hammering-sound screenings, structural analysis, final repair, and/or reinforcement.

Author Contributions: All authors made equal contributions to all the aspects of the paper. All authors have read and agreed to the published version of the manuscript.

Funding: This research was partially funded by Cross-ministerial Innovation Promotion Program—Infrastructure Maintenance Renovation and Management-(2015–2019), Cabinet Office of Japan.

Institutional Review Board Statement: Not applicable.

Informed Consent Statement: Not applicable.

Data Availability Statement: Not applicable.

Conflicts of Interest: The authors declare no conflict of interest.

References

1. Japan Ministry of Land, Infrastructure, Transport and Tourism. Status Quo of Social Infrastructure Inspection and Maintenance. Available online: https://www.mlit.go.jp/sogoseisaku/maintenance/02research/02_01.html (accessed on 30 March 2021).
2. Japan Prestressed Concrete Contractors Associate. *Design of PC Bridges*, 5th ed.; Japan Prestressed Concrete Contractors Associate: Tokyo, Japan, 2019; Volume 1, pp. 15–25.
3. Anton, O.; Komárková, T.; Heřmánková, V. An optimal method of determining the position of bends on shear reinforcement as part of the diagnosis of reinforced concrete beam structures. *IOP Conf. Ser. Mater. Sci. Eng.* **2019**, *549*, 012015.
4. Available online: https://home.hiroshima-u.ac.jp/bridge2/mb/mb-po.html (accessed on 15 March 2017).
5. Abdel-Jaber, H.; Glisic, B. Monitoring of prestressing forces in prestressed concrete structures—An overview. *Struct. Control Health Monit.* **2019**, *26*, e2374. [CrossRef]
6. Bonopera, M.; Chang, K.-C.; Lee, Z.-K. State-of-the-art review on determining prestress losses in prestressed concrete girders. *Appl. Sci.* **2020**, *10*, 7257. [CrossRef]
7. Yang, J.; Guo, T.; Li, A. Experimental investigation on long-term behavior of prestressed concrete beams under coupled effect of sustained load and corrosion. *Adv. Struct. Eng.* **2020**, *23*, 2587–2596. [CrossRef]
8. Huang, H.; Huang, S.-S.; Pilakoutas, K. Modeling for assessment of long-term behavior of prestressed concrete box-girder bridges. *J. Bridge Eng.* **2018**, *23*, 1–15. [CrossRef]
9. Meng, Q.; Zhu, J.; Wang, T. Numerical prediction of long-term deformation for prestressed concrete bridges under random heavy traffic loads. *J. Bridge Eng.* **2019**, *24*, 1–19. [CrossRef]
10. SCREENING EAGLE. Available online: https://www.proceq.com/product/pundit-pd8000/?utm_content=buffer3af4c&utm_medium=social&utm_source=facebook.com&utm_campaign=buffer (accessed on 21 March 2017).
11. Konica Minolta "Magnetic Stream Method" Enters Infrastructre Non-Destructive Inspection. 2018. Available online: https://www.konicaminolta.hk/hk/zh-hk/home.php?gclid=EAIaIQobChMI8LbJnPqO8QIVATdgCh2Q6QIvEAMYASAAEgLNevD_BwE (accessed on 19 March 2017).
12. Available online: https://xtech.nikkei.com/atcl/nxt/column/18/00107/00033/ (accessed on 23 March 2017).
13. Available online: http://www.kokusai-se.co.jp/technology/technology03.html (accessed on 25 March 2017).
14. Uesaka, M.; Mitsuya, Y.; Hashimoto, E.; Dobashi, K.; Yano, R.; Takeuchi, H.; Bereder, J.M.; Kusano, J.; Tanabe, E.; Maruyama, N.; et al. Onsite Non-destructive Inspection of the Actual Bridge using the 950 keV X-band Electron Linac X-ray Source. *J. Disaster Res.* **2017**, *12*, 578–584. [CrossRef]

15. Uesaka, M.; Mitsuya, Y.; Dobashi, K.; Kusano, J.; Yoshida, E.; Oshima, Y.; Ishida, M. On-Site Bridge Inspection by 950 keV/3.95 MeV Portable X-band Linac X-ray Sources. In *Bridge Optimization—Inspection and Condition Monitoring*; InTech Open: London, UK, 2018. [CrossRef]
16. Uesaka, M.; Dobashi, K.; Mitsuya, Y.; Yang, J.; Kusano, J. Highway Bridge Inspection by 3.95 MeV X-ray/Neutron Source. In *Computational Optimization Techniques and Applications*; InTech Open: London, UK, 2021. [CrossRef]
17. Maekawa, K.; Pimanmas, A.; Okamura, H. *Nonlinear Mechanics of Reinforced Concrete*; Spon Press: London, UK, 2003.
18. NRA JAPAN. Available online: https://www.nsr.go.jp/english/library/RI.html (accessed on 21 March 2017).
19. Pei, C.; Wu, W.; Uesaka, M. Image enhancement for onsite X-ray nondestructive inspection of reinforced concrete structures. *J. X-ray Sci. Technol.* **2016**, *24*, 797–805. [CrossRef] [PubMed]

Article

Influence of HFM Thermal Contact on the Accuracy of In Situ Measurements of Façades' U-Value in Operational Stage

Katia Gaspar *, Miquel Casals and Marta Gangolells

Group of Construction Research and Innovation (GRIC), Universitat Politècnica de Catalunya·BarcelonaTech (UPC), C/Colom, 11, Ed. TR5, 08222 Terrassa, Spain; miquel.casals@upc.edu (M.C.); marta.gangolells@upc.edu (M.G.)
* Correspondence: katia.gaspar@upc.edu

Abstract: Accurate information on the actual thermal transmittance of walls is vital to select appropriate energy-saving measures in existing buildings to meet the commitments of the European Green Deal. To obtain accurate results using the heat flow meter (HFM) method, good thermal contact must be made between the heat flow meter plate and the wall surface. This paper aimed to assess the influence of the non-perfect thermal contact of heat flow meter plates on the accuracy of in situ measurement of the façades' U-value when a film was applied to avoid damage to the wall surface. Given the fact that to avoid harm to the wall surface, the laying of a film is a usual procedure in the installation of equipment during the building's operational stage. The findings show that deviations between measured U-values when an HFM was installed directly on the wall surface and when an HFM was installed with a PVC film were found to differ significantly from the theoretical effect of including a PVC film during the monitoring process.

Keywords: non-destructive test; monitoring; housing; buildings; façade; thermal transmittance; HFM method

Citation: Gaspar, K.; Casals, M.; Gangolells, M. Influence of HFM Thermal Contact on the Accuracy of In Situ Measurements of Façades' U-Value in Operational Stage. *Appl. Sci.* **2021**, *11*, 979. https://doi.org/10.3390/app11030979

Academic Editor: Jerzy Hoła
Received: 30 November 2020
Accepted: 18 January 2021
Published: 22 January 2021

Publisher's Note: MDPI stays neutral with regard to jurisdictional claims in published maps and institutional affiliations.

Copyright: © 2021 by the authors. Licensee MDPI, Basel, Switzerland. This article is an open access article distributed under the terms and conditions of the Creative Commons Attribution (CC BY) license (https://creativecommons.org/licenses/by/4.0/).

1. Introduction

Efficiency in terms of energy and resources is required by the European Green Deal when building and renovating the European building stock [1] for achieving the European Union's energy efficiency targets [2]. To meet the 2050 long-term climate and energy targets, 97% of the residential building stock, which is responsible for around 36% of CO_2 emissions in the European Union [3], needs to be upgraded to become highly energy efficient and obtain the Energy Performance Certificate (EPC) label A [4]. This high percentage is due partly to the average age of European residential building stock: more than 80% of residential buildings are over 25 years old (Table 1) and, therefore, most of them were built without thermal regulation. Construction is considered the sector with the most potential for energy saving. Buildings represent 40% of the energy used in the European Union [5]. Almost 26% of Europe's total final energy consumption is used in residential buildings [6]. Space heating represents a large amount of energy consumption in the housing sector [7,8] and, therefore, has a significant capacity for energy saving.

Table 1. Distribution of residential floor space by year of construction in the European Union 27, Switzerland and Norway [9].

Region	Floor Space Distribution	Average Age of Residential Floor Space		
		Pre 1960	1961–1990	1991–2010
North and West	50%	42%	39%	19%
Central and East	14%	35%	48%	17%
South	36%	37%	49%	14%

According to the IEA EBC Annex 71-project [10], measuring the actual energy performances discloses a notable gap between actual and expected energy performances of buildings. The actual thermal behaviour of building envelopes could contribute to this energy performance gap in buildings. Deviations between the predicted and actual behaviour of envelopes are related to aspects of the design and construction stages [11]. Moreover, assumptions about energy efficiency improvements resulting from building refurbishment are not always met [12,13]. Consequently, accurate in situ measurements of the actual U-value of façades are necessary.

The thermal performance of building façades is a fundamental parameter that should be evaluated to obtain an accurate energy diagnosis of buildings [14–21]. Successful decision-making during energy renovation processes of existing buildings requires precise characterization of thermal properties of building components [22,23]. Classifying the opaque part of the façades is essential to precisely analyse the thermal performance of façades in the housing sector.

In recent years, the thermal behaviour of the materials that compose façades had been studied in depth by several authors such as Laaroussi et al. [24] measuring the thermal properties of brick materials, Kuman and Suman [25] or Björk and Enochsson [26] measuring the properties of thermal insulation materials. However, when analysing the thermal transmittance of walls, assumed U-values have been a meaningful source of error in estimations of energy savings and carbon emissions [27]. Evidence suggests that assumptions concerning heat loss from a dwelling pre-retrofit and post-retrofit are not correct [12]. Therefore, accurate on-site measurements are required to provide information on the actual thermal transmittance of façades. To effectively quantify the actual performance of buildings, optimized on-site measurements combined with dynamic data analysis techniques are needed [28].

Several methods can be used for the in situ measurement of U-value of existing buildings' façades [29,30]. One of the most common is the heat flow meter method, standardised by ISO 9869-1:2014 [31]. This method obtains the thermal transmittance by measuring the heat flow rate that passes through a wall and the inside and outside environmental temperatures. However, difficulties can arise in on-site measurements of walls U-value in the existing building stock, leading to inaccuracies [32]. These difficulties can be classified into three groups, according to the IEA EBC Annex 58-project [33]: errors related to the measurement accuracy, errors related to the analysis of data and errors related to the boundary conditions of the in situ measurement. Difficulties related to the measurement accuracy include factors like reading and calibration of heat flow meters and temperature sensors and had been deeply analysed by authors such as Ficco et al. [17], Trethowen [34] and Meng et al. [35]. Difficulties related to the analysis of data were recently analysed in depth [32,36–39]. Difficulties related to the boundary conditions include factors as an imbalance of the heat flow, edge heat loss and accuracy on the position of sensors, which were highly analysed by Peng and Wu [32], Meng et al. [35], Cesaratto et al. [40], Ahmad et al. [41] and Guattari et al. [42]. Delving into boundary monitoring conditions, the factor of contact between the wall surface and the heat flow meter was analysed by simulations but has not been analysed in experimental tests [40], so as for the present study focuses on this aspect.

The IEA EBC Annex 58-project [33] and authors such as Cesaratto et al. [40], Tadeu et al. [43] and Gori and Elwell [44] highlight the importance of ensuring good thermal contact between the heat flow meter plate and the wall to be measured to obtain accurate results. However, conducting HFM in situ measurements during the operational stage of dwellings is challenging due to the need for avoiding damage to wall surfaces. In this sense, the usual practice is placing a PVC film to preserve the wall surface [27,45].

Considering all the aspects aforementioned, the study aimed to analyse the influence of the non-perfect thermal contact of heat flow meter plates on the accuracy of in situ measurement of the façades' U-value when a film was applied to avoid damage to the wall surface, as a usual equipment installation procedure during the building's operational stage. This research provides valuable additional evidence on the accuracy of in situ monitoring of the actual U-value of existing buildings' façades and will therefore assist practitioners in pre-retrofit diagnosis.

The method used to analyse the influence of the non-perfect thermal contact of heat flow meter plates on the accuracy of in situ measurement of the façades' U-value is explained in the second section following this introduction. The third section presents the results. Finally, the discussion and conclusions are given in the fourth section.

2. Materials and Methods

2.1. Methodology

The method for assessing the impact of the heat flow meter thermal non-perfect contact on the accuracy of in situ measurement of walls U-value consists of three steps (Figure 1):

- First, two façades with a range of theoretical thermal transmittance values were selected.
- Second, in situ measurements of walls' U-value were conducted with two heat flow meters. One heat flow meter was installed by applying a layer of interface material between the heat flow meter and the wall surface, and the other heat flow meter was installed by applying a PVC film between the layer of interface material and the wall surface. During the monitoring process, recommendations on apparatus and environmental conditions were considered [31]. Then, data were analysed using the dynamic method.
- Third, the variability of results was analysed by comparing the differences between the measured thermal transmittances obtained from the two heat flow meters with the theoretical effect of including a PVC film during the monitoring process, for the two case studies.

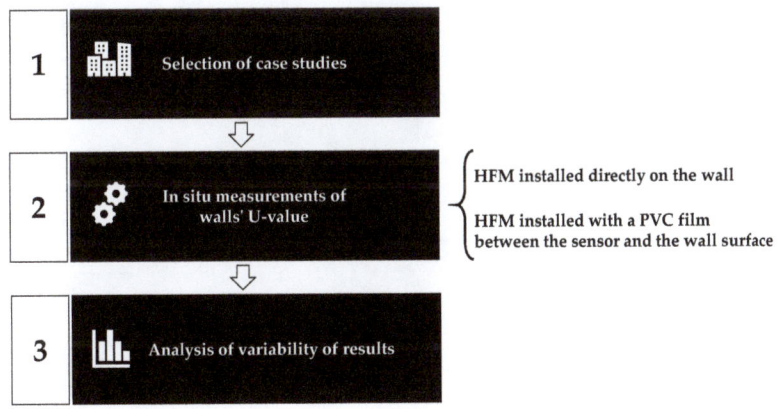

Figure 1. Research methodology.

2.2. Case Studies

To analyse the implications of using a protective film during the installation process of heat flow meters on the accuracy of in situ measurements of the façades' thermal transmittance, two brick masonry walls with varying theoretical thermal transmittances were selected as case studies (Case studies 1 and 2).

Case study 1 was built in 1960 and according to Gaspar et al. [46] can be defined as a single-skin wall with no air cavity or insulation. This case study is located between an interior habitable zone and an interior non-habitable zone. Case study 2 was built in 2005, it is a double-skin façade with internal insulation but no air cavities finished with continuous covering [46]. This case study is located between an indoor habitable zone and an outdoor zone.

The theoretical total thermal resistance (RT) of the two multilayer walls can be calculated as follows [47]:

$$R_T \left(\frac{m^2 \cdot K}{W} \right) = \frac{1}{U_t} = R_{si} + R_1 + R_2 + \ldots + R_N + R_{se} \quad (1)$$

where U_t is the theoretical thermal transmittance of the wall, R_{si} and R_{se} are the internal and external superficial resistance values (0.13 m²·K/W and 0.04 m²·K/W, respectively [47]) and $R_1 + R_2 + \ldots + R_N$ are the design thermal resistance values for layers from 1 to N.

To calculate the design thermal resistance (R) of a uniform layer, the following expression was used:

$$R \left(\frac{m^2 \cdot K}{W} \right) = \frac{d}{\lambda} \quad (2)$$

where d is the thickness of the layer and λ is its design thermal conductivity.

Design data to calculate the designed thermal transmittance was obtained from the buildings' executive projects and reports. This information was corroborated by the facility managers of the buildings. Subsequently, the thickness of the wall was measured on-site. The theoretical U-values of the two case studies were determined according to Equations (1) and (2), following guidelines of ISO 6949:2007 [47] and the Spanish Technical Building Code's Catalogue of Building Elements [48]. Even though the theoretical U-values were not absolutely certain were taken as reference values. Table 2 provides a summary of the thickness, thermal conductivity and thermal resistance of each layer and the theoretical thermal transmittance for the case studies.

Table 2. Composition of the case studies.

Case Study	No. Layer	Material Layer (Inside-Outside)	Thickness (m)	Thermal Conductivity (W/m·K)	Thermal Resistance (m²·K/W)	Total Thickness (m)	Theoretical U-Value (W/m²·K)
					0.13		
Case study 1	1	Mortar plaster	0.01	0.570	0.018	0.12	2.20
	2	Hollow brick wall	0.10		0.160		
	3	Mortar plaster	0.01	0.570	0.018		
					0.13		
					0.13		
Case study 2	1	Mortar plaster	0.02	1.300	0.015	0.34	0.36
	2	Hollow brick wall	0.10		0.160		
	3	Polyurethane insulation	0.06	0.028	2.143		
	4	Perforated brick wall	0.14		0.210		
	5	Single-layer mortar plaster	0.02	0.340	0.059		
					0.04		

2.3. In Situ Measurement of Façades' Thermal Transmittance

This section specifies the instrumentation and data collection process and the subsequent analysis of data.

2.3.1. Instrumentation

Proper instrumentation was carefully selected for the in situ measurement of the actual thermal transmittance of the walls. The apparatus consisted of an internal acquisition system, to which two heat flow plates and an internal environmental temperature sensor were connected, and an external environmental temperature sensor with its acquisition system. Table 3 summarises the main specifications and a priori accuracy of the calibrated instrumentation.

Table 3. Main specifications of the instrumentation.

Type of Equipment	Model and Manufacturer	Range	A Priori Accuracy
Heat flow meter plates	HFP01, Hukseflux	±2000 W/m^2	±5%
Inside environmental temperature sensor	T107, Campbell Scientific, Inc.	−35° to +50 °C	±0.5 °C
Inside acquisition system	CR850, Campbell Scientific, Inc.	Input ±5 Vdc	±0.06% of reading
Outside environmental temperature sensor and its acquisition system	175T1, Instrumentos Testo, SA	−35° to +50 °C	±0.5 °C

The monitoring process was designed following ISO guidelines [31]. Taking into account the considerations of Asdrubali et al. [14], Evangelisti et al. [16], Ahmad et al. [41], Tejedor et al. [49,50], Barreira et al. [51] and Nardi et al. [52], the placement of equipment was examined with an infrared thermographic camera (FLIR E60bx Infrared Camera). Proximity to defects, joints and borders of the wall, direct solar radiation and direct impact of heating or cooling devices were avoided, as recommended by Guattari et al. [42] and Evangelisti et al. [53].

The two heat flow meter plates were placed on the inner side of the wall due to it is the location where the temperature was most stable. Proper thermal contact was ensured between the entire area of one heat flow meter plate and the wall surface by carefully applying a layer of thermal interface material. The other heat flow meter plate was meticulously installed by applying a film between the layer of thermal interface material and the wall (Figure 2), which is usual procedure to protect the wall surface, as described in Section 1.

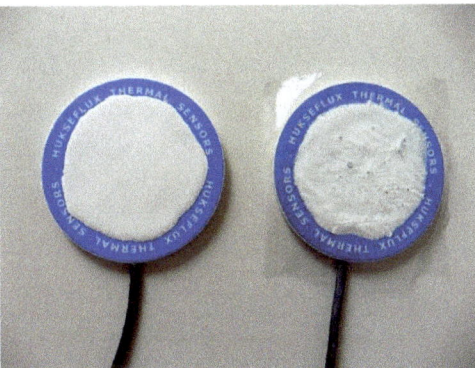

Figure 2. View of two heat flow meter plates, without and with a film between the layer of thermal interface material and the wall surface.

During the experimental campaign, climatic conditions were observed. The monitoring process took into consideration that the internal environmental temperature always exceeded the external environmental temperature, ensuring a stable heat flow direction. Alternating this flow direction could lead to inaccurate measurements, as described by Tadeu et al. [43], and could greatly influence the minimum test duration and the variability of the results [39]. Data were sampled every 1 s and recorded every 30-min averaged data in both dataloggers.

Data obtained from process monitoring is depicted in Figure 3, where *Tin* is the inside environmental temperature, *Tout* is the outside environmental temperature, *qNF* is the heat flow measured with a heat flow meter plate installed with direct contact on the wall surface and *qF* is the heat flow measured with a heat flow meter plate installed by applying a PVC film between the layer of thermal interface material and the wall surface. The experimental campaign of Case study 1 was conducted from 10–17 June 2016 (from 12:00 a.m. to 12:00 a.m.) and of Case study 2 from 24–30 October 2016 (from 12:30 p.m. to 12:30 p.m.).

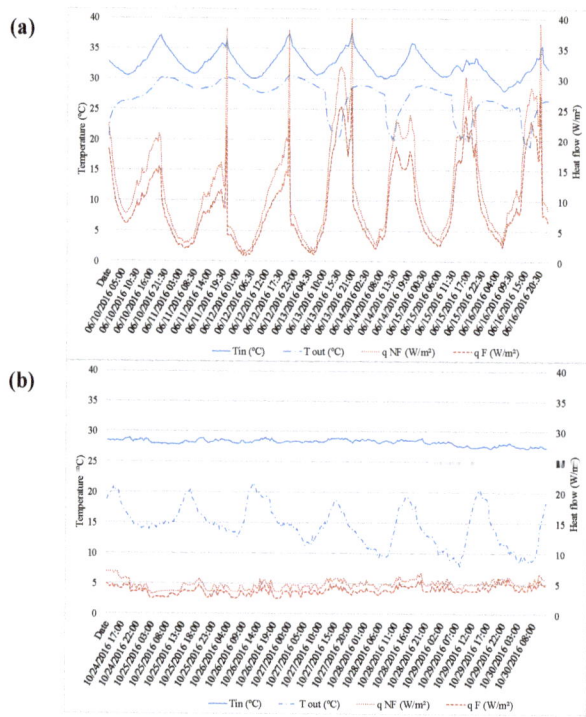

Figure 3. Indoor temperature (*Tin*), outdoor temperature (*Tout*), heat flow measured with an heat flow meter (HFM) installed with a layer of thermal interface material directly on the wall surface (*qNF*) and heat flow measured with an HFM installed with a PVC film between the layer of thermal interface material and the wall surface (*qF*) for (**a**) Case study 1 and (**b**) Case study 2.

2.3.2. Data Analysis

The measured thermal transmittance was determined according to the standardised dynamic method described by ISO 9869-1:2014 [31] and extensively detailed in Gaspar et al. [54]. To apply the dynamic analysis, a spreadsheet was programmed following the recommendations of Gaspar et al. [54]. The best estimate of the thermal transmittance was obtained for each cycle of 24 h. A confidence interval of 95% was adopted in the study to evaluate the quality of the thermal transmittance estimation results [31,55].

The duration of the test was evaluated considering the requirements established in the ISO 9869-1:2014 standard [31,56]. The three requirements are summarised in the following equations (Equations (3)–(5)):

$$D_T \ (days) \geq 3 \qquad (3)$$

$$\left| \frac{U_{mi} - U_{mi-1}}{U_{mi-1}} \times 100 \right| \leq 5\% \qquad (4)$$

$$\left| \frac{U_{m_{i=1}}^{INT(2\times\frac{DT}{3})} - U_{m_{i=DT-INT(2\times\frac{DT}{3})+1}}^{DT}}{U_{m_{i=DT-INT(2\times\frac{DT}{3})+1}}^{DT}} \right| \times 100 \leq 5\% \qquad (5)$$

where DT is the length of the test in days, U_m is the measured U-value of the wall, index i indicates the number of the cycle and INT is the integer part.

2.4. Analysis of Variability of Results

The introduction of a PVC film during the monitoring process changes the theoretical U-value of the wall due to its inherent thermal resistance plus the thermal contact resistance between the film and the wall surface. The theoretical maximum thermal resistance associated with the inclusion of a PVC film is estimated according to Equations (1) and (2), as follows:

$$R_{film\_max}\left(\frac{m^2 \cdot K}{W}\right) = 0.0005 + \frac{15 \cdot 10^{-6}}{0.12} = 6.25 \cdot 10^{-4} \qquad (6)$$

where the thermal contact resistance is usually between 0.000005 to 0.00005 m²·K/W according to Çengel and Ghajar [57], the thickness of the film is usually between 7×10^{-6} to 15×10^{-6} m, the thermal conductivity of the material is between 0.12 to 0.25 W/m·K and R_{film_max} is the maximum thermal resistance associated to the inclusion of a PVC film, considering the most insulation case of the film inclusion.

The theoretical effect of the inclusion of a PVC film on the thermal transmittance is calculated according to Equation (1) for the two case studies and shown in Table 4. The deviations between the theoretical thermal transmittance (NF) and the thermal transmittance when using a PVC film during the installation of the heat flow meter (F) are practically imperceptible values, 0.14% in Case study 1 and 0.02% in Case study 2.

Table 4. Theoretical impact of the inclusion of a PVC film on the thermal transmittance in the two case studies.

Case Study	Theoretical U-Value NF (W/m²·K)	Theoretical U-Value F (W/m²·K)	Deviation between U-Values(%)
Case study 1	2.20	2.19	0.14
Case study 2	0.36	0.36	0.02

To assess the use of a protective film during the installation of heat flow meters, two values of the measured thermal transmittance were calculated for each case study. The first value was obtained using a heat flow meter plate installed with a layer of thermal interface material directly on the wall surface (U_{mNF}) and the second value was obtained using a heat flow meter plate installed with a PVC film between the layer of thermal interface material and the wall (U_{mF}). To check the adjustment between both measured U-values, the absolute value of the relative difference between measured U-values were calculated using the following expression:

$$Absolute\ value\ of\ the\ relative\ difference\ U_{m\ NF} - U_{m\ F}(\%) = \left|\frac{(U_{m\ NF} - U_{m\ F})}{U_{m\ NF}}\right| \times 100 \qquad (7)$$

where U_{mNF} is the measured U-value of the wall using the dynamic method using an HFM installed without film and U_{mF} is the measured U-value of the wall using the dynamic method using an HFM installed with a PVC film between the layer of thermal interface material and the wall.

The variability of results was analysed by comparing the relative difference between the measured thermal transmittances obtained according to Equation (7), with the theoretical effect of including a PVC during the monitoring process shown in Table 4.

3. Results

The data acquisition process was conducted following the indications in Section 2, with a sampling duration of 168 h in Case study 1 (337 readings) and of 144 h in Case study 2 (289 readings). Two measured thermal transmittances and its confidence interval were calculated, for 24-h test cycles using the dynamic method in the two case studies [54]. One value was calculated using data from the heat flow meter plate installed with a layer of thermal interface material directly on the wall surface and the other value was calculated using data from the heat flow meter installed with a PVC film between the layer of thermal interface material and the wall.

The minimum test duration was checked considering the ISO standard [31]. The first requirement is that the sampling duration must be an integer of 24 h and at least 72 h (Equation (3)). Consequently, the second and third requirements were verified from the third day onwards. The second requirement for test completion is that the value of thermal transmittance obtained at the end of the sampling duration shall not deviate more than 5% from the value obtained 24 h before (Equation (4)). In accordance with this condition, in Case studies 1 and 2 the test could be ended after 72 h, as the requirement was met for all cycles with both HFM installation methods. The last requirement for ending the test is that the U-value obtained by analysing data from the first 2/3 of the sampling duration shall not deviate more than 5% from the value obtained from the data for the last period of the same length (Equation (5)). According to this condition, in Case study 1 the monitoring process could be ended in 96 h using an HFM installed directly on the wall surface and in 120 h when an HFM was used with a PVC film between the layer of thermal interface material and the wall surface. In Case study 2, the test could be stopped at 72 h when the heat flow meter was installed directly on the wall surface and in 120 h when the heat flow meter was installed with a PVC film between the layer of thermal interface material and the wall surface. The minimum test duration when using an HFM installed directly on the wall surface was found to be shorter than when using an HFM installed with a PVC film between the layer of thermal interface material and the wall surface.

The results of the measured thermal transmittance for the two case studies are shown in Table 5 and depicted in Figure 4, where $U_{m\text{-}Dyn} \pm I_{95\%}$ is the measured U-value using the dynamic method, HFM_{NF} indicates the use of a heat flow meter plate installed with a layer of thermal interface material directly on the wall surface and HFM_F indicates the use of a heat flow meter installed with a PVC film between the layer of thermal interface material and the wall.

Table 5. Measured thermal transmittance values using the dynamic method in 24-h cycles, using an HFM installed with a layer of thermal interface material directly on the wall surface (HFM_{NF}) and an HFM installed with a PVC film between the layer of thermal interface material and the wall surface (HFM_F).

Duration of the Test (h)	Case Study 1 $U_{m\text{-}Dyn} \pm I_{95\%}$ (W/m²·K)		Case Study 2 $U_{m\text{-}Dyn} \pm I_{95\%}$ (W/m²·K)	
	HFM_{NF}	HFM_F	HFM_{NF}	HFM_F
24 h	2.19 ± 0.12	2.03 ± 0.10	0.34 ± 0.02	0.27 ± 0.02
48 h	2.21 ± 0.10	1.62 ± 0.05	0.35 ± 0.01	0.28 ± 0.01
72 h	2.25 ± 0.09	1.64 ± 0.05	0.35 ± 0.01	0.27 ± 0.01
96 h	2.24 ± 0.06	1.69 ± 0.03	0.34 ± 0.01	0.27 ± 0.01
120 h	2.24 ± 0.05	1.70 ± 0.03	0.35 ± 0.01	0.28 ± 0.01
144 h	2.23 ± 0.04	1.70 ± 0.02	0.35 ± 0.01	0.28 ± 0.01
168 h	2.24 ± 0.04	1.70 ± 0.02	-	-

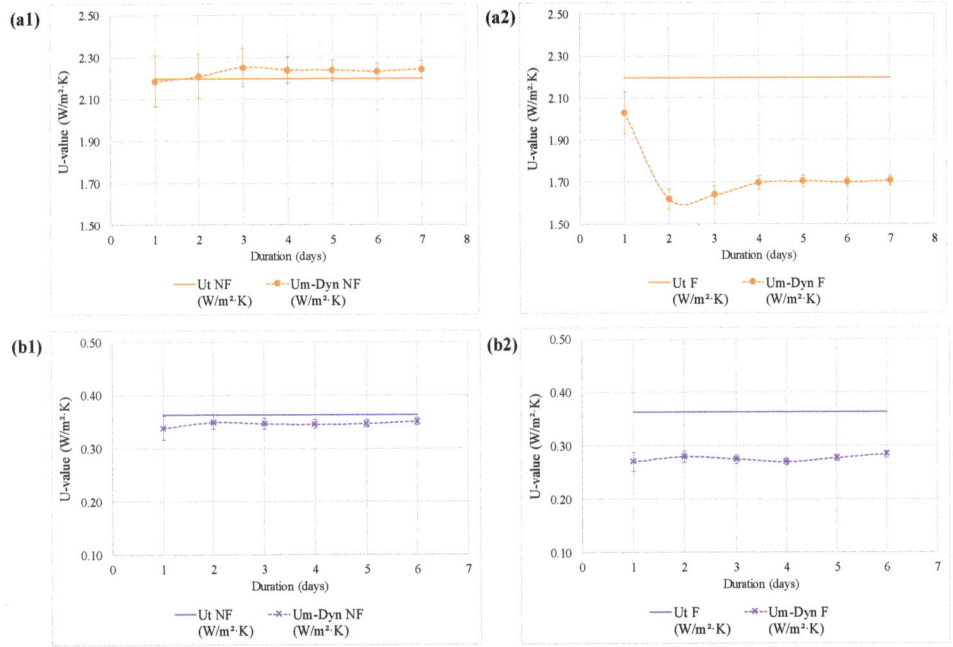

Figure 4. Theoretical and measured thermal transmittance values, (**1**) using an HFM installed with a layer of thermal interface material directly on the wall surface (NF) and, (**2**) an HFM installed with a PVC film between the layer of thermal interface material and the wall surface (F) for (**a**) Case study 1 and (**b**) Case study 2.

As tests lengthened, the confidence intervals were reduced. These findings are aligned with those analysed in the existing literature, in which the length of the test that was too short led to greater confidence intervals [14,37].

Deviations between the two measured thermal transmittance values were calculated following Equation (7), using an HFM installed with a layer of thermal interface material directly on the wall surface and an HFM installed with a PVC film between the layer of thermal interface material and the wall surface. These relative differences between measured U-values using both heat flow meters for the two case studies are depicted in Figure 5 and summarised in Table 6.

Table 6. Influence of using a PVC film between the layer of thermal interface material and the wall surface on the in situ measurement of the thermal transmittance in the two case studies.

Duration of the Test (h)	$\left\lvert \frac{(U_{qNF} - U_{qF})}{U_{qNF}} \right\rvert$ (%)	
	Case Study 1	Case Study 2
24 h	7.27	20.17
48 h	26.81	20.01
72 h	27.27	20.70
96 h	24.33	21.74
120 h	23.94	20.13
144 h	23.91	19.09
168 h	23.96	-

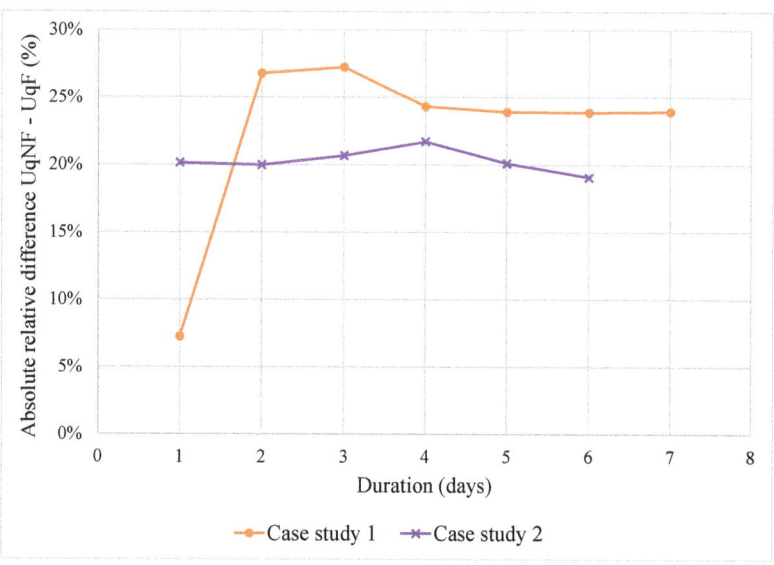

Figure 5. Deviation between the measured thermal transmittances for the two case studies, using an HFM installed with a layer of thermal interface material directly on the wall surface and an HFM installed with a PVC film between the layer of thermal interface material and the wall surface.

4. Discussion and Conclusions

This research assesses the influence of the heat flow meter plate non-perfect thermal contact on the accuracy of the in situ measurement of walls' thermal transmittance using the standardised heat flow meter method. The assessment considers usual practices in the installation of heat flow meters, consisting of the application of a PVC film to avoid damage to the wall in occupied buildings, in two case studies with different theoretical thermal transmittance values. The experimental campaign was designed to isolate measurement errors related to non-perfect thermal contact from those related to the measurement accuracy and the analysis of data. In situ measurements were conducted in the same conditions and equipment, with two heat flow meters: one was carefully installed with a layer of thermal interface material directly on the wall surface and the other was carefully installed by applying a film between the thermal interface material and the wall. Data were analysed in 24-h periods using the dynamic method. Finally, the variability of results was analysed by comparing the deviation between the measured thermal transmittances obtained from the two heat flow meters with the theoretical effect of including a PVC film during the monitoring process.

Test completion results indicate that the minimum duration of experimental campaigns was influenced by the installation of the heat flow meter plate. Generally, when the heat flow meter was installed by applying a PVC film between the thermal interface material and the wall to protect its surface, the minimum duration of the test was longer than when the heat flow meter was directly installed on the wall surface without a PVC film. In Case study 1, the test lasted 24 h more and in Case study 2, the test lasted 48 h more.

The findings show that the use of a PVC film hinders the installation of sensors, worsening the thermal contact between sensor and wall. Relative differences between the measured U-values were found to be greater than the expected by the theoretical calculation and also by the ISO standard quantification in both case studies. Deviations between the two measured thermal transmittance values, using an HFM installed with a layer of thermal interface material directly on the wall surface and an HFM installed with a PVC film between the layer of thermal interface material and the wall surface, were found to be significantly different than those expected resulting from the calculation of the theoretical effect of including a PVC during the monitoring process summarised in Table 4. In the theoretical estimation, the use of a PVC film between the layer of thermal interface material and the wall surface had an expected deviation of 0.14% in Case study 1 and 0.02% in Case study 2. However, in the experimental tests results from the third cycle onwards showed that for Case study 1, the relative differences were around 24% to 27%, and in Case study 2 around 19% to 21%. On the other hand, random variations caused by a non-perfect thermal contact between the sensors and the surface are quantified by 5% according to ISO 9869-1:2014 [31]. The experimental results also differ significantly from those quantified by ISO standard [31]. These considerable differences between the expected differences and the measured ones might be due to collateral effects when installing an HFM with a PVC film between the layer of thermal interface material and the wall surface that hampers a good thermal contact, as Li et al. [27] suggested. Including a PVC film to protect the wall may complicate the sensor installation process and consequently, the quality of the thermal contact might be hampered. Additionally, the theoretical calculation of including a PVC film on the thermal transmittance does not consider random variations in the thermal contact of the HFM with the surface of the wall tested. Moreover, extending the duration of the test did not seem to reduce the differences between the measured U-values. Therefore, it is highly recommended to avoid the use of a PVC film between the layer of thermal interface material and the wall surface in the in situ measurement of the thermal transmittance of existing buildings' façades when accurate results are required.

This research could be useful for practitioners when they conduct energy audits. It was found that when HFMs were installed with a PVC film between the layer of thermal interface material and the wall surface, the measured thermal transmittance was around 19 to 27% lower than when HFMs were installed with a layer of thermal interface material directly on the wall surface. Therefore, depending on the installation of heat flow meter plates, technical staff could take into account these possible deviations in the in situ measurement of the façades' actual thermal transmittance during building's operational stage and, consequently, propose appropriate energy retrofitting strategies.

Author Contributions: Conceptualization, K.G., M.C. and M.G.; methodology, K.G., M.C. and M.G.; investigation, K.G., M.C. and M.G.; data curation, K.G.; writing—original draft preparation, K.G.; writing—review and editing, K.G., M.C. and M.G.; visualization, K.G.; supervision, M.C. and M.G.; project administration, M.C.; funding acquisition, M.C. All authors have read and agreed to the published version of the manuscript.

Funding: This research was partially funded by the Government of Catalonia, Research Grant 2017—SGR—227.

Data Availability Statement: Not applicable.

Conflicts of Interest: The authors declare no conflict of interest. The funders had no role in the design of the study; in the collection, analyses, or interpretation of data; in the writing of the manuscript, or in the decision to publish the results.

Abbreviations

d	thickness of a layer [m]
HFM $_{NF}$	heat flow meter plate installed without a film between the sensor and the wall surface
HFM $_F$	heat flow meter plate installed with a film between the sensor and the wall surface
I	confidence interval [%]
q $_{NF}$	heat flux using a heat flow meter plate installed without a film between the sensor and the wall surface [W/m^2]
q $_F$	heat flux using a heat flow meter plate installed with a film between the sensor and the wall surface [W/m^2]
R	theoretical thermal resistance of a uniform layer [m^2·K/W]
R$_T$	theoretical total thermal resistance of an element [m^2·K/W]
T$_{in}$	internal environmental temperature [°C]
T$_{out}$	external environmental temperature [°C]
U$_{m\text{-}Dyn}$	measured thermal transmittance using the dynamic calculation method [W/m^2·K]
Ut	theoretical thermal transmittance [W/m^2·K]
λ	design thermal conductivity of a material [W/m·K]

References

1. Martynov, A.; Sushama, L.; Laprise, R. Simulation of temperate freezing lakes by one-dimensional lake models: Performance assessment for interactive coupling with regional climate models. *Boreal Environ. Res.* **2010**, *15*, 143–164. [CrossRef]
2. Pedregosa, F.; Varoquaux, G.; Gramfort, A.; Michel, V.; Thirion, B.; Grisel, O.; Vanderplas, J. Scikit-learn: Machine Learning in Python. *J. Mach. Learn. Res.* **2012**, *12*, 2825–2830. [CrossRef]
3. European Union. Directive (EU) 2018/844 of the European Parliament and of the Council of 30 May 2018 Amending Directive 2010/31/EU on the Energy Performance of Buildings and Directive 2012/27/EU on Energy Efficiency. 2018. Available online: https://eur-lex.europa.eu/legal-content/EN/TXT/PDF/?uri=CELEX:32018L0844&from=EN (accessed on 24 October 2019).
4. Buildings Performance Institute Europe (BPIE). 97% of Buildings in the EU Need To Be Upgraded. 2017, Available online: http://bpie.eu/publication/97-of-buildings-in-the-eu-need-to-be-upgraded/ (accessed on 3 September 2020).
5. European Commission. Communication from the Commission to the European Parliament, the Council, the European Economic and Social Committee and the Committee of the Regions. A Renovation Wave for Europe—Greening Our Buildings, Creating Jobs, Improving Lives. COM(2020) 662 Final. Available online: https://eur-lex.europa.eu/resource.html?uri=cellar:0638aa1d-0f02-11eb-bc07-01aa75ed71a1.0003.02/DOC_1&format=PDF (accessed on 21 October 2020).
6. European Union. *EU Energy in Figures*; Publications Office of the European Union: Luxembourg, 2020.
7. Government of Spain. *2017–2020 National Energy Efficiency Action Plan*; Government of Spain: Madrid, Spain, 2017; pp. 1–216.
8. Gangolells, M.; Casals, M.; Forcada, N.; Macarulla, M.; Cuerva, E. Energy mapping of existing building stock in Spain. *J. Clean. Prod.* **2016**, *112*, 3895–3904. [CrossRef]
9. Buildings Performance Institute Europe (BPIE). *Europe's Buildings Under the Microscope*; A Country-by-Country Review of the Energy Performance of Buildings; Buildings Performance Institute Europe (BPIE): Brussels, Belgium, 2011.
10. Roels, S. *IEA EBC Annex 71: Building Energy Performance Assessment Based on In-Situ Measurements*; EBC: Birmingham, UK, 2017.
11. Kampelis, N.; Gobakis, K.; Vagias, V.; Kolokotsa, D.; Standardi, L.; Isidori, D.; Cristalli, C.; Montagino, F.M.; Paredes, F.; Muratore, P.; et al. Evaluation of the performance gap in industrial, residential & tertiary near-Zero energy buildings. *Energy Build.* **2017**, *148*, 58–73. [CrossRef]
12. Farmer, D.; Gorse, C.; Swan, W.; Fitton, R.; Brooke-Peat, M.; Miles-Shenton, D.; Johnston, D. Measuring thermal performance in steady-state conditions at each stage of a full fabric retrofit to a solid wall dwelling. *Energy Build.* **2017**, *156*, 404–414. [CrossRef]
13. Evangelisti, L.; Guattari, C.; Gori, P.; Asdrubali, F. Assessment of equivalent thermal properties of multilayer building walls coupling simulations and experimental measurements. *Build. Environ.* **2018**, *127*, 77–85. [CrossRef]
14. Asdrubali, F.; D'Alessandro, F.; Baldinelli, G.; Bianchi, F. Evaluating in situ thermal transmittance of green buildings masonries—A case study. *Case Stud. Constr. Mater.* **2014**, *1*, 53–59. [CrossRef]
15. Desogus, G.; Mura, S.; Ricciu, R. Comparing different approaches to in situ measurement of building components thermal resistance. *Energy Build.* **2011**, *43*, 2613–2620. [CrossRef]
16. Evangelisti, L.; Guattari, C.; Gori, P.; Vollaro, R.D.L. In Situ Thermal Transmittance Measurements for Investigating Differences between Wall Models and Actual Building Performance. *Sustainability* **2015**, *7*, 10388–10398. [CrossRef]
17. Ficco, G.; Iannetta, F.; Ianniello, E.; Alfano, F.R.D.; Dell'Isola, M. U-value in situ measurement for energy diagnosis of existing buildings. *Energy Build.* **2015**, *104*, 108–121. [CrossRef]
18. Hughes, M.; Palmer, J.; Cheng, V.; Shipworth, D. Sensitivity and uncertainty analysis of England's housing energy model. *Build. Res. Inf.* **2013**, *41*, 156–167. [CrossRef]
19. Sunikka-Blank, M.; Galvin, R. Introducing the prebound effect: The gap between performance and actual energy consumption. *Build. Res. Inf.* **2012**, *40*, 260–273. [CrossRef]

20. Symonds, P.; Taylor, J.; Mavrogianni, A.; Davies, M.; Shrubsole, C.; Hamilton, I.; Chalabi, Z. Overheating in English dwellings: Comparing modelled and monitored large-scale datasets. *Build. Res. Inf.* **2016**, *45*, 195–208. [CrossRef]
21. Zheng, K.; Cho, Y.K.; Wang, C.; Li, H. Noninvasive Residential Building Envelope R-Value Measurement Method Based on Interfacial Thermal Resistance. *J. Arch. Eng.* **2016**, *22*, A4015002. [CrossRef]
22. Lucchi, E. Thermal transmittance of historical brick masonries: A comparison among standard data, analytical calculation procedures, and in situ heat flow meter measurements. *Energy Build.* **2017**, *134*, 171–184. [CrossRef]
23. Sassine, E. A practical method for in-situ thermal characterization of walls. *Case Stud. Therm. Eng.* **2016**, *8*, 84–93. [CrossRef]
24. Laaroussi, N.; Lauriat, G.; Garoum, M.; Cherki, A.; Jannot, Y. Measurement of thermal properties of brick materials based on clay mixtures. *Constr. Build. Mater.* **2014**, *70*, 351–361. [CrossRef]
25. Kumar, A.; Suman, B. Experimental evaluation of insulation materials for walls and roofs and their impact on indoor thermal comfort under composite climate. *Build. Environ.* **2013**, *59*, 635–643. [CrossRef]
26. Björk, F.; Enochsson, T. Properties of thermal insulation materials during extreme environment changes. *Constr. Build. Mater.* **2009**, *23*, 2189–2195. [CrossRef]
27. Li, F.G.N.; Smith, A.; Biddulph, P.; Hamilton, I.; Lowe, R.J.; Mavrogianni, A.; Oikonomou, E.; Raslan, R.; Stamp, S.; Stone, A.; et al. Solid-wall U-values: Heat flux measurements compared with standard assumptions. *Build. Res. Inf.* **2015**, *43*, 238–252. [CrossRef]
28. Roels, S. *IEA EBC Annex 71: Building Energy Performance Assessment Based on Optimized In-Situ Measurements*; EBC: Birmingham, UK, 2018.
29. Bienvenido-Huertas, D.; Moyano, J.; Marín, D.; Fresco-Contreras, R.; Marín-García, D. Review of in situ methods for assessing the thermal transmittance of walls. *Renew. Sustain. Energy Rev.* **2019**, *102*, 356–371. [CrossRef]
30. Soares, N.; Martins, C.; Gonçalves, M.; Santos, P.; Da Silva, L.S.; Costa, J.J. Laboratory and in-situ non-destructive methods to evaluate the thermal transmittance and behavior of walls, windows, and construction elements with innovative materials: A review. *Energy Build.* **2019**, *182*, 88–110. [CrossRef]
31. International Organization for Standardization (ISO). *Thermal Insulation—Building Elements—In-Situ Measurement of Thermal Resistance and Thermal Transmittance—Part 1: Heat Flow Meter Method*; ISO Standard: Geneva, Switzerland, 2014.
32. Peng, C.; Wu, Z. In situ measuring and evaluating the thermal resistance of building construction. *Energy Build.* **2008**, *40*, 2076–2082. [CrossRef]
33. Janssens, A. *IEA EBC Annex 58: Reliable Building Energy Performance Characterisation Based on Full Scale Dynamic Measurements*; KU Leuven: Leuven, Belgium, 2016.
34. Trethowen, H. Measurement errors with surface-mounted heat flux sensors. *Build. Environ.* **1986**, *21*, 41–56. [CrossRef]
35. Meng, X.; Yan, B.; Gao, Y.; Wang, J.; Zhang, W.; Long, E. Factors affecting the in situ measurement accuracy of the wall heat transfer coefficient using the heat flow meter method. *Energy Build.* **2015**, *86*, 754–765. [CrossRef]
36. Albatici, R.; Tonelli, A.M.; Chiogna, M. A comprehensive experimental approach for the validation of quantitative infrared thermography in the evaluation of building thermal transmittance. *Appl. Energy* **2015**, *141*, 218–228. [CrossRef]
37. Nardi, I.; Ambrosini, D.; De Rubeis, T.; Sfarra, S.; Perilli, S.; Pasqualoni, G. A comparison between thermographic and flow-meter methods for the evaluation of thermal transmittance of different wall constructions. *J. Phys. Conf. Ser.* **2015**, *655*, 012007. [CrossRef]
38. Deconinck, A.-H.; Roels, S. Comparison of characterisation methods determining the thermal resistance of building components from onsite measurements. *Energy Build.* **2016**, *130*, 309–320. [CrossRef]
39. Atsonios, I.A.; Mandilaras, I.; Kontogeorgos, D.A.; Founti, M.A. A comparative assessment of the standardized methods for the in–situ measurement of the thermal resistance of building walls. *Energy Build.* **2017**, *154*, 198–206. [CrossRef]
40. Cesaratto, P.G.; De Carli, M.; Marinetti, S. Effect of different parameters on the in situ thermal conductance evaluation. *Energy Build.* **2011**, *43*, 1792–1801. [CrossRef]
41. Ahmad, A.; Maslehuddin, M.; Al-Hadhrami, L.M. In situ measurement of thermal transmittance and thermal resistance of hollow reinforced precast concrete walls. *Energy Build.* **2014**, *84*, 132–141. [CrossRef]
42. Guattari, C.; Evangelisti, L.; Gori, P.; Asdrubali, F. Influence of internal heat sources on thermal resistance evaluation through the heat flow meter method. *Energy Build.* **2017**, *135*, 187–200. [CrossRef]
43. Tadeu, A.; Simões, N.; Simões, I.; Pedro, F.; Škerget, L. In-Situ Thermal Resistance Evaluation of Walls Using an Iterative Dynamic Model. *Numer. Heat Transf. Part A Appl.* **2014**, *67*, 33–51. [CrossRef]
44. Gori, V.; Elwell, C.A. Estimation of thermophysical properties from in-situ measurements in all seasons: Quantifying and reducing errors using dynamic grey-box methods. *Energy Build.* **2018**, *167*, 290–300. [CrossRef]
45. Biddulph, P.; Gori, V.; Elwell, C.A.; Scott, C.; Rye, C.; Lowe, R.J.; Oreszczyn, T.; Oreszczyn, T. Inferring the thermal resistance and effective thermal mass of a wall using frequent temperature and heat flux measurements. *Energy Build.* **2014**, *78*, 10–16. [CrossRef]
46. Gaspar, K.; Casals, M.; Gangolells, M. Classifying System for Façades and Anomalies. *J. Perform. Constr. Facil.* **2016**, *30*, 04014187. [CrossRef]
47. International Organization for Standardization. *Building Components and Building Elements—Thermal Resistance and Thermal Transmittance—Calculation Method*; ISO: Geneva, Switzerland, 2007.
48. Spanish Ministry of Housing; Instituto de Ciencias de la Construcción Eduardo Torroja (IETcc). Constructive Elements Catalogue of Technical Building Code. 2010. Available online: https://www.codigotecnico.org/pdf/Programas/CEC/CAT-EC-v06.3_marzo_10.pdf (accessed on 21 April 2020).

49. Tejedor, B.; Casals, M.; Macarulla, M.; Giretti, A. U-value time series analyses: Evaluating the feasibility of in-situ short-lasting IRT tests for heavy multi-leaf walls. *Build. Environ.* **2019**, *159*, 106123. [CrossRef]
50. Tejedor, B.; Casals, M.; Gangolells, M. Assessing the influence of operating conditions and thermophysical properties on the accuracy of in-situ measured U -values using quantitative internal infrared thermography. *Energy Build.* **2018**, *171*, 64–75. [CrossRef]
51. Barreira, E.; Almeida, R.M.; Delgado, J.M.P.Q. Infrared thermography for assessing moisture related phenomena in building components. *Constr. Build. Mater.* **2016**, *110*, 251–269. [CrossRef]
52. Nardi, I.; Lucchi, E.; De Rubeis, T.; Ambrosini, D. Quantification of heat energy losses through the building envelope: A state-of-the-art analysis with critical and comprehensive review on infrared thermography. *Build. Environ.* **2018**, *146*, 190–205. [CrossRef]
53. Evangelisti, L.; Guattari, C.; Asdrubali, F. Influence of heating systems on thermal transmittance evaluations: Simulations, experimental measurements and data post-processing. *Energy Build.* **2018**, *168*, 180–190. [CrossRef]
54. Gaspar, K.; Casals, M.; Gangolells, M. A comparison of standardized calculation methods for in situ measurements of façades U-value. *Energy Build.* **2016**, *130*, 592–599. [CrossRef]
55. Roulet, C.; Gass, J.; Markus, I. In-Situ U-Value Measurement: Reliable Results in Shorter Time by Dynamic Interpretation of Measured Data. 1985. Available online: https://web.ornl.gov/sci/buildings/conf-archive/1985B3papers/057.pdf (accessed on 23 April 2020).
56. Gaspar, K.; Casals, M.; Gangolells, M. Review of criteria for determining HFM minimum test duration. *Energy Build.* **2018**, *176*, 360–370. [CrossRef]
57. Çengel, Y.A.; Ghajar, A.J. *Heat and Mass Transfer: Fundamentals and Applications*, 5th ed.; McGraw-Hill Education: New York, NY, USA, 2015.

Article

Non-Destructive Techniques for Building Evaluation in Urban Areas: The Case Study of the Redesigning Project of Eleftheria Square (Nicosia, Cyprus)

Marilena Cozzolino [1,*], Vincenzo Gentile [1], Paolo Mauriello [1] and Agni Peditrou [2]

1. Department of Human, Social and Educational Science, University of Molise, Via De Sanctis, 86100 Campobasso, Italy; vincenzo.gentile86@gmail.com (V.G.); mauriello@unimol.it (P.M.)
2. Architectural Planning Office, Nicosia Municipality, Eptanisou 11, 1016 Nicosia, Cyprus; Agni.Petridou@nicosiamunicipality.org.cy
* Correspondence: marilena.cozzolino@unimol.it

Received: 1 June 2020; Accepted: 16 June 2020; Published: 23 June 2020

Abstract: This paper deals with the application of non-destructive geophysical techniques of investigation in the urban environment of the city of Nicosia (Cyprus). The main aim of the research was, in the frame of the Eleftheria Square redesign project, to image subsurface properties in order to reduce the impact of hazards on the old buildings (therefore preserving the cultural heritage of the place), and on the new infrastructure under construction. Since 2008, electrical resistivity tomography (ERT), ground penetrating radar (GPR) and induced electromagnetic method (EMI) were employed during the different phases of the project to provide an understanding of geological stratigraphy, the detection of buried objects (archaeological structures and underground utilities) and the solution of unexpected events (such as water infiltration in the course of works). The geophysical results proved the efficiency of the adopted methods, adding scientific value to the knowledge of the studied area. The new gathered information helped the public administration technicians to plan direct and targeted interventions and to modify the original design of the project according to the discovery of archaeological findings.

Keywords: applied geophysics; urban geophysics; Eleftheria Square; ERT; probability-based ERT inversion (PERTI) method; GPR; EMI

1. Introduction

Applied geophysics, through the employment of different methodologies, represents a useful tool to map subsurface features in a non-invasive way, and up to date, successful documented researches in rural settings dealing with different research objectives are available in the literature. Geological investigations for characterizing faults [1–3], landslides [4,5], paleo-morphologies [6,7], litho-stratigraphies [8,9], acquifers [10], sinkoles [11,12], and seepage detection [13] are generally carried out through electrical resistivity tomography (ERT) [1–3,5,7,9], ground penetrating radar (GPR) [8,10], seismic [2,4–6], magnetic [14–17] and gravity [18] methods. Archae-geophysical prospections are particularly common in applications for the detection of buried structures, tombs, and channels through the application of electromagnetic methods [19,20], magnetometry [21], ERTs [22,23], GPR [24,25] or the combination of them. The diagnostics and the monitoring of buildings in architectural and engineering surveys are often reached by implementing linear variable displacement transducer [26], sonic tomography [27], infrared thermography [27–29] and GPR [30–32] because of the high resolution of results.

In contrast, the geophysical exploration in urban environments for different purposes is currently in constant development. The consistent noise, the low signal-to-noise ratio, the reduced spaces to operate

Appl. Sci. 2020, *10*, 4296; doi:10.3390/app10124296   www.mdpi.com/journal/applsci

in, and the presence of paved surfaces, asphalt or complex layers disturbed by anthropological buried shallow artifacts and metallic structures represent a problematic difficulty in geophysical prospections that can significantly influence data quality. The rising interest of those aspects are evidenced by the inclusion of the matter in ad-hoc sessions of international workshops and conferences [33–36] and by the increment of special issues of international journals aiming to provide an enrichment of showcases for urban geophysics [37–39]. Nevertheless, some applications can be found in current literature, providing information on cavity individuation [40], geological mapping [41,42], archaeological investigations [43–45] and underground pipe detection [46,47].

In this work, ground-sensing prospections were used to characterize the subsurface in the areas involved in the redesign project of Eleftheria Square (Nicosia, Cyprus) (Figure 1).

Figure 1. Location of Nicosia on a Google Earth™ satellite image (2020) of Cyprus (**a**), the historical center of Nicosia (Cyprus) with the indication of the three gates, Eleftheria Square and the surrounding area, the Green Line (the border line that separates the island and the city in two parts) and the main Ledras Street (**b**) and an ancient map enhancing the Venetian walls [48] (**c**).

The activities lasted more than ten years, from 2008 until today; the geophysical studies were an integral part of the project and here we report the various steps. The action was performed in the framework of a collaboration between University of Molise and the Nicosia Master Plan and was partly supported by the Ministry of Foreign Affairs. In particular, the research was in the frame of a scientific archaeological mission that had not only the function of scientific study but also represented a valuable tool for intercultural dialogue and development policies in the host country. The archeo-geophysical surveys performed at Famagusta Gate, Castellionissa Hall, Phafos Gate, the monumental complex of the Bedestan and Plateia Di Marche are within this project [49,50]. The choice of Eleftheria Square as a study case is justified by the majesty and prestige of the construction project. In addition, as it provided for a distortion of the original setting with an enormous impact on its soils, the landscape and the urban system, there was the real need to acquire information before direct interventions.

In this case, a careful evaluation of the best method to implement to solve a particular problem was carried out and the ERT, GPR and induced electromagnetic (EMI) techniques were used in this contest. Sometimes investigations were conducted in an emergency situation by providing quick answers directly on the investigation site for problem solving.

According to the architectural work being planned, the surveys were organized during the executive phases of the intervention with the following aims:

- Evaluation of the potential risk of presence of archaeological remains, and consequently their hazard of destruction, at Eleftheria Square (Survey 1 in Figure 2) and Solomon Square (Survey 2 in Figure 2) in 2008, before the beginning of the construction of the new structure.
- Delineation of the geological stratigraphy of the soil proving the absence of cavities or archaeological structures in function of the construction of the underground car parking along Amirou Avenue in 2009 (Survey 3 in Figure 2).
- Detection of underground utilities with the aim to facilitate technicians' work during excavation and avoid unexpected damages to city supplies along Amirou Avenue in 2017 (Survey 4 in Figure 2).
- Comprehension of the reason for an abundant flow of water that occurred inside the new underground electrical substation positioned under the new Eleftheria Square during excavations at the intersection of Evagorou Avenue and Omirou Avenue in 2009 (Survey 5 in Figure 2). The most reliable hypothesis of the cause of this phenomenon was thought to be the failure of water pipes south of Evagorou Avenue.

From a scientific point of view, we wanted to show, through the chosen case study, the efficiency of this type of investigation in very disturbed environments such as an urban context where stratigraphies are often disturbed by recent or ancient buried anthropogenic elements. In particular, for ERT data processing, this was the opportunity to test the probability-based ERT inversion (PERTI) method [51], applied here for the first time to model near-surface urban geoelectrical data. The algorithm was developed contextually with needs that the case study presented, having, among its main features, a good filter capacity and the possibility of performing real-time inversion directly in the field, thus allowing for fast modifications of the survey plan to better focus on the expected targets [51].

In the following sections, a brief description of the case study and its cultural value is given, then the methodological approach to the research is explained and the results are reported and discussed.

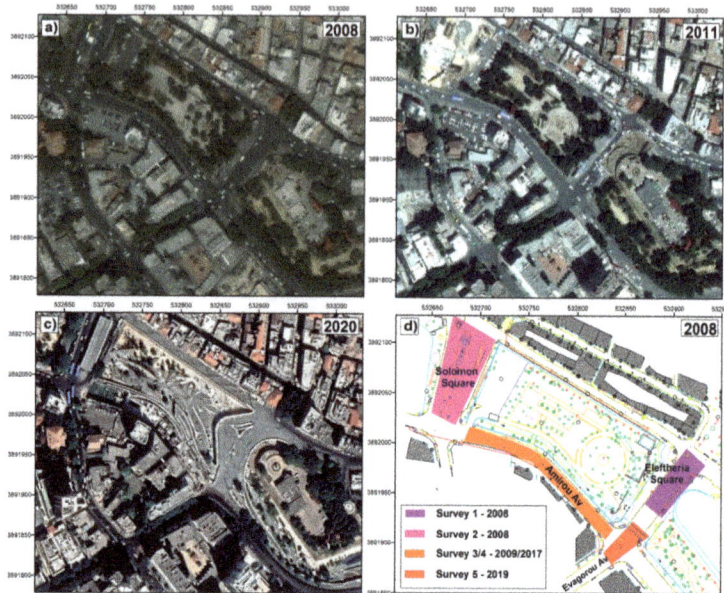

Figure 2. Google Earth™ satellite images (2008, 2011, 2020) of Eleftheria Square and surrounding areas (**a–c**) and location of geophysical surveys (**d**).

2. Case Study

Nicosia has been the capital of Cyprus since the 10th century. During the Middle Ages, walls larger than the actual fortification were built to protect the city. The engineer Giulio Savorgnano designed the Venetian walls that subsist today during the Venetian period (1489–1570) for defense from Ottoman attacks. They have a circular shape containing eleven pentagonal bastions and incorporate specific innovative military architecture (Figure 1). The fortification had three gates known today as Famagusta Gate, Paphos Gate and Kerynia Gate (Figure 1a). At the end of the 19th century, the city began to expand towards the southwest direction outside of the walls and new openings were realized by constructing bridges over the moat connecting the old town with the surrounding areas. The main square of Eleftheria Square today is located at the site of one of the first openings that was created on the walls towards the end of the 19th century in order to facilitate the direct connection of the old city with the newly developed areas around the walls (Figure 3). In 1882, a bridge in wood was built but later it was substituted with an in-filled solid bridge with stone walls. In 1930 during the English regime, the passage was widened while demolishing part of the bastion of the Venetian walls. The square has been a timeless gathering place for citizens linked to important historical moments of the country and the major celebrations of the capital city.

In 1974, the Turks invaded the island, occupying the 37% of it. The so-called "Green Line" (Figure 1a) separates the island in two parts; crossing the city of Nicosia actually represents today the unique divided capital in Europe. Due to this division, the historic city center suffered from depopulation, significant physical decay, loss of commercial activities and employment, and high concentration of social problems [52,53]. Since April 2003, it was possible to cross the Green Line and move between the two regions. In 2008, the first passage was unlocked inside the historical center of the capital through Ledra Street (Figure 1a).

In 1979, a significant agreement was reached between the Turkish and Cypriot community in order to prepare a bi-communal master plan with the aim to improve current and future living conditions for all residents of Nicosia. The Nicosia Master Plan was put under the auspices and the financial backing of the United Nations Development Programme (UNDP), incorporating the following objectives [53,54]:

- Social rehabilitation of old residential quarters, community development and population increase;
- Economic revitalization and increase of employment activities;
- Restoration of monuments and buildings, preserving culture heritage and increasing the potential tourism attraction;
- Overall urban planning for a harmonic and balanced development of different sectors of the historic center.

Figure 3. Photographs of Eleftheria Square (Metaxa Square) in 1945 (Photo captured by J. Arthur Dixon) (**a,b**), 1957 (Photo captured by Felix Giaxis) (**c**), 1950 (**d**) and 1969 (**e**).

Thus, in the last decades the town has witnessed the growth of projects focusing on restoring historic buildings, preserving cultural heritage, and calling for new constructions in the periphery and along the historic fortification walls. An example is testified by the renovation of Elefteria Square, an important link between the old town and the modern city outside the walls. In 2005, an architectural competition assigned the project to Zaha Hadid Architects, an architectural office of international reputation, that proposed a modern solution which was blending the Venetian walls, the ditch and the limits between the two parts of the city [55] (Figure 4). The project, directed by Christos Passas, transformed the moat into a green park in combination with the restored Venetian walls, offering recreational facilities, spaces to be used for sport and cultural activities, and art exhibitions, while the upper part became the central core with a modern pedestrianized plaza. The eastern Solomon square was also restored with a new bus station (today completed) and a car park, which is currently under implementation under Omirou Avenue.

Figure 4. Eleftheria Square project: photographs of the work in progress captured in December 2019.

3. Methodological Approach, Data Acquisition and Processing

As a premise to the methodological approach used, it should be noted that the project of Eleftheria Square is a long-term plan that began in 2005 with its design and, today, its implementation is still ongoing. Furthermore, during the years the geophysical support was needed for the various purposes previously introduced, which cannot be separated from a case-by-case evaluation of the methodology to be adopted. Thus, the following workflow was considered:

- Analysis of the needs of the investigation in collaboration with the technicians involved in the project, taking into consideration the typology of searched targets (archaeological remains, underground utilities, geological layers, cavities, water), their supposed depth (from centimeters to tens of meters), their constitutive material (stones, landfills, metal, plastic, fluids), their physical properties (density, conductivity, susceptibility) and their geometry (punctual targets, stretched objects, layers).
- Evaluation of the geological stratigraphy of the area. It is composed by anthropogenic soils and fluvial deposits of different thickness overlying deep marl and sandy marl of the Nicosia Formation [56].
- Evaluation of the environmental noise of the study area considering the presence of metals, buried iron grids, anthropic buried artifacts and the occurrence of recent mechanical interventions.
- Analysis of the typology of surfaces on which to operate (asphalt, concrete, paved areas, land) and the spaces available to work in.
- Consideration of the differences in the operating principles and the applicability of the various geophysical survey techniques, taking into account the previous points.
- Choice of the proper method that is suitable for solving the goal of the research. ERT, EMI and GPR were preferred over other methods. ERT, although having slower acquisition times and a small invasiveness due to the insertion of electrodes in the surfaces, was applied during surveys 1, 2 and 3 in order to obtain a good compromise between depth of investigation and resolution of results, features not achievable with EMI, GPR, seismic, magnetic and gravity methods. In addition, the presence of metals, energy lines and buried grids under paved surfaces during surveys 1 and 2 excluded the use of the expeditious technique such as the magnetic method and the GPR that have an extreme sensitivity to these environmental noises. GPR was used during survey 4 and 5 with the aim to produce high-resolution maps of shallow targets such as buried pipelines. Other methods, in this urban context, would not have led to a proper representation of them. EMI prospections were applied during survey 5 as they are very suitable for the identification of high conductivity bodies in a fast way as required by the research question.
- Choice of the geophysical instruments to use during data acquisition. ERT was carried out using a multi-electrode resistivity meter, the A3000E (M.A.E. s.r.l., Frosolone, Italy). GPR was implemented through an IDS Georadar (IDS GeoRadar s.r.l., Pisa, Italy), equipped with a multi-frequency TRMF antenna (200–600 MHz), and a MALA X3M Ramac Georadar (Guideline Geo AB, Skolgatan, Sweden), equipped with a shielded 500 MHz antenna frequency. EMI prospections were conducted utilizing the GSSI Profiler EMP-400 [57].
- Definition of the parameters of acquisition taking into consideration the required depth of investigation and the resolution required by the research question. For ERT, profiles had different lengths, taking into account the available spaces in the different survey areas and the electrode spacing was in any case set to 1.2 m. Regarding archaeological prospections, the selected electrode spacing has been considered proper to reach supposed large deep walls. A dipole-dipole (DD) configuration was used. For GPR investigation in both cases, all radar reflections were recorded digitally in the field as 16-bit data, 512 (for IDS) and 628 (for MALA) samples per radar scan at 25 scan s^{-1} for IDS (1 scan approximately corresponds to 0.025 m) and 24 scan s-1 for MALA (1 scan approximately corresponds to 0.024 m). Half meter equally spaced GPR profiles were acquired in grids adapted to the available areas. EMI measurements were collected in stationary

(point-to-point) mode using frequencies in the range 5–16 kHz with vertically oriented dipoles. The investigation area was covered by profiles spaced 0.5 m inserted in a regular grid.

- Choice of the techniques of data processing for the selected methodologies:

    - For ERT, the measured apparent resistivity datasets were processed in order to remove dragging effects that are typical of DD array and model the survey targets converting the values in real electrical resistivity values displayed as a function of depth below surface. To this end, a probability tomography approach has been applied for imaging the sources of anomalies into the analyzed grounds. The theory was first stated for the self-potential method [58], and then adapted to the resistivity method [59–61]. The primary approach, even if capable of distinguishing resistivity highs and lows in the field datasets considering a reference background resistivity, precluded the estimate of the intrinsic resistivities of the source bodies. The method was successfully used to delineate the geometry of the self-potential sources in the central volcanic area of Vesuvius (Naples, Italy) [62], to imaging through the resistivity method buried archaeological structures in the archeological site of Pompei (Naples, Italy) [63], at the Castle of Zena (Carpeneto Piacentino, Italy) [64], the prehistoric sites of Checua (Cundinamarca, Colombia) [65] and Grotta Reali (Rocchetta a Volturno, Italy) [66], the Archaeological Park of Aeclanum [67] and for fault detection in a site on the Matese Mountain (Italy) [3]. Successively, a data-adaptive probability-based ERT inversion (PERTI) method [51] was directly derived from the principles of the probability tomography in order to estimate the true resistivities. From a probabilistic point of view, the algorithm, being a non-linear approach, identifies, inside the set of possible solutions the most probable one, compatibly with the dataset acquisition scheme. The main features of the PERTI method, as reported in [51] are: (i) unnecessity of a priori information; (ii) full, unconstrained adaptability to any kind of dataset, including the case of non-flat topography; (iii) drastic reduction of computing time of even two orders of magnitude, with respect to the previous methods in complex 3D cases using the same computer; (iv) real-time inversion directly in the field, thus allowing for fast modifications of the survey plan to better focus the expected targets; (v) full independence from data acquisition techniques and spatial regularity, (vi) possibility to be used as an optimum starting model in standard iterative inversion processes in order to speed up convergence. As the PERTI method does not require a priori information and iterative processes, the computation of the route mean square (RMS) error between measured and modelled apparent resistivity values is useless. In fact, the resulting RMS error, whatever it is, can be lowered within the PERTI scheme. Many applications of the PERTI approach in near surface prospections are available in literature for solving archaeological research questions [49,68–71], for defining faults in Crete [72] and for imaging of the near-surface structure of the Solfatara crater, Campi Flegrei (Naples, Italy) [73]. In [49,51,68,72] the PERTI routine was tested using well-known commercial software of inversion of geoelectrical data and the comparison put in evidence for the coherence between the obtained results and the better filter capacity and great versatility of the PERTI algorithm. Here, for the first time, the method is applied to process field datasets acquired in urban environments, as its main features fit perfectly with the needs of the case study presented.

    - For GPR, as the interpretation of each section can lead to underestimation or overestimation of the reflected signals and makes it not easy to identify the effect of lateral bodies present in the subsoil, all sections were processed together using standard methodological approaches in order to obtain 2D horizontal maps at a different range of depth (GPR-SLICE 7.0 software) [74]. Data were converted by subtracting out the dc-drift (wobble) in the data, and at the same time adding a gain with time of 20. A time-zero correction was determined to designate the starting point of the wave and the center frequency of the antenna was matched. Then the bandpass filter and the background removal were respectively applied to reduce noise from oscillating

components that had a regular frequency cycle in the frequency domain and to remove striation noises that occurred at the same time. Processed radargrams were subsequently corrected with an automatic gain function applied to each trace based on the difference between the mean amplitude of the signal in the time window and the maximum amplitude of the trace. Thus, horizontal sections (time slices) were processed considering the whole dataset. Data were gridded using the inverse distance algorithm, which includes a search of all data within a fixed radius of 0.75 m of the desired point to be interpolated on the grid and a smoothing factor of two. Grid cell size was set to 0.01 m to produce high resolution images.
- For EMI, during data elaboration the measured values of conductivity were transformed in values of resistivity and visualized in 2D maps through a contouring software.

4. Results

4.1. Survey 1

In 2008, five ERT profiles, 36 m long, were conducted at Eleftheria Square with the aim to cross out possible remains of ancient walls and traces of portion of D'Avila Bastion (Figure 5). It clearly appears to have an asymmetrical shape, with the uncertainty that they were destroyed during the construction of the openings in the circuit at the beginning of the 19th century. The distance between the close profiles 2–4 was 1 m. The maximum reached depth was about 12 m at the center of the sections.

At 1 m in depth at the northeastern side of ERT 1 and ERT 5, two resistivity highs (about 300 Ohm m), vertically elongated, were detected (Figure 6a,d). They are precisely located were the portion of Venetian walls should be as pointed out with yellow dots in Figure 7a.

ERT profiles 2–5 highlight coupled medium resistive spots (about 100 Ohm) spaced 4 m; the location is reported with blue circles in Figure 6b–d. They were interpreted as the walls of the bastion. The conductive zones between them can be associated with the soil filling the fortress. In ERT 1 and ERT 5, the section of the ditch is represented by low resistivity values from the surface until a maximum depth of 8 m. It rests on a resistive layer, the probable deep bedrock.

Figure 5. Google Earth™ satellite images (2008) with the indication of gaps in Venetian wall (in blue) research subjects of Surveys 1 and 2.

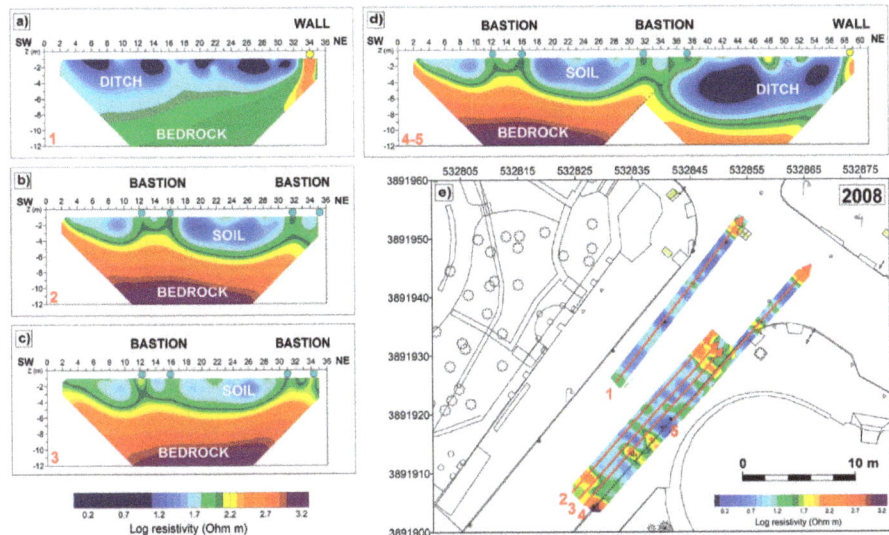

Figure 6. Eleftheria Square: ERT profiles 1–5 (**a**–**d**) and the modelled horizontal resistivity map relative to 1.2 m in depth (**e**).

Figure 6e shows a horizontal resistivity map in which the resistivity highs are located on a technical plan of the area. In order to better highlight the medium resistive spots, the limits of color scale have been slightly modified.

An archaeological excavation performed in 2009 proved the existence of the Venetian walls in the point indicated at the end of ERT 5 and the bastion was entirely brought to light in the exact points indicated by geophysical prospections (Figures 6 and 7a). The ditch was also partially excavated. Subsequently, the original project of the new square needed to be modified, maintaining a distance of 3 m between the new construction and the northwestern side of the bastion as regulated by the International Charter for the Conservation of Monuments and Sites. The new construction is now completed, and it incorporates the ancient bastion (Figure 7b).

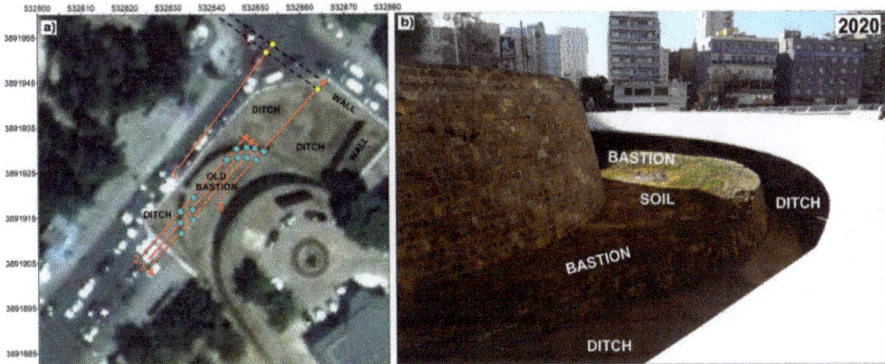

Figure 7. Eleftheria Square: Location of ERT profiles 1–5 and geophysical anomalies on a Google Earth™ satellite images (2011) after archaeological excavations at Eleftheria Square (**a**) and a picture of the excavated ancient bastion inserted in the new square (**b**).

Appl. Sci. **2020**, *10*, 4296

4.2. Survey 2

Survey 2 was carried in 2008 and concerned Solomos Square, where the city's bus station is, at the intersection of Rigenis Street and Omirou Avenue (Figures 2 and 8a). Here, the presence of buried walls and archaeological remains were the target of the investigation. Five ERT profiles (6–10) with different lengths where acquired. ERT profiles 7–9 (Figure 8d–f), located at the center of the square, clearly put forth evidence of the filling of the ditch and the deep resistive substratum. With the exception of rare near-surface high-resistivity nuclei probably due to modern alteration, the unique dishomogeneities that can be attributable to ancient walls are the high-resistive features located at the right side of ERT profiles 6 and 10 (yellow dots in Figure 8a,c,f). In 2010, the square was remodeled, and direct archaeological excavations were not performed to verify geophysical results.

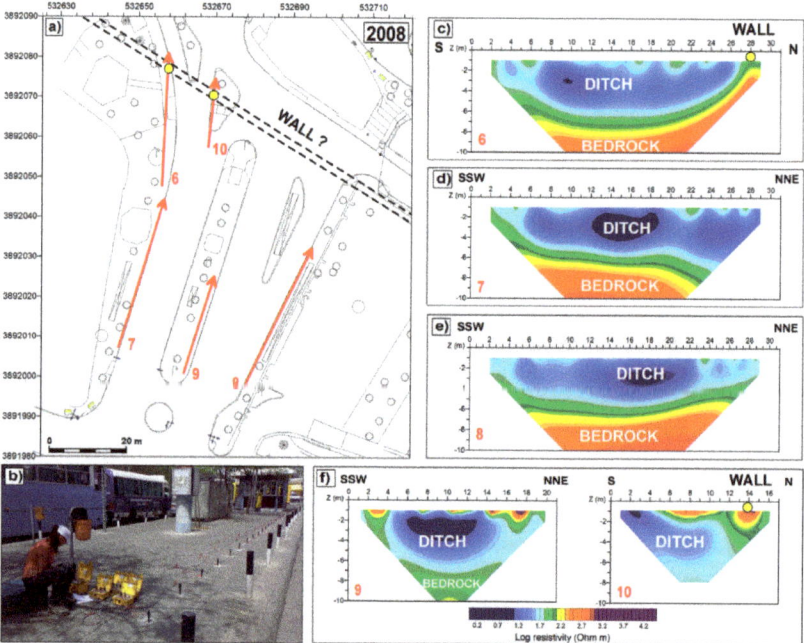

Figure 8. Solomon Square: location of ERT profiles 6–10 (**a**), data acquisition (**b**) and geophysical results (**c**–**f**).

4.3. Surveys 3–4

In 2009, twelve ERT profiles were carried out along Omirou Avenue with the aim of imaging the stratigraphy of the soil and to verify the presence of archaeological structures into the subsoil (ERT profiles 11–22 in Figure 9). As the planning of the survey was influenced by the necessity to ensure the traffic of cars, two main fragmented lines of investigation were conducted (blue and red lines in Figure 9). Profiles had variable lengths and the depth reached for all was 10 m. Figure 10 shows two tridimensional perspective views of the imaged ERT profiles. In general, a uniform stratigraphy is detected and characterized by materials with low resistivity values. In some cases, an increase of resistivity can be attested at about 8 m in depth. High resistivity features that can be associated to huge empty cavities were not detected, as well as it was not possible to distinguish small cavities filled up with loose conductive sediments.

Figure 9. Omirou Avenue: location of ERT profiles 11–22. Blue and red lines indicate the two main fragmented lines of investigation.

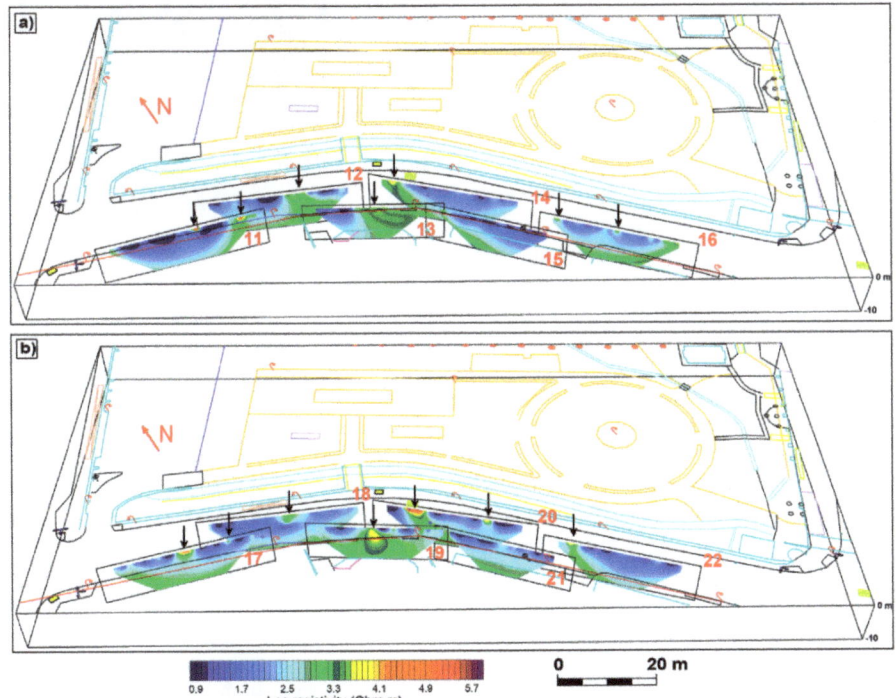

Figure 10. Omirou Avenue: 3D perspective views of ERT profiles 11–16 (**a**) and 17–22 (**b**).

Resistivity features with well-defined shapes similar to the ones that were detected on the ERT profiles of surveys 1 and 2 are not attested here, such as to hypothesize the presence of archaeological remains under the road plane. However, isolated shallow high resistivity values were highlighted in each section (marked with black arrows in Figure 10) that can suggest the existence of nuclei with a non-vanishing probability to find ancient remains. Noteworthy are the two high resistive nuclei that repeat on the ERT profiles 13, 14, 19 and 20, which seem to be on the same route.

In 2017, before the removal of the asphalt layer from the road, a GPR survey was performed. Figure 11a displays two examples of radargrams in which punctual targets, probable manholes, and horizontal discontinuities are marked with coloured arrows. The same marks are reported on the horizontal slice relative to the time window 10–15 ns, the one that better enhances the presence of targets into the subsoil (Figure 11b). The map shows the presence of different linear high amplitude values of the electromagnetic signal that probably refer to pipes buried into the shallow ground. Particularly evident is the presence of a high amplitude feature flanking the south side of the road attributable to the underground sewer pipes (with dotted lines in Figure 11b). Furthermore, the map clearly highlights the presence of manholes indicated with crosses in Figure 11b.

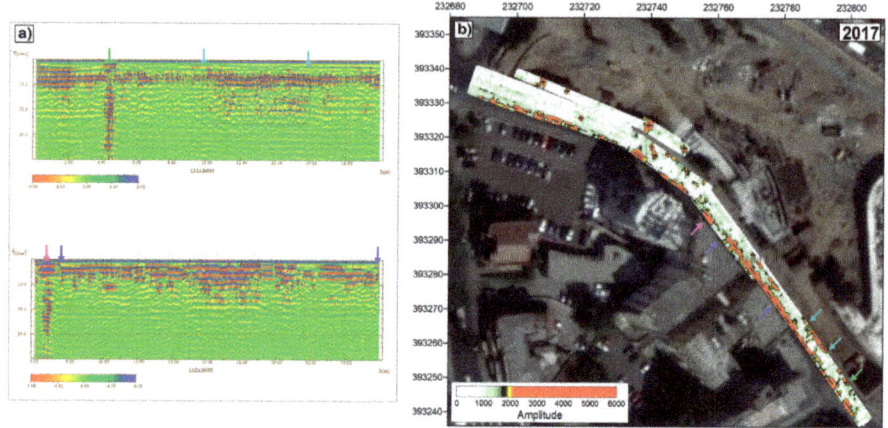

Figure 11. Omirou Avenue: two examples of radargrams (**a**) and GPR slice relative to the time window 10–15 ns. (**b**) Dotted lines and crosses indicate respectively underground pipes and manholes; the colored arrows mark the same anomalies on radargrams and slice.

4.4. Survey 5

In 2019, an unexpected problem occurred during the works: inside the underground substation where electrical cables arranged in six plastic pipes converge, water began to flow in a very abundant cascade directly from the pipes. To try to find out the cause, an excavation trench was carried out south of the cabin (Figure 12a), partially bringing to light the cables that were embedded inside a layer of reinforced concrete. The action was not conclusive as the leak could not be detected. The most reliable hypothesis of the origin of the water was the breakdown of the sewage pipes detected by the GPR in 2017 that pass to the center of the road junction, but no evidence was identified by direct verification. To this point, GPR and EMI investigations were performed in the very few spaces available (Figure 12b).

The resistivity maps obtained from the EMI survey allowed for locating the most probable wet areas. Figure 13 shows all the horizontal maps relating to the frequencies from 16 kHz to 5 kHz. At the edge of the sidewalk there is a conductive band that at lower frequencies (which give us information at greater depths) gradually flags and highlights the retaining wall of the underground rooms in the square. The slice that showed the highest conductivity is that related to the 11 kHz frequency

(Figure 14b). The maps relative to lower frequencies put in evidence of a decreasing of the conductive values and highlight the resistive piles of reinforcement.

Figure 12. Corner between Evagorou Avenue and Omirou Avenue: the excavated trench (**a**), GPR and EMI data acquisition (**b**).

As expected, the GPR survey was not a solution in wet areas because the water completely absorbed the electromagnetic signal (Figure 14a). However, the buried electric cabin was precisely reported on the map and two nuclei with high amplitude of the electromagnetic signals, possible manholes, were detected.

A targeted excavation verification at that point confirmed the presence of conspicuous water into the soil. Although the origin of the water has not been identified, the investigation has averted the southern origin of the leak as hypothesized before the investigation. The most probable theory is that the problem should originate to the west of the trench where the pipeline of the city aqueduct is located. The issue is still being analyzed through inexpensive and non-invasive solutions.

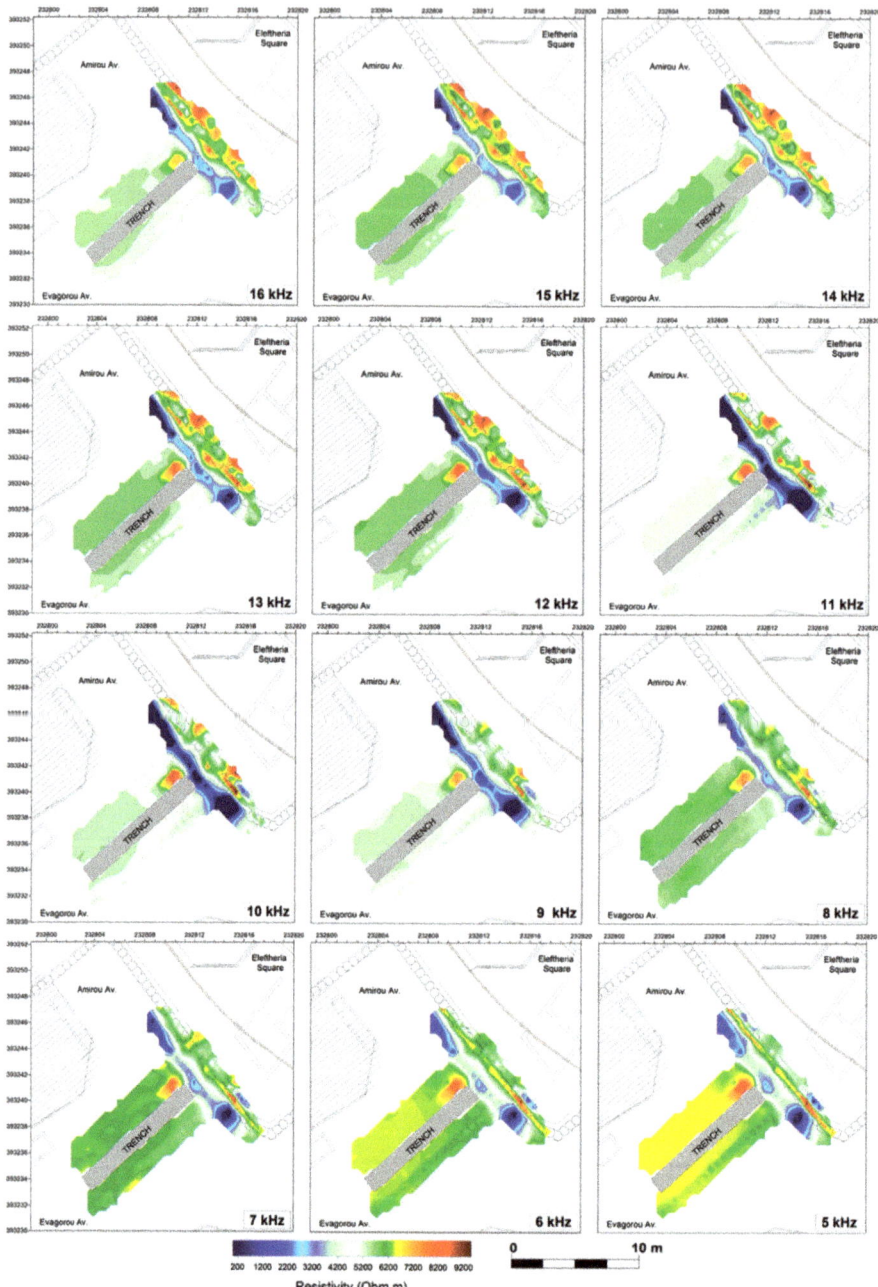

Figure 13. EMI resistivity maps relative to the frequencies in the range 16–5 kHz.

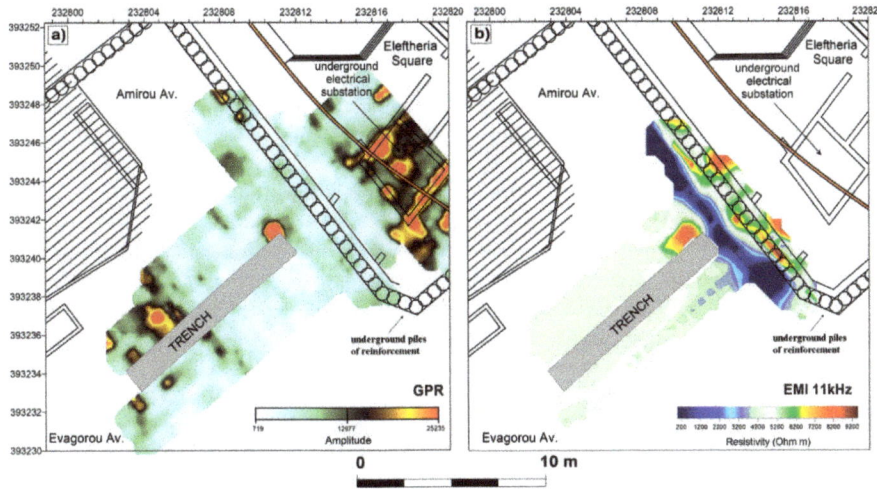

Figure 14. GPR slice relative to the time window 14–18 ns (**a**), EMI resistivity map relative to the frequency of 11 kHz (**b**).

5. Conclusions

This paper concerned the contribution of non-destructive investigations within the redesigning project of Eleftheria Square and its cultural value protection according to its important role in the life of the city of Nicosia.

As a premise which forms an integral part of the discussion of the reported results, it is important to emphasize that the unique limitation of the methodological approach used in this work is in the general principle of geophysical diagnostics as it is an indirect methodology aimed at the spatial definition of anomalies present in the subsoil. Only a direct verification in the field can give certainty of the nature of the buried objects and in the case of ancient finds can define their age and the real archaeological value of the anomalies found. On the contrary, the strength of preventive investigations is the non-invasiveness of the survey, the speed of data acquisition and the possibility of being able guide direct investigations in a targeted and precise manner.

The Elephteria Square redesign project is an important case study for its majesty and prestige and for the historical and cultural context within which it is inserted. The activities lasted more than ten years, from 2008 until today, and the geophysical researches were an integral part of the project. Geophysical research combined with this type of initiative is important as it can be easily replicated in one of the many historical centers scattered around Cyprus or in other parts of the world interested in changes in new urban planning that can have an impact on modern and ancient structures.

The obtained results provided a precise answer to the research questions. In particular, the evaluation of the potential archaeological risk at Eleftheria Square and Solomon Square in 2008, before the beginning of the construction of the new structure, was successfully reached through the application of ERTs: the highs of resistivity interpreted as buried archaeological constructions were proved to be, after excavation, part of the Venetian wall and portions of D'Avila Bastion. Such a result allowed for preserving pieces of the local culture, avoiding its destruction. Furthermore, once discovered, they were inserted within the new complex of the square, adopting a modification to the project in order to highlight the archaeological findings.

The ERT surveys along Amirou Avenue in 2009 have permitted to delineate the geological stratigraphy of the soils, clarifying the absence of huge empty cavities and suggesting the existence of near surface resistive nuclei that probably can be attributable to ancient remains.

Regarding the methodological approach used to process electrical resistivity datasets, the data-adaptive probability-based ERT inversion (PERTI) method was proven to be appropriate for urban measurements because of its ability to remove dragging effects due to the dipole-dipole configuration used and to filter, sometimes, very noisy raw data. In general, within the first application of the probability tomography in urban settings, a good performance in drawing sections is proved.

The GPR surveys, using standard technique of data acquisition and processing, along Amirou Avenue in 2017 allowed the precise detection of underground utilities facilitating the technicians during works of excavation and avoiding unexpected damages to city supplies.

Finally, GPR and EMI investigations were carried out in order to understand the reason for an abundant flow of water that occurred inside the new underground electrical substation positioned under the new Eleftheria Square during excavations at the intersection of Evagorou Avenue and Omirou Avenue in 2009. As expected, the GPR survey was not able to overcome the limit of the presence of water but it was useful to report the precise location of the buried electric cabin on the map. The resistivity maps obtained from the EMI survey allowed the precise locating of the most probable wet areas. A targeted excavation verification at that point confirmed the presence of conspicuous water into the soil, averting the southern origin of the leak as hypothesized before the investigation.

In general, the geophysical results proved the efficiency of the adopted methods, adding scientific value to the status of knowledge of the subject area of study. In some cases, the verification of the supposed buried features was proved by excavation and a perfect correspondence between the geophysical previsions and the structures found in the subsoil was attested. In addition, the presented research supported the actions of the technicians involved in each phase of the project and often, in emergency, the results were provided in real time to resolve the problems encountered. A result of all the various urban activities is, in fact, the collaboration between geophysicists and urban decision makers, which should be encouraged in cases similar to that presented for an optimal city management.

Author Contributions: Conceptualization, M.C. and P.M.; methodology, M.C. and P.M.; software, M.C. and P.M.; validation, M.C. and P.M.; formal analysis, M.C. and P.M.; investigation, M.C., V.G. and P.M.; data curation, M.C., V.G. and P.M.; writing—original draft preparation, M.C., P.M. and A.P.; writing—review and editing, M.C. and P.M.; supervision, P.M. and A.P.; project administration, P.M. and A.P.; funding acquisition, P.M. and A.P. All authors have read and agreed to the published version of the manuscript.

Funding: This research received no external funding.

Acknowledgments: The Ministry of Foreign Affairs of Italy, in the frame of the Archaeological Missions, and Nicosia Master Plan funded part of this research. We thank the Embassy of Italy and the Ambassador Andrea Cavallari for the continuous support for the activities and His Excellency Gherardo La Francesca for being the promoters of this Italian-Cypriot collaboration. Special thanks to the technicians of Nicosia Municipality for administrative and technical support during the research activities.

Conflicts of Interest: The authors declare no conflict of interest.

References

1. Nguyen, F.; Garambois, S.; Chardon, D.; Hermitte, D.; Bellier, O.; Jongmans, D. Subsurface electrical imaging of anisotropic formations affected by a slow active reverse fault, Provence, France. *J. Appl. Geophys.* **2007**, *62*, 338–353. [CrossRef]
2. Moisidi, M.; Vallianatos, F.; Soupios, P.; Kershaw, S. Spatial spectral variations of microtremors and electrical resistivity tomography surveys for fault determination in southwestern Crete, Greece. *J. Geophys. Eng.* **2012**, *9*, 261–270. [CrossRef]
3. Valente, E.; Ascione, A.; Ciotoli, G.; Cozzolino, M.; Porfido, S.; Sciarra, A. Do moderate magnitude earthquakes generate seismically induced ground effects? The case study of the Mw = 5.16, 29th December 2013 Matese earthquake (southern Apennines, Italy). *Int. J. Earth Sci.* **2013**, *107*, 517–537. [CrossRef]
4. Bichler, A.; Bobrowsky, P.; Best, M.; Douma, M.; Hunter, J.; Calvert, T.; Burns, R. Three-dimensional mapping of a landslide using a multi-geophysical approach: The Quesnel Forks landslide. *Landslides* **2004**, *1*, 29–40. [CrossRef]

5. Jäger, D.; Sandmeier, C.; Schwindt, D.; Terhorst, B. Geomorphological and geophysical analyses in a landslide area near Ebermannstadt, Northern Bavaria. *E G Quat. Sci. J.* **2013**, *62*, 150–161. [CrossRef]
6. Grippa, A.; Bianca, M.; Tropeano, M.; Cilumbriello, A.; Mucciarelli, M.; Sabato, L.; Scalo, T. Use of the HVSR method to detect buried paleomorphologies (filled incised-valleys) below a coastal plain: The case of the metaponto plain (Basilicata, Southern Italy). *Boll. Geofis. Teor. Appl.* **2011**, *52*, 225–240.
7. Green, R.T.; Klar, R.V.; Prikryl, J.D. Use of Integrated Geophysics to Characterize Paleo-Fluvial Deposits. In Proceedings of the Site Characterization and Modeling, Geo-Frontiers Congress, Austin, TX, USA, 24–26 January 2015; American Society of Civil Engineers: Reston, VA, USA, 2015; pp. 1–13.
8. Davis, J.L.; Annan, A.P. Ground-penetrating radar for high-resolution mapping of soil and rock stratigraphy 1. *Geophys. Prospect.* **1989**, *37*, 531–551. [CrossRef]
9. Darwin, R.L.; Ferring, C.R.; Ellwood, B.B. Geoelectric stratigraphy and subsurface evaluation of quaternary stream sediments at the Cooper Basin, NE Texas. *Geoarchaeology* **1990**, *5*, 53–79. [CrossRef]
10. Martel, R.; Castellazzi, P.; Gloaguen, E.; Trépanier, L.; Garfias, J. ERT, GPR, InSAR, and tracer tests to characterize karst aquifer systems under urban areas: The case of Quebec City. *Geomorphology* **2018**, *310*, 45–56. [CrossRef]
11. Zini, L.; Calligaris, C.; Forte, E.; Petronio, L.; Zavagno, E.; Boccali, C.; Cucchi, F. A multidisciplinary approach in sinkhole analysis: The Quinis village case study (NE-Italy). *Eng. Geol.* **2015**, *197*, 132–144. [CrossRef]
12. Kaufmann, G.; Romanov, D.; Tippelt, T.; Vienken, T.; Werban, U.; Dietrich, P.; Mai, F.; Börner, F. Mapping and modelling of collapse sinkholes in soluble rock: The Münsterdorf site, northern Germany. *J. Appl. Geophys.* **2018**, *154*, 64–80. [CrossRef]
13. Titov, K.; Loukhmanov, V.; Potapov, A. Monitoring of water seepage from a reservoir using resistivity and self polarization methods: Case history of the Petergoph fountain water supply system. *First Break* **2000**, *18*, 431–435. [CrossRef]
14. Nabighian, M.N.; Grauch, V.J.S.; Hansen, R.O.; LaFehr, T.R.; Li, Y.; Peirce, J.W.; Phillips, J.D.; Ruder, M.E. The historical development of the magnetic method in exploration. *Geophysics* **2005**, *70*, 33ND–61ND. [CrossRef]
15. Gavazzi, B.; Bertrand, L.; Munschy, M.; de Lépinay, J.M.; Diraison, M.; Géraud, Y. On the Use of Aeromagnetism for Geological Interpretation: 1. Comparison of Scalar and Vector Magnetometers for Aeromagnetic Surveys and an Equivalent Source Interpolator for Combining, Gridding, and Transforming Fixed Altitude and Draping Data Sets. *J. Geophys. Res. Solid Earth* **2020**, *125*, 018870. [CrossRef]
16. Bertrand, L.; Gavazzi, B.; de Lépinay, J.M.; Diraison, M.; Géraud, Y.; Munschy, M. On the Use of Aeromagnetism for Geological Interpretation: 2. A Case Study on Structural and Lithological Features in the Northern Vosges. *J. Geophys. Res. Solid Earth* **2020**, *125*, 017688. [CrossRef]
17. Gavazzi, B.; le Maire, P.; de Lépinay, J.M.; Calou, P.; Munschy, M. Fluxgate three-component magnetometers for cost-effective ground, UAV and airborne magnetic surveys for industrial and academic geoscience applications and comparison with current industrial standards through case studies. *Geomech. Energy Environ.* **2019**, *20*, 100117. [CrossRef]
18. Nabighian, M.N.; Ander, M.E.; Grauch, V.J.S.; Hansen, R.O.; la Fehr, T.R.; Li, Y.; Ruder, M.E. Historical development of the gravity method in exploration Historical Development of Gravity Method. *Geophysics* **2005**, *70*, 63ND–89ND. [CrossRef]
19. Scollar, I. Electromagnetic Prospecting Methods in Archaeology. *Archaeometry* **1962**, *5*, 146–153. [CrossRef]
20. Bozzo, E.; Merlanti, F.; Ranieri, G.; Sambuelli, L.; Finzi, E. EM-VLF soundings on the eastern hill of the archaeological site of Selinunte. *Boll. Geofis. Teor. Appl.* **1991**, *34*, 132–140.
21. Fassbinder, J.W.E.; Reindel, M. Magnetometer prospection as research for pre-Spanish cultures at Nasca and Palpa, Perù. In Proceedings of the 6th International Archaeological Prospection Conference, CNR, Rome, Italy, 14–17 September 2005; Piro, S., Ed.; Institute of Technologies Applied to Cultural Heritage: Rome, Italy, 2005; pp. 6–10.
22. Mol, L.; Preston, P.R. The writing's in the wall: A review of new preliminary applications of electrical resistivity tomography within archaeology. *Archaeometry* **2010**, *52*, 1079–1095. [CrossRef]
23. Tsokas, G.N.; Tsourlos, P.I.; Kim, J.-H.; Yi, M.-J.; Vargemezis, G.; Lefantzis, M.; Fikos, E.; Peristeri, K. ERT imaging of the interior of the huge tumulus of Kastas in Amphipolis (northern Greece). *Archaeol. Prospect.* **2018**, *25*, 347–361. [CrossRef]

24. Urban, T.M.; Leon, J.F.; Manning, S.W.; Fisher, K.D. High resolution GPR mapping of Late Bronze Age architecture at Kalavasos-Ayios Dhimitrios, Cyprus. *J. Appl. Geophys.* **2014**, *107*, 129–136. [CrossRef]
25. Cozzolino, M.; Gentile, V.; Giordano, C.; Mauriello, P. Imaging Buried Archaeological Features through Ground Penetrating Radar: The Case of the Ancient Saepinum (Campobasso, Italy). *Geosciences* **2020**, *10*, 225. [CrossRef]
26. Knudson, D.L.; Rempe, J. Linear variable differential transformer (LVDT)-based elongation measurements in Advanced Test Reactor high temperature irradiation testing. *Meas. Sci. Technol.* **2012**, *23*, 025604. [CrossRef]
27. Burrows, S.E.; Rashed, A.; Almond, D.P.; Dixon, S. Combined laser spot imaging thermography and ultrasonic measurements for crack detection. *Nondestruct. Test. Eval.* **2007**, *22*, 217–227. [CrossRef]
28. Candoré, J.C.; Bodnar, J.-L.; Detalle, V.; Grossel, P. Non destructive testing in situ, of works of art by stimulated infra-red thermography. *J. Phys. Conf. Ser.* **2010**, *214*, 012068. [CrossRef]
29. Costanzo, A.; Minasi, M.; Casula, G.; Musacchio, M.; Buongiorno, M.F. Combined Use of Terrestrial Laser Scanning and IR Thermography Applied to a Historical Building. *Sensors* **2014**, *15*, 194–213. [CrossRef]
30. Arias, P.; Armesto, J.; di Capua, D.; Gonzalez-Drigo, R.; Lorenzo, H.; Perez-Gracia, V. Digital photogrammetry, GPR and computational analysis of structural damages in a mediaeval bridge. *Eng. Fail. Anal.* **2007**, *14*, 1444–1457. [CrossRef]
31. Cozzolino, M.; Gabrielli, R.; Galata', P.; Gentile, V.; Greco, G.; Scopinaro, E. Combined use of 3D metric surveys and non-invasive geophysical surveys for the determination of the state of conservation of the Stylite Tower (Umm ar-Rasas, Jordan). *Ann. Geophys.* **2019**, *61*, 72. [CrossRef]
32. Cozzolino, M.; di Meo, A.; Gentile, V. The contribution of indirect topographic surveys (photogrammetry and laser scanner) and GPR investigations in the study of the vulnerability of the Abbey of Santa Maria a Mare, Tremiti Islands (Italy). *Ann. Geophys.* **2019**, *61*, 71. [CrossRef]
33. Available online: https://seg.org/Events/Distributed-Sensing-for-Geophysics (accessed on 25 May 2020).
34. Available online: https://www.eage.org/event/?eventid=1419&evp=18807 (accessed on 25 May 2020).
35. Available online: https://www.eegs.org/sageep-2018-special-session--geophysics-for-urban-underground-space-development-i (accessed on 25 May 2020).
36. Available online: https://www.science-community.org/en/node/177491 (accessed on 25 May 2020).
37. Available online: https://www.sciencedirect.com/journal/engineering-geology/special-issue/10WK1QM13L7 (accessed on 25 May 2020).
38. Available online: https://www.mdpi.com/journal/geosciences/special_issues/urban_geophysics (accessed on 25 May 2020).
39. Miller, R. Introduction to this special section: Urban geophysics. *Geophysics* **2013**, *32*, 248–249. [CrossRef]
40. Piscitelli, S.; Rizzo, E.; Cristallo, F.; Lapenna, V.; Crocco, L.; Persico, R.; Soldovieri, F. GPR and microwave tomography for detecting shallow cavities in the historical area of "Sassi of Matera" (southern Italy). *Near Surf. Geophys.* **2007**, *5*, 275–284. [CrossRef]
41. Gabàs, A.; Macau, A.; Benjumea, B.; Bellmunt, F.; Figueras, S.; Vila, M. Combination of Geophysical Methods to Support Urban Geological Mapping. *Surv. Geophys.* **2013**, *35*, 983–1002. [CrossRef]
42. Rizzo, E.; Capozzoli, L.; de Martino, G.; Grimaldi, S. Urban geophysical approach to characterize the subsoil of the main square in San Benedetto del Tronto town (Italy). *Eng. Geol.* **2019**, *257*, 105133. [CrossRef]
43. Lück, E.; Eisenreich, M.; Spangenberg, U.; Christl, G. A note on geophysical prospection of archaeological structures in urban contexts in Potsdam (Germany). *Archaeol. Prospect.* **1997**, *4*, 231–238. [CrossRef]
44. Amato, V.; Cozzolino, M.; de Benedittis, G.; di Paola, G.; Gentile, V.; Giordano, C.; Marino, P.; Rosskpof, C.M.; Valente, E. An integrated quantitative approach to assess the archaeological heritage in highly anthropized areas: The case study of Aesernia (southern Italy). *Acta IMEKO* **2016**, *5*, 33. [CrossRef]
45. Paz-Arellano, P.; Tejero-Andrade, A.; Argote-Espino, D. 2D-ERT Survey for the Identification of Archaeological and Historical Structures beneath the Plaza of Santo Domingo, Mexico City, Mexico. *Archaeol. Prospect.* **2016**, *24*, 183–194. [CrossRef]
46. Sagnard, F.; Norgeot, C.; Dérobert, X.; Baltazart, V.; Merliot, E.; Derkx, F.; Lebental, B. Utility detection and positioning on the urban site Sense-City using Ground-Penetrating Radar systems. *Measurement* **2016**, *88*, 318–330. [CrossRef]
47. Sărăcin, A. Using georadar systems for mapping underground utility networks. *Procedia Eng.* **2017**, *209*, 216–223. [CrossRef]

48. Camocio, G.F.; Bertelli, D.; Zenoi, D.; Rota, M.; Furlani, P. *Isole Famose Porti, Fortezze, e Terre Maritime Sottoposte Alla Ser.ma Sig.ria di Venetia, ad Altri Principi Christiani, et al. Sig.or Turco, Nouame[n]te Poste in Luce*; Alla libraria del segno di S. Marco: Venice, Italy, 1570–1573; pp. 97–100.
49. Cozzolino, M.; Mauriello, P.; Patella, D. Resistivity Tomography Imaging of the Substratum of the Bedestan Monumental Complex at Nicosia, Cyprus. *Archaeometry* **2013**, *56*, 331–350. [CrossRef]
50. Cozzolino, M.; di Giovanni, E.; Mauriello, P. Geophysical prospections applied to historical centres. In Proceedings of the 19th International Conference on Cultural Heritage and New Technologies 2014 (CHNT 19, 2014), Vienna, Austria, 3–5 November 2014; pp. 1–14.
51. Mauriello, P.; Patella, D. A data-adaptive probability-based fast ERT inversion method. *Prog. Electromagn. Res.* **2009**, *97*, 275–290. [CrossRef]
52. Oktay, D. An Analysis and Review of the Divided City of Nicosia, Cyprus, and New Perspectives. *Geography* **2007**, *3*, 234–236.
53. Petridou, A. *Nicosia Master Plan: Perspectives for Urban Rehabilitation—Building Bridges between the Two Communities of the Divided City of Nicosia*; EU Partnership for the Future Programme; Nicosia Municipality: Nicosia, Cyprus, 2010.
54. Petridou, A. *Rehabilitating Traditional Mediterranean Architecture. The Nicosia Rehabilitation Project: An Integrated Plan*; Monumenta, Nicosia Municipality: Nicosia, Cyprus, 2007.
55. Available online: www.zaha-hadid.com/masterplans/eleftheria-square (accessed on 24 April 2020).
56. Harrison, R.W.; Newell, W.L.; Panayides, I.; Stone, B.; Tsiolakis, E.; Necdet, M.; Batihanli, H.; Ozhur, A.; Lord, A.; Berksoy, O.; et al. *Bedrock Geologic Map of the Greater Lefkosia Area, Cyprus*; US Geological Survey: Reston, VA, USA, 2008.
57. Available online: www.geophysical.com (accessed on 24 April 2020).
58. Patella, D. Introduction to ground surface self-potential tomography. *Geophys. Prospect.* **1997**, *45*, 653–681. [CrossRef]
59. Mauriello, P.; Patella, D. Resistivity anomaly imaging by probability tomography. *Geophys. Prospect.* **1999**, *47*, 411–429. [CrossRef]
60. Mauriello, P.; Patella, D. Resistivity tensor probability tomography. *Prog. Electromagn. Res. B* **2008**, *8*, 129–146. [CrossRef]
61. Mauriello, P.; Patella, D. Geoelectrical anomalies imaged by polar and dipolar probability tomography. *Prog. Electromagn. Res.* **2008**, *87*, 63–88. [CrossRef]
62. Alaia, R.; Patella, D.; Mauriello, P. Application of geoelectrical 3D probability tomography in a test-site of the archaeological park of Pompei (Naples, Italy). *J. Geophys. Eng.* **2007**, *5*, 67–76. [CrossRef]
63. Alaia, R.; Mauriello, P.; Patella, D. Imaging multipole self- potential sources by 3D probability tomography. *Prog. Electromagn. Res. B* **2009**, *14*, 311–339. [CrossRef]
64. Compare, V.; Cozzolino, M.; Mauriello, P.; Patella, D. Resistivity Probability Tomography Imaging at the Castle of Zena, Italy. *EURASIP J. Image Video Process.* **2009**, 1–9. [CrossRef]
65. Minelli, A.; Cozzolino, M.; di Nucci, A.; Guglielmi, S.; Giannantonio, M.; D'Amore, D.; Pittoni, E.; Groot, A.M. The prehistory of the Colombian territory: The results of the Italian archaeological investigation on the Checua site (Municipality of Nemocòn, Cundinamarca Department). *J. Biol. Res.* **2012**, *85*, 94–97. [CrossRef]
66. Compare, V.; Cozzolino, M.; Mauriello, P.; Patella, D. 3D Resistivity probability tomography at the prehistoric site of Grotta Reali (Molise, Italy). *Archaeol. Prospect.* **2009**, *16*, 53–63. [CrossRef]
67. Compare, V.; Cozzolino, M.; di Giovanni, E.; Mauriello, P. Examples of Resistivity Tomography for Cultural Heritage Management. In *Near Surface 2010—16th EAGE European Meeting of Environmental and Engineering Geophysics*; European Association of Geoscientists and Engineers (EAGE): Utrecht, Holland, 2010.
68. Cozzolino, M.; di Giovanni, E.; Mauriello, P.; Desideri, A.V.; Patella, D. Resistivity Tomography in the Park of Pratolino at Vaglia (Florence, Italy). *Archaeol. Prospect.* **2012**, *19*, 253–260. [CrossRef]
69. Cozzolino, M.; Longo, F.; Pizzano, N.; Rizzo, M.L.; Voza, O.; Amato, V. A Multidisciplinary Approach to the Study of the Temple of Athena in Poseidonia-Paestum (Southern Italy): New Geomorphological, Geophysical and Archaeological Data. *Geosciences* **2019**, *9*, 324. [CrossRef]
70. Cozzolino, M.; Caliò, L.M.; Gentile, V.; Mauriello, P.; Di Meo, A. The Discovery of the Theater of Akragas (Valley of Temples, Agrigento, Italy): An Archaeological Confirmation of the Supposed Buried Structures from a Geophysical Survey. *Geosciences* **2020**, *10*, 161. [CrossRef]

71. Cozzolino, M.; Baković, M.; Borovinić, N.; Galli, G.; Gentile, V.; Jabučanin, M.; Mauriello, P.; Merola, P.; Živanović, M. The Contribution of Geophysics to the Knowledge of the Hidden Archaeological Heritage of Montenegro. *Geosciences* **2020**, *10*, 187. [CrossRef]
72. Cozzolino, M.; Mauriello, P.; Moisidi, M.; Vallianatos, F. A Probability Electrical Resistivity Tomography Imaging of complex tectonic features in the Kissamos and Paleohora urban areas, Western Crete (Greece). *Ann. Geophys.* **2019**, *62*, 13. [CrossRef]
73. Di Giuseppe, M.G.; Troiano, A.; Fedele, A.; Caputo, T.; Patella, D.; Troise, C.; de Natale, G. Electrical resistivity tomography imaging of the near-surface structure of the Solfatara crater, Campi Flegrei (Naples, Italy). *Bull. Volcanol.* **2015**, *77*, 27. [CrossRef]
74. Goodman, D. *GPR-SLICE. Ground Penetrating Radar Imaging Software, User's Manual*; Geophysical Archaeometry Laboratory: Los Angeles, CA, USA, 2004.

© 2020 by the authors. Licensee MDPI, Basel, Switzerland. This article is an open access article distributed under the terms and conditions of the Creative Commons Attribution (CC BY) license (http://creativecommons.org/licenses/by/4.0/).

MDPI
St. Alban-Anlage 66
4052 Basel
Switzerland
Tel. +41 61 683 77 34
Fax +41 61 302 89 18
www.mdpi.com

Applied Sciences Editorial Office
E-mail: applsci@mdpi.com
www.mdpi.com/journal/applsci